The Geological Society of America
Memoir 168

The Cretaceous System of Southern South America

A. C. Riccardi
División Paleozoología Invertebrados
Museo de Ciencias Naturales
1900 La Plata, Argentina

1988

Published by The Geological Society of America, Inc.
3300 Penrose Place, P.O. Box 9140, Boulder, Colorado 80301

Printed in U.S.A.

GSA Books Science Editor Campbell Craddock

Library of Congress Cataloging-in-Publication Data

Riccardi, A. C.
    The cretaceous system of southern South America / A. C. Riccardi.
        p.   cm.—(Memoir / Geological Society of America ; 168)
    Bibliography: p.
    Includes index.
    ISBN 0-8137-1168-1
    1. Geology, Stratigraphic—Cretaceous.   2. Geology—South America.
I. Title.   II. Series: Memoir (Geological Society of America) ; 168.
QE685.R53   1988
551.7′7′0982—dc19   87-32081
                                                       CIP

Contents

Acknowledgments

For lending fossil types, for the gift of plastotypes, and/or for providing photographs of ammonites, I am indebted to Dr. M. B. Aguirre Urreta of the Universidad Nacional de Buenos Aires; Dr. B. Baird, Dr. R. H. Hansman, and Mrs. P. Hasson of Princeton University; Dr. G. Blasco de Nullo of the Servicio Geológico Nacional, Buenos Aires; Dr. M. Hünicken of the Universidad Nacional de Córdoba, Argentina; Dr. R. W. Imlay of the U.S. Geological Survey, Washington, D.C.; Dr. J. Sornay of the Museum National d'Histoire Naturelle, Paris; Professor V. Standish Mallory of the University of Washington, Seattle; and Dr. G. Thorbecke of the Geologisch-Paläontologisches Institut der Universität, Freiburg.

Specific information, advice, and/or corrections were kindly given by: Dr. S. Ballent, Universidad Nacional de La Plata, on micropaleontology; Dr. C. A. Cione and Dr. Z. B. de Gasparini, Universidad Nacional de La Plata, on fossil vertebrates; Lic. C. Gulisano, Yacimientos Petrolíferos Fiscales, on the Cretaceous stratigraphy of Neuquén and Mendoza; Dr. M. Morbelli, Universidad Nacional de La Plata, on microfloras; Dr. J. Salfity, Universidad Nacional de Salta, Argentina, on the Cretaceous stratigraphy of the Aimará Basin; Dr. M. Uliana, Astra C.A.P.S.A., on Cretaceous changes of sea-level and paleogeographic evolution of the area. Three unidentified reviewers and Campbell Craddock, Books Science Editor of the Geological Society of America, helped to improve the final version of this book.

Geological Society of America
Memoir 168
1988

The Cretaceous System of
Southern South America

ABSTRACT

The initial history of all Cretaceous sedimentary basins of southern South America is related to the early breakup of Gondwanaland. Along the subduction zone of the Pacific coast, Late Jurassic–Early Cretaceous marine basins were initiated in central Chile and west-central Argentina (the Andean Basin) and in southern Patagonia (the Austral Basin), with the development of volcanic arcs and ensialic troughs. South of 50° south latitude an intra-arc or back-arc marginal basin was formed. At about the same time, several isolated rift basins, formed on the continent and along its Atlantic margin, were the site of continental volcanism and sedimentation. In Barremian time the magmatic arc of the Andean Basin underwent an important eastward migration, resulting in the initial uplift of the Cordillera Principal of Argentina and Chile. This uplift caused a reversal of the regional slope, and continental basins developed. South of 50° south latitude, closure of the marginal basin that began in Late Albian time was accompanied by uplift, resulting in a foreland stage of development of the Austral Basin with deposition of turbidites within a generally regressive pattern.

In Maastrichtian time, sedimentation was discontinued in the continental basins of west-central Argentina and central and northern Chile. The marine area of the Austral Basin had become more restricted. Large regions of northern Patagonia, northeastern Argentina, and Bolivia were covered by a shallow sea, and some restricted fore-arc marine basins were developed on the coast of central Chile. Marine sedimentation was also continuous throughout most of Late Cretaceous time in the Atlantic basins of southern Brazil and central Argentina. Continental deposition was restricted to central Patagonia and the intracratonic Paraná Basin. There was a correlation between diastrophism, igneous activity, and the global cycles of sea-level changes. The transgressive-regressive pattern of the Andean Basin appears to have been controlled by regional tectonics in an area in which local vertical movements were greater than global sea-level changes. In the Austral Basin a different transgressive-regressive pattern emerged, as local tectonic movements could not completely overprint the record of global changes of sea level. Widespread marine sedimentation in a series of Atlantic and Pacific basins during late Late Cretaceous time was coincident with a worldwide transgressive peak.

The Cretaceous flora is known mostly from Patagonia. Lower Cretaceous invertebrates are known mainly from west-central Argentina, whereas Upper Cretaceous invertebrates are known almost exclusively from southern Patagonia. Vertebrates are most common in continental Upper Cretaceous strata of central-northern Patagonia and southern Brazil. Changes in the diversity, endemism, and evolutionary rates of marine faunas appear to be related to transgressive-regressive pulses. The Cretaceous climate north of 43° south latitude was characterized by warm and extremely arid conditions. To the south of this arid belt was an area with temperate and humid climate.

INTRODUCTION

PREVIOUS WORK

The scientific geological study of the Cretaceous system in southern South America goes back to d'Orbigny (1842) and Darwin (1846) and the study of their fossil collections. It was, however, during the last decades of the 19th century and the beginning of the 20th century that the most important contributions to the stratigraphy and paleontology of the Cretaceous system here were made.

In west-central Argentina the work of Bodenbender (1892), whose fossil collections were studied by Behrendsen (1891–92), and that of Burckhardt (1900a,b, 1903) outlined the general features of the Cretaceous stratigraphy of the area. In southern Patagonia, several expeditions resulted in publications including general information on the Cretaceous of the area, e.g., Hatcher (1897, 1900), Hauthal (in Wilckens, 1907b), Quensel (1911), and Halle (1913). The subject was also addressed by Ameghino (1906). A number of important observations on the Mesozoic rocks of the western margin of South America came from Steinmann, whose studies ranged from Patagonia (1883) to central Bolivia (1906).

In the first half of the present century, the stratigraphy of the Cretaceous system of west-central Argentina was thoroughly studied by Groeber (1929, 1933, 1946, 1953), Windhausen (1918b, 1931), Gerth (1914), Lahee (1927), Weaver (1931), and Fossa Mancini (1937). At the same time, Wichmann (1924, 1927a), Kranck (1932), Feruglio (1938, 1944), Piatnitzky (1938), Heim (1940), Brandmayr (1945), and Thomas (1949) made important contributions to our knowledge of the Patagonian Cretaceous rocks south of 39° to 40° south latitude. In northern Argentina the work of Brackebush (1883, 1891) was followed by the more specialized studies of Bonarelli (1914, 1921, 1927). A number of paleontological papers were also published (see under Paleontology), followed by the detailed reviews of Feruglio (1949–50) and Groeber (1953, 1959). Most geological studies made in Chile during the first half of the present century were economically oriented. Information about the Cretaceous system was therefore distributed in diverse publications, although summaries were made by Brüggen (1950), Muñoz Cristi (1956), and Hoffstetter et al. (1957). With respect to southern Brazil and Uruguay, general accounts on the lithostratigraphic units presently assigned to the Cretaceous were produced by Gonzaga de Campos (1889), Walther (1927, 1930), Washburne (1930a,b), Falconer (1931), Lambert (1941), Maack (1947), and Gordon (1947). By comparison, the Cretaceous sequences in Bolivia, had barely been mentioned at that time in a few isolated papers, e.g., Fritzsche (1924), Heald and Mather (1922), and Schlagintweit (1941).

The last 30 years have been characterized by the publication of mostly specialized studies that have improved and refined our knowledge of the Cretaceous period of southern South America.

An outline on the Cretaceous stratigraphy of Argentina, based on those studies, has already been published by Malumián and Báez (1978).

General accounts, covering part or the whole Cretaceous of this region, can also be found in Muñoz Cristi (1956), Hoffstetter et al. (1957), Zeil (1964), and Ruiz et al. (1965) for Chile; Caorsi and Goñi (1958), Goñi and Hoffstetter (1964), Bossi (1966), and Sprechmann et al. (1981) for Uruguay; Hoffstetter and Ahlfeld (1958), Eckel (1959), and Putzer (1962) for Paraguay; Ahlfeld (1946, 1956), Ahlfeld and Braniša (1960), YPFB (1972), Cherroni (1977), and Pareja et al. (1978) for Bolivia; Beurlen (1970), Schneider et al. (1974), Soares (1981), Petri and Fúlfaro (1983), Schobbenhaus et al. (1984), and Petri (1987) for Brazil; and Malumián et al. (1983), and Riccardi (1987) for Argentina, Chile, Paraguay, and Uruguay.

SCOPE AND PURPOSE

The work herein contains a synthesis of the existing knowledge of the Cretaceous system of southern South America; its purpose is aimed to give a general account of the stratigraphy, magmatism, tectonism, paleontology, and paleogeography, and to provide access to more detailed information. In the first part, a description of the Cretaceous rocks is provided on the basis of generalized sections of the most important areas or basins known to date, i.e., (1) Austral Basin, (2) Deseado Massif, (3) Chubut or San Jorge Basin, (4) Concepción-Chubut Dorsal, (5) Andean Basin, (6) Colorado Basin, (7) Macachín Basin, (8) Salado Basin, (9) Santa Lucía–Laguna Merín Basins, (10) Pelotas-Santos Basins, (11) Paraná Basin, (12) Chaco-Pampean Plain, (13) Central Argentina, (14) Aimará Basin, and (15) Pacific Basins. The second part contains information about diastrophic phases, igneous activity, and paleontology. The last part includes the geological history and paleogeographical synthesis. The bibliography is not exhaustive, but an effort has been made to include the most important relevant literature.

This synthesis follows the same general plan previously used to deal with the Jurassic of Argentina and Chile (Riccardi, 1983a), where the reader is directed for some supplementary information.

Cretaceous rocks in southern South America crop out over extensive areas (Fig. 1) and comprise a great variety of marine and continental facies. Most outcrops of the marine Cretaceous are restricted to the Andean area of Argentina and Chile. Lower Cretaceous marine strata extend from about Cape Horn, in the south, to the latitude of Arica in northern Chile, although between 38° and 47° and 18° and 26° south latitude, the records are sparse and/or the Cretaceous is mostly continental. South of 47° south latitude (Fig. 2), Cretaceous rocks are mostly present along the boundary between Argentina and Chile, and the marine succession, which extends into the Upper Cretaceous, has been used as the basis for a sequence of local stages (Natland et al., 1974). North of 32° south latitude (Fig. 10), the marine Lower Cretaceous is restricted to Chile, where it is found on the western

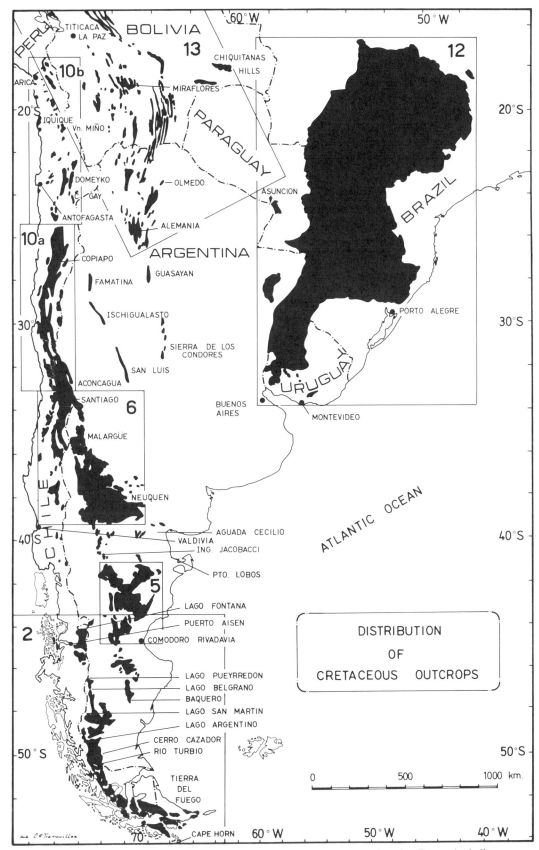

Figure 1. Distribution of outcrops of Cretaceous rocks in southern South America. Rectangles indicate locations of Figures 2, 5, 6, 10, 12, and 13.

monoclinal (Coastal Range) and folded eastern limb (Andean Range) of a NNW–trending synclinorium. To the west the Cretaceous consists of a thick volcanic succession with some minor marine intercalations, while eastward it is developed as a relatively thinner volcanoclastic sequence with important Upper Jurassic–Lower Cretaceous marine intercalations.

In northeastern Argentina, Uruguay, Paraguay, and southern Brazil (Fig. 12), the Cretaceous is represented by extensive basaltic flows and continental sandstones. Similar continental sequences, in some instances interbedded with volcanics, are present in a series of basins developed along the Atlantic margin of Argentina, Uruguay, and southern Brazil, although the uppermost Cretaceous is often of marine origin.

In northern Argentina and Bolivia (Fig. 13) there is an important sedimentary succession, mainly of Late Cretaceous age, that includes continental beds with some minor shallow marine intercalations. Other relatively small basins filled by continental sediments are also present in central Argentina (Fig. 1).

Within the Cretaceous of west-central Argentina and central Chile, there is a marine sedimentary succession that extends from the Tithonian to the Albian, which is followed by Cenomanian-Campanian continental volcanics and sediments and Maastrichtian–Lower Tertiary marine sediments. The whole sequence was divided by Groeber (1946, 1953) into three sedimentary "cycles" (and several "subcycles"), in ascending order, the Andean, Neuquenian, and Malalhueyan. These names were applied to both the sedimentary events and the resulting depositional units, and were understood as South American chronostratigraphic units differing in scope from those of the Standard European Chronostratigraphic Scale. The rank of these units was not clearly stated, although in Groeber's (1946) scheme they were ranked higher than stages and treated as equivalent to series or systems. Subsequently, almost all the names of these "cycles" (and "subcycles") were also applied to the lithostratigraphic classification of the units of west-central Argentina on which Groeber's (1946) scheme had been based.

Groeber's (1946) "cycles" and "subcycles" are, however, still in use as they refer to conformable successions of genetically related strata bounded by unconformities (or disconformities) of different magnitudes (Groeber, 1953; Stipanicic and Rodrigo, 1970; Stipanicic, 1983). The Andean "cycle" is said (Digregorio and Uliana, 1980) to comprise a transgressive-regressive sedimentary succession laid down in a seaway of Pacific origin, that began in the Tithonian, reached its maximum expansion in the Hauterivian, and receded during the Barremian-Albian. The Neuquenian includes continental sediments in the east and thick volcanic sequences in the west. The Malalhueyan consists of Maastrichtian marine sediments deposited during a transgression from the Atlantic. Some of these "cycles" (such as the Andean cycle) refer to transgressive-regressive sequences or cyclothems, and are, therefore, related to global sea-level changes partially controlled by local tectonics. Groeber's (1946) "cycles/subcycles" are closely related, if not equal, to unconformity-bounded or seismic stratigraphic units, as defined by Chang (1975) and

Mitchum et al. (1977), respectively. The Andean "Cycle" has in fact been divided into several depositional sequences (Gulisano et al., 1984), and could be ranked as a "supersequence" (Mitchum et al., 1976). The concept of "sequence" as defined by Sloss et al. (1949) and Krumbein and Sloss (1963, p. 34–35), and that of "cyclothem" as first proposed by Weller (in Wanless and Weller, 1932), could also apply to some of these units.

It seems evident that the various sedimentary "cycles/subcycles" recognized by Groeber (1946) could have different genetic meanings and could be the result of diverse events or any combination of them. Use of this nomenclature could be justified only on the previous definition of all units within a consistent scheme of classification. Furthermore, these divisions are not developed in other areas of southern South America. Thus, in southern Patagonia, marine sedimentation extended throughout Cretaceous time, whereas in all other basins the successions usually consist of continental Upper Cretaceous rocks.

REPOSITORIES

The specimens shown in Plates 1 through 18 are deposited in the following collections under the catalog numbers listed in the plate explanations.

FCEN: Facultad de Ciencias Exactas y Naturales (Departamento de Geología, Cátedra de Paleontología), Universidad de Buenos Aires, Argentina

GPIF: Geologisch-Paläontologischen Institut der Universität, D-78 Freiburg i. Br., Germany

GPIG: Geologisch-Paläontologisches Institut, Georg-August Universität, Göttingen, Germany

MLP: Museo de Ciencias Naturales (División de Paleozoología Invertebrados), Universidad Nacional de La Plata, Paseo del Bosque, La Plata, Argentina

MNHNP: Muséum National d'Histoire Naturelle, Institut de Paléontologie, Paris-V, France

PU: Princeton University (Department of Geological and Geophysical Sciences), Princeton, New Jersey 08540, USA

SGN: Servicio Geológico Nacional (Paleontología), Avenida Santa Fé 1548, Buenos Aires, Argentina

UNC: Universidad Nacional de Córdoba, Facultad de Ciencias Exactas, Físicas y Naturales, Paleozoología, Avenida Vélez Sarsfield 299, Córdoba, Argentina

USNM: United States National Museum, U.S. Geological Survey, Washington, D.C., USA

GENERAL STRATIGRAPHY

AUSTRAL BASIN

The Patagonian Cordillera (Figs. 1, 2) is located on the western border of the structurally depressed Magallanes or Austral Basin. The latter extends over most of southern Patagonia south of 47° south latitude and was formed in Late Jurassic time. The older units recognized in this area include sedimentary and

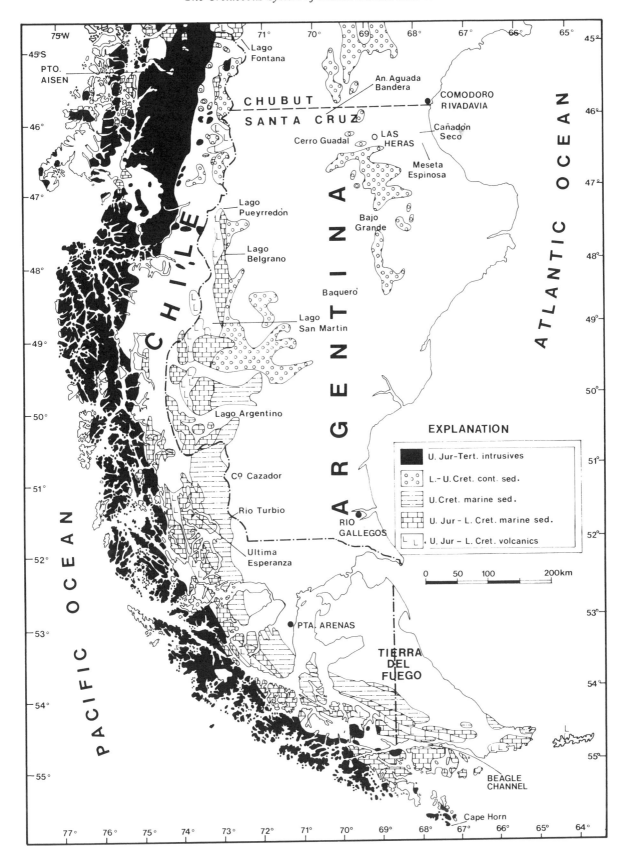

Figure 2. Distribution of major facies within the Cretaceous rocks of southern Patagonia.

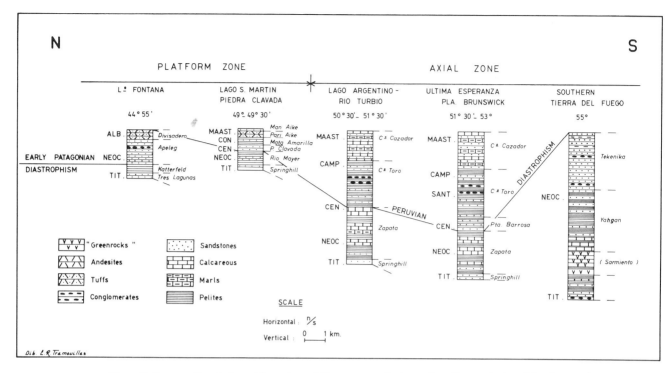

Figure 3. Comparative stratigraphic sections of Cretaceous rocks in southern Patagonia (modified from Riccardi and Rolleri, 1980).

metamorphic rocks of early and late Paleozoic age and the Middle to Upper Jurassic volcanics of the El Quemado Complex (or Tobífera Formation). The basin was filled with Upper Jurassic–Tertiary sediments, chiefly marine (see Leanza, 1972; Riccardi and Rolleri, 1980).

The Cretaceous sequence (Figs. 3, 4; Table 17, in pocket inside back cover) is quite well known from a number of outcrops in the Cordilleran area, and borehole data in the east. In the northeastern part of the basin (Figs. 3, 4), the platform area, the weathering of the volcanic rocks resulted in the oil-bearing continental to marine sandstones of the Springhill Formation which disconformably overlies the Jurassic El Quemado Volcanic Complex, and is generally restricted to the depressions of its irregular surface (Robles, 1982, 1984; Hinterwimmer et al., 1984). The Springhill Formation has an average thickness of 30 to 40 m, with a maximum of 130 to 150 m. Regionally it consists of at least three backstepping but individually prograding sandstone intervals, which may or may not be in contact with each other (Biddle et al., 1986). Locally, the basal part or even the entire formation may be missing. In different wells and outcrops the Springhill Formation has yielded fossil remains. The plants (Table 1) and palynomorphs (Table 3), probably of Tithonian age, have been described by Archangelsky (1977a), Baldoni (1979), Baldoni and Archangelsky (1983), and Baldoni and Taylor (1985), whereas the invertebrates, of Tithonian-Berriasian

age, have been illustrated by Riccardi (1976, 1977) and Blasco et al. (1979). The Springhill Formation has also been dated on the basis of microfauna as Oxfordian-Kimmeridgian (Sigal et al., 1970; Natland et al., 1974) and Valanginian (Kielbowicz et al., 1984). The wide range is attributed to diachronism by some authors (Cecioni and Charrier, 1974; but also see Riccardi, 1976).

The volcanic rocks of the El Quemado Complex and the psammites of the Springhill Formation are overlain by marine pelites, which are widely distributed throughout the basin. Toward the north in the basin platform (Figs. 3, 4), the lower Cretaceous pelites are only 700 m thick, are more calcareous and fossiliferous, and have been ascribed to the Río Mayer Formation. In the northern extreme of the basin (see Ramos, 1976; Marshall et al., 1984; Niemeyer et al., 1984), these pelites were named the Katterfeld or Coyhaique Formation. Eastward, the Katterfeld Formation grades into shallow marine and continental medium- to coarse-grained brown sandstones of the Apeleg Formation, which includes some pyroclastic and marine intercalations (Archangelsky and Seiler, 1980; Archangelsky et al., 1981; Marshall et al., 1984), and is conformably covered by pyroclastic and volcanic rocks, the Divisadero Formation. Westward, the Katterfeld Formation interfingers with the andesitic volcanics of the Carrenleufú Formation (Ramos and Palma, 1983).

In the southwest (Figs. 3, 4; Table 17), the upper levels of the Jurassic volcanics are interbedded with 1,000 to 1,200 m of rather uniform, dark-colored, commonly banded, hard marine pelites, siltstones, and thin turbidites of the Zapata Formation (Katz, 1963), which have yielded Tithonian-Albian invertebrate faunas (Feruglio, 1936–1937, 1944, 1949; Leanza, 1968; Riccardi, 1970). Farther south along the Pacific side of the basin, the Zapata Formation (Katz, 1963) is replaced by the Yahgan Formation (Fig. 3), consisting of more than 3,000 m of alternating volcanoclastic turbidites and shales with intercalations of chert and intermediate pyroclastic rocks. It contains the upfaulted gabbros, dolerites, and basalts of the Tortuga Complex, and is underlain and interbedded with basic volcanics, commonly known as "rocas verdes" (green rocks) or the Sarmiento Formation. The Yahgan Formation exhibits continuous southward coarsening (? including the Tekenika Beds of Halle, 1913) (see Suárez and Pettigrew, 1976; Dott et al., 1977; Suárez, 1978; Suárez et al., 1985a). It seems to interfinger with rhyolitic flows, basalts, and pyroclastic rocks of the Upper Jurassic–Lower Cretaceous Hardy Formation (Suárez, 1978).

On the Argentinian coast of Tierra del Fuego, north of the Beagle Channel (Fig. 2; Table 17; see Caminos, 1980), the Cretaceous begins with at least 500 m of graywackes and slates intercalated with gray sandstones and conglomerates, the Beauvoir and Vicuña formations; these rest conformably on the Jurassic volcanics. This unit includes Upper Aptian–Lower Albian fossils of the Hito XIX fauna (Camacho, 1949; Macellari, 1979), and is equivalent to the subsurface Pampa Rincón and Nueva Argentina formations (Flores et al., 1973) found in the northern part of Tierra del Fuego. The Pampa Rincón Formation rests conformably on the Springhill Formation.

The lower Cretaceous of the Austral Basin bears fossil remains (Tables 1, 3, 6, 8, 12); especially in its lower and upper part. The ammonites (Table 15) and belemnites have been described by Stanton (1901), Favre (1908), Stolley (1912, 1928), Bonarelli and Nágera (1921), Piatnitzky (1938), Cecioni (1955a), Wetzel (1960), Leanza (1963, 1970), Riccardi (1968, 1970, 1977, 1984a,b), Blasco et al. (1980), Aguirre Urreta and Ramos (1981a), Aguirre Urreta and Suárez (1985), Olivero (1983), Aguirre Urreta (1985a, 1986), Aguirre Urreta and Klinger (1986), Leanza H. (1986) and Riccardi et al. (1987). On that basis the Río Mayer Formation and its equivalents have been ascribed to Tithonian-Cenomanian age. Radiometric age dates (Pbα and K-Ar) of plutonites intruding the Divisadero Formation gives values between 83 and 115 ± 7 Ma (Skarmeta and Charrier 1976; Skarmeta, 1978; Ramos, 1979). The Divisadero Formation has yielded K-Ar ages of 97 ± 3 and 111 ± 2 Ma (Charrier et al., 1979; Busteros and Lapido, 1984).

The Río Mayer Formation is conformably overlain by 70 to 300 m of shallow marine to continental whitish, greenish and yellowish sandstones with intercalated conglomerates, the Río Belgrano, Río Tarde, Kachaike, and Piedra Clavada Formations (Figs. 3, 4; Table 17). These grade upward into 250 to 500 m of variegated volcanic and pyroclastic rocks of the Pari Aike or Cardiel Formation, which may include as much as 350 m of intercalated parallic gray-greenish argillites, siltstones, and sandstones, the Mata Amarilla Formation. These formations have yielded invertebrates (Tables 6, 7) and rare ammonites (Tables 15, 16; Bonarelli and Nágera, 1921; Piatnitzky, 1938; Leanza, 1969b; Nullo G. et al., 1981; Nullo F. et al., 1981a,b), vertebrates (Table 12), and plants (Table 2), which attest to a Barremian-Aptian age for the Río Belgrano and Río Tarde formations, an Albian-Cenomanian age for the Kachaike and Piedra Clavada formations, and a Coniacian age for the Mata Amarilla Formation.

Farther southwest (Figs. 3, 4), in the internal parts of the basin, the Zapata Formation is conformably(?) overlain by 250 to 400 m of alternating coarse-grained sandstones and shales of Cenomanian age, the Punta Barrosa Formation. This formation grades eastward and upward into a flysch-like sequence (Cecioni, 1957a; Vilela and Csaky, 1968; Scott, 1966; Winn and Dott, 1977, 1978; Arbe and Hechem, 1984, 1985) consisting of 2,000 m of dark pelites, which commonly exhibit rhythmic alternations with thin-bedded and fine-grained psammites. This unit, named the Cerro Toro Formation (Katz, 1963), includes conglomerates and has yielded ammonites (Paulcke, 1907; Favre, 1908; Cecioni, 1956a; Leanza, 1963, 1965, 1967a; Riccardi, 1979, 1983b; Nullo G. et al., 1981; Nullo F. et al., 1981b; Riccardi and Rolleri, 1984) of Middle Cenomanian and Late Santonian–Early Campanian age (Table 16), other invertebrates (Table 7), and microfauna (Table 8).

The Cerro Toro Formation thins northward (Fig. 4), where it interfingers with the Piedra Clavada, Mata Amarilla, and El Alamo Formations. South of Lake Argentino (Fig. 2) it is conformably overlain by 350 to 1,200 m of yellowish and green marine sandstones that increase in thickness from the north to the south, where they have, respectively, been named La Anita (including *pars* Río Guanaco Formation of Nullo F. et al., 1981a) and Cerro Cazador (including the Tres Pasos Formation of Katz, 1963) formations. Farther south, near Magallanes Strait (Table 17), a similar sequence is represented in upward succession by the gray shales of the Estero La Pera Complex, the pelites of the Balcarcel Formation, the sandstones of the Rosa Formation, the shales with intercalated sandstones of the Fuentes Formation (or Santa Ana and Blanco Formations), and finally the glauconitic sandstones of the Rocallosa Formation (Charrier and Lahsen, 1968; Macellari, 1985a). These units have yielded ammonites (Table 16; Paulcke, 1907; Feruglio, 1936–1937; Cecioni, 1955b, 1956b; Leanza, 1963, 1967a; Lahsen and Charrier, 1972; Hünicken et al., 1975, 1980), other invertebrates (Table 7), and dinoflagellates (Table 4), indicating a Late Campanian–Early Maastrichtian age (Table 17).

The Cerro Cazador Formation (Feruglio, 1938, 1944, 1949) grades upward (Fig. 4; Table 17) into 300 to 500 m of gray, green, and brown shallow marine sandstones with intercalated conglomerates, the Cerro Dorotea Formation, which grades toward the top into continental beds with coal seams. The Cerro Dorotea Formation, an approximate equivalent of the Rocallosa Formation farther south, has yielded marine bivalves, gastropods,

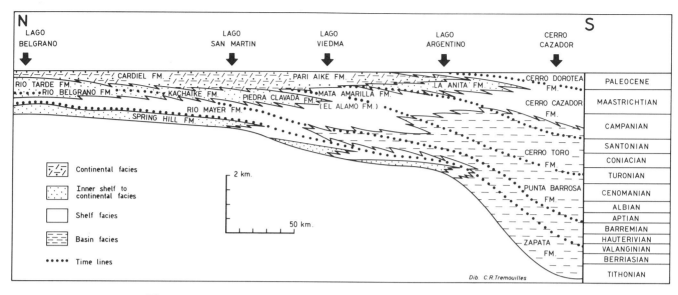

Figure 4. Diagram interpreting relations in the Cretaceous rocks of southern Patagonia.

brachiopods (Table 7), microflora (Table 4), and some plants (Table 2) and vertebrates (Table 12) of Maastrichtian to Paleocene (Freile, 1972) and perhaps early Eocene (Hünicken, 1967) age.

In the subsurface of the basin, the Cretaceous sequence is mostly pelitic and has been named the Palermo Aike Formation. It has also yielded invertebrates (García and Camacho, 1965) and microfossils (Malumián, 1968; Malumián et al., 1972; Rossi de García, 1979).

In central Tierra del Fuego, Chile (Figs. 1, 2; Table 17), the Upper Cretaceous consists of 1,000 to 1,500 m of shales and sandstones, the Cerro Matrero or Río García formations, bearing Campanian-Maastrichtian ammonites (Hünicken et al., 1975; Macellari, 1985a). The Upper Cretaceous sequence of eastern Tierra del Fuego, Argentina consists of (Table 17)—at the base of greenish to brownish gray fine- to medium-grained sandstones with intercalated conglomerates—the Policarpo Formation (including the Bahía Tethis Formation), with Late? Campanian marine invertebrates. In turn, it is conformably overlain by gray, greenish, bluish, fine-grained sandstones, i.e., the Leticia Formation, with marine Maastrichtian fossils. These Cretaceous units are unconformably covered by marine Paleocene (Furque and Camacho, 1949; Bernasconi, 1954; Furque, 1966; Borrello, 1972; Caminos, 1980). In the subsurface (Table 17), the Upper Cretaceous is represented by the pelites of the Cabeza de León Formation (Flores et al., 1973), bearing Early Turonian–Early Campanian microfauna (Table 8; Malumián and Masiuk, 1976b, 1978), microplankton, and pollen (Table 4; Menéndez, 1965a; Menéndez and Caccavari, 1975).

DESEADO MASSIF

North of the Austral Basin, extra-Andean Patagonia can be divided into four structural units. From the south to the north, they are the Deseado Massif, the Chubut Basin, the Somuncurá Massif, and the Colorado Basin (Figs. 17, 18).

The Deseado Massif comprises a large area that has remained structurally positive throughout most of its geologic history. It is covered by Middle to Upper Jurassic volcanics and pyroclastic rocks. Older units are concealed or restricted to small and rare outcrops. Cretaceous rocks are also poorly represented. They constitute the southernmost exposures of a continental sedimentary sequence that is quite well represented farther north in the Chubut Basin (see below).

The oldest Cretaceous rocks in the Deseado Massif (Table 17) are the Bajo Grande Formation (see Lesta and Ferello, 1972; De Giusto et al., 1980); these unconformably overlie the Middle–Upper Jurassic volcanic and pyroclastic rocks of the Bahía Laura Group, within which the Bajo Grande Formation was previously included (Archangelsky, 1967a). The lower part of the Bajo Grande Formation consists of whitish, reddish, and greenish psephites and psammites of continental origin, which grade upward into tuffs. The thickness is about 1,000 m, and several levels have yielded silicified tree trunks. Stratigraphic relationships indicate Tithonian-Berriasian ages (De Giusto et al., 1980).

The Bajo Grande Formation is unconformably overlain (Table 17) by 95 to 247 m of continental conglomerates, gray and red sandstones, carbonaceous mudstones, and whitish to reddish tuffs, which have been ascribed to the Baqueró Formation (Stipanicic, 1957). This unit has yielded abundant micro- and megafloras (Archangelsky, 1963a-d, 1964, 1965, 1967a,b, 1977b; Archangelsky and Baldoni, 1972a,b; Archangelsky and Gamerro, 1964, 1966a-c, 1967; Gamerro, 1965a,b; Menéndez, 1965b, 1966a, 1973; Traverso, 1966, 1968; Martínez, 1968; Herbst, 1971a; Baldoni, 1975, 1977a,b, 1981) (Tables 1, 3), and dated as Barremian-Aptian. The Baqueró Formation is unconformably overlain (Table 17) by Tertiary marine sediments of the Salamanca Formation.

CHUBUT OR SAN JORGE BASIN

The Chubut or San Jorge Basin is an extensional basin developed in central Patagonia, north of the Deseado Massif and south of the Somuncurá Massif (Figs. 2, 5, 17, 18) during Early Cretaceous time. It is a structurally depressed area that was closed to the east (see Lesta et al., 1973; Urien and Zambrano, 1973) but probably had an Early Cretaceous (Hauterivian) western connection with the Austral Basin (Fig. 17). It contains a thick continental sequence that mostly overlies Middle–Upper Jurassic volcanic and pyroclastic rocks of the Lonco Trapial Group. The Cretaceous exhibits a wide range of intercalated facies and unconformities of doubtful significance. Opinions have therefore differed about the definition of the lithostratigraphic units, their nomenclature, and relationships.

The Cretaceous sequence (Table 17) begins with 2,500 m of shales and tuffs, the Las Heras Group (Lesta et al., 1980). Most of the Las Heras Group consists of early rift lacustrine and pyroclastic continental deposits (Brown et al., 1982) restricted to the structurally deepest areas of the basin, and known only through subsurface data. It has been divided into three formations: the Pozo Anticlinal Aguada Bandera-l Formation, the Pozo Cerro Guadal Formation, and the Pozo D-129 Formation (Table 17; Figs. 2, 5).

The Anticlinal Aguada Bandera-1 Formation (Lesta et al., 1980) consists of 1,700 m of black shales, dark siltstones and mudstones, which, on the basis of unpublished palynological studies have been assigned to the ?Tithonian-Berriasian. This is unconformably overlain by a 560-m-thick continental sequence of hard quartzose sandstones intercalated with tuffaceous siltstones, light-colored tuffs, and black silicified shales, the Cerro Guadal Formation (Ferello and Lesta, 1973). The upper unconformable part of the Las Heras Group consists of lacustrine brown oolitic limestones, argillaceous tuffs, siltstones, and sandstone lenses of the Pozo D-129 Formation (Lesta, 1968), with a maximum thickness of about 1,500 m.

In western Chubut, sediments equivalent to the Pozo D-129 Formation have been included in the Paso Río Mayo Formation (Archangelsky et al., 1984; Seiler and Moroni, 1984). The Cerro Guadal and D-129 formations represent the first coarse clastic sediments deposited in the San Jorge Basin by braided fluvial and fan-delta systems (Brown et al., 1982). The Pozo D-129 Formation grades laterally (Table 17) into the fluviatile facies of the Matasiete Formation (Sciutto, 1981; Barcat et al., 1984b). The Matasiete Formation (Lesta and Ferello, 1972; Lesta et al., 1980) has a maximum thickness of 820 m at its type locality; it grades upward from green to brown conglomerates and coarse, cross-bedded sandstones that are intercalated with greenish and pinkish tuffs to whitish tuffs and gray sandstones and marls with fresh- to brackish-water gastropods and bivalves, *Diplodon* sp. and *Corbicula* sp.

On the basis of seismic stratigraphic analyses, the Las Heras Group has been divided (Barcat et al., 1984) into two depositional sequences. The oldest is equivalent to the Pozo Anticlinal Aguada Bandera and Pozo Cerro Guadal formations, and the youngest to the Pozo D-129 and Matasiete formations.

Micropaleontological studies on borehole samples from western Chubut have shown the presence of marine microplankton (Table 3), including dinoflagellates and microforaminifera (Seiler, 1979; Archangelsky and Seiler, 1980; Lesta et al., 1980; Archangelsky et al., 1981, 1983, 1984; Seiler and Moroni, 1984), dating strata ascribed to the Las Heras Group in this area as being of Tithonian-Aptian age. This unit is intruded by granites with a radiometric age of 107 ± 15 Ma (Stipanicic and Linares, 1969). Furthermore, it has been assumed that the strata of the Paso Río Mayo Formation in western Chubut interfinger to the west with sedimentary units of the Austral Basin (Lesta et al., 1980; Riccardi and Rolleri, 1980).

Unconformably overlying the Las Heras Group is a sequence of continental tuffs and tuffites with intercalated sandstones and conglomerates called the Chubut Group (Lesta, 1968, 1969; Lesta and Ferello, 1972; Lesta et al., 1980). This has a maximum thickness of 1,500 m and has been divided into at least three different sets of formations in three different areas.

In the southeastern part of the basin (Figs. 2, 5), the Chubut Group includes the following subsurface units (Table 17): the Mina El Carmen, Comodoro Rivadavia, and Cañadón Seco formations (Lesta, 1968). The Mina El Carmen Formation (Aptian-Albian), which unconformably overlies the Pozo D-129 Formation, consists of 200 to 1,200 m of black shales and light-colored tuffs and tuffites of lacustrine origin. It is conformably overlain by a 200- to 1,200-m succession of Santonian-Campanian oil-bearing sandstones intercalated with tuffaceous argillites, which, on the basis of the variable fraction of sandstones from the north to the south, have been named the Comodoro Rivadavia and Cañadón Seco formations, respectively (Lesta, 1968). Most parts of the Comodoro Rivadavia and Cañadón Seco formations consist of a repetition of fan-delta progradational and supperposed aggradational facies (Brown et al., 1982).

In the central part of southern Chubut (Fig. 5), the Chubut Group is quite well exposed in the area of San Bernardo and Castillo Hills located on the western and northern margins of the Musters Lake (Sciutto, 1981). There, the Chubut Group comprises the Castillo, Bajo Barreal, and Laguna Palacios formations (Table 17). The Castillo Formation (Lesta and Ferello, 1972; Lesta et al., 1980), which unconformably overlies the Las Heras Group, reaches a maximum thickness of 460 m and includes greenish lithic tuffs and tuffaceous sandstones that grade upward into whitish argillites and tuffs. It represents a proximal fluvial and fan-delta system (Brown et al., 1982) and has yielded plant remains (Table 2) that have been studied by Hicken (*in* Frenguelli, 1930b), Menéndez (1959), and Romero and Arguijo, (1981), as well as vertebrates (see Bonaparte, 1978). The Castillo Formation grades upward into 200 to 400 m of whitish tuffs and tuffaceous and argillaceous sandstones and siltstones of alluvial

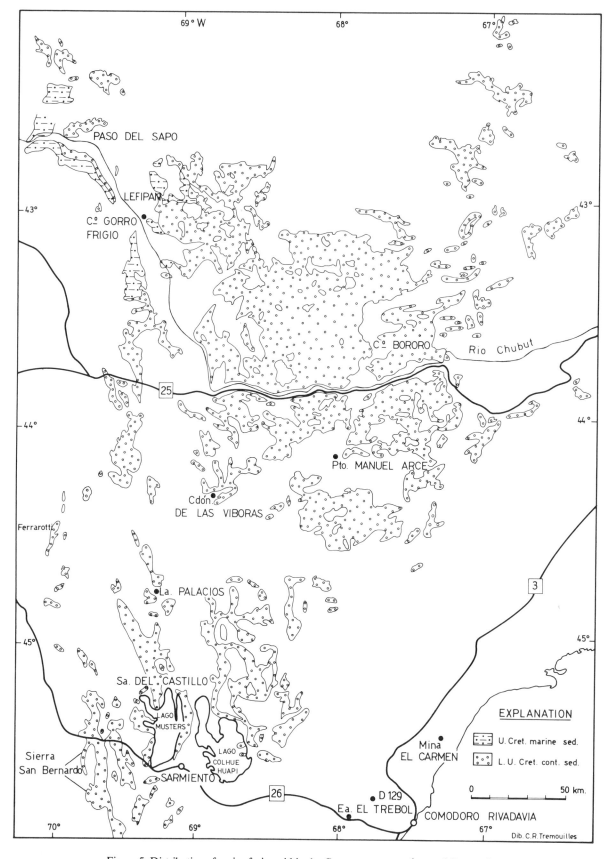

Figure 5. Distribution of major facies within the Cretaceous strata of central Patagonia.

origin that have been included in the Bajo Barreal Formation (Lesta and Ferello, 1972; Lesta et al., 1980). This unit is, in turn, conformably overlain by 130 to 240 m of yellowish, pinkish, and brownish tuffs and conglomeratic sandstones ascribed to the Laguna Palacios Formation. These formations have yielded fossil vertebrates (Table 12) (see Bonaparte and Gasparini, 1979; Martínez et al., 1986). On the basis of seismic stratigraphic analysis, the Chubut Group has been divided (Barcat et al., 1984b) for this area into two depositional sequences. The oldest is equivalent to the Castillo Formation, and the youngest to the Bajo Barreal and Laguna Palacios formations.

In north-central Chubut (Figs. 1, 5) the Chubut Group, also known locally as Cerro Fortín Formation (Fernández Garrasino, 1977; or the Puesto Albornoz Formation of Cortiñas and Arbe, 1981), rests unconformably on continental Jurassic rocks and has been divided into the Gorro Frigio, Cañadón de las Víboras, and the Puesto Manuel Arce formations (Table 17; Chebli et al., 1976; see also Codignotto et al., 1979; Lesta et al., 1980). The Gorro Frigio Formation has a maximum thickness of 1,664 m and consists of brownish to grayish coarse conglomerates and sandstones intercalated with gray silty tuffites and tuffs. These grade upward into sandy tuffs, reddish to brownish argillites and siltstones, and coarse- to fine-grained sandstones and conglomerates interbedded with tuffites, argillites, and siltstones. In this area the basal part of the Chubut Group has yielded spores (Gamerro, 1975a, 1977; Cortiñas and Arbe, 1981), charophytes (Table 3), ostracods (Table 9) (Musacchio, 1972b; Musacchio and Chebli, 1975), and vertebrates (Table 12; del Corro, 1966, 1975; Bonaparte, 1978), which suggests an early Cretaceous (probably Hauterivian-Aptian) age, and thus is equivalent to the Pozo D-129 Formation.

The Gorro Frigio Formation is conformably overlain by a 276-m-thick sequence of red, green, brown, and whitish tuffaceous sandstones, included in the Cañadón de las Víboras Formation. This is, in turn, overlain by a 98-m-thick succession of light-colored conglomerates, sandstones, and argillites bearing charophytes, ostracods, and megaspores, which has been named the Puesto Manuel Arce Formation (Chebli et al., 1976).

Sediments equivalent to the Chubut Group of northern and western Chubut, including those of the middle valley of the Chubut River, have also been ascribed to the Los Adobes Formation (Stipanicic et al., 1968; Stipanicic and Methol, 1972; Tasch and Volkheimer, 1970; Turner, 1979). This unit contains fishes (Table 12; Bordas, 1943; Bocchino, 1979) and sponges (Table 5; Ott and Volkheimer, 1972; Volkheimer and Ott, 1973), and has been ascribed to the Valanginian-Albian period.

In most areas the Chubut Group is unconformably overlain by basalts of Tertiary age or by the marine Salamanca Formation of ?Maastrichtian-Paleocene age (Table 17). In the southeastern part of the basin, however, subsurface data indicate that a 600 to 650-m-thick succession of variegated clays unconformably lies between sediments included in the Chubut Group and the Salamanca Formation. These clays are interbedded with sandstones representing a north-south progradational episode (Brown et al.,

1982). They have been assigned to the Meseta Espinoza and Yacimiento Trébol formations, respectively (Lesta, 1968; Lesta and Ferello, 1972; Lesta et al., 1980; Table 17), and dated as Campanian-Maastrichtian (Brown et al., 1982).

In north-central Chubut Province (Figs. 1, 5), the Chubut Group is unconformably overlain by a Cretaceous sequence consisting at the base of 300 m of brown, reddish, and yellowish conglomerates and sandstones, the Paso del Sapo Formation. This is, in turn, overlain by 180 m of sandstones and gray to green pelites, the Lefipán Formation (Piatnitzky, 1936; Petersen, 1946; Feruglio, 1949; Lesta and Ferello, 1972; Pesce, 1979b; Lesta et al., 1980). The last unit has yielded a rich marine invertebrate fauna (Table 7), mostly studied by Feruglio (1936– 1937) and Camacho (1967b), which includes *Eubaculites* sp., indicating a Maastrichtian age (Camacho, 1967b; Riccardi, 1975). These beds have also yielded vertebrates (Table 12) such as *Aristonectes parvidens* Cabrera (1941), and ostracods (Table 10) of Late ?Campanian-Maastrichtian age (Rossi de García and Proserpio, 1980). Similar strata also seem to occur in south-central Chubut (Lesta et al., 1980). These represent the southernmost deposits of the Maastrichtian sea (Fig. 20) that was widely distributed farther north, i.e., the Jagüel Formation (see under Neuquén). Some authors, however, have suggested that this Maastrichtian sea had connections with southern Patagonia and the Pacific Ocean (see Camacho, 1967a, 1971; Bertels, 1970b; Zambrano, 1986).

Campanian-Maastrichtian continental and marine strata are also present (Fig. 1) in northeastern Chubut near Puerto Lobos (Cortes, 1980), and southeastern Río Negro (Bonaparte and Soria, 1985). In the last area, Campanian–Lower Maastrichtian strata, included in the Los Alamitos Formation, have yielded mammalian remains and other vertebrates (Albino, 1986; Bonaparte et al., 1985; Bonaparte, 1986a,b; Bonaparte and Pascual, 1987). Some Cretaceous units widely represented within the Chubut Basin have correlative units to the south in the Deseado Massif (see above), and to the north near the Somuncurá Massif. In the latter area, however, most Cretaceous formations are formally included within the stratigraphical nomenclatorial systems of the Chubut Basin or the Neuquén Embayment, and are discussed under those headings in the text.

THE CONCEPCIÓN-CHUBUT DORSAL

Between about 43° and 38° south latitude is a north-northwest–trending structural high, the Concepción-Chubut Dorsal, or High, separating the southern Austral (see Austral Basin) from the northern Andean Basin (see Andean Basin) (Aubouin and Borrello, 1966; Cecioni, 1970; Aubouin et al., 1973). In that region mostly volcanic and continental Cretaceous rocks have been recorded. About 1,000 m of andesites, tuffs, and breccias of probable "Neocomian age" (Chotin, 1977), resting unconformably on Lias strata, occur at Lonquimay (Fig. 6). Southward, in Aluminé (Fig. 6), an equivalent sequence consists of 500 m of yellowish sandstones and conglomerates, the Colo Colo Formation (Turner, 1965). These rocks are unconformably covered by andesitic flows and tuffs of Eocene age.

Between 39° and 41°30′ south latitude, east of Valdivia, Chile (Fig. 1), there is a continental sequence of volcanic and clastic rocks, the Curarrehue Formation (Aguirre and Levi, 1964; Hervé et al., 1974; Moreno and Parada, 1976), which contains Upper Cretaceous–Lower Tertiary plants. This sequence rests unconformably on Upper Triassic rocks and is unconformably overlain by Upper Tertiary–Quaternary volcanics. Farther east, within Argentina, conglomeratic sandstones and andesitic rocks of ?Late Cretaceous age have been ascribed to the Llancamil (Ljungner, 1930) and Auca Pan formations (Turner, 1965), although according to González Díaz and Nullo (1980), the Cretaceous is represented only by the granites of the Los Machis Formation.

In the Jacobacci area (Fig. 1), strata equivalent to the Chubut Group (see Chubut or San Jorge Basin), have been named the Angostura Colorada Formation (Volkheimer, 1973); these rest unconformably on Jurassic volcanics and are conformably covered by the marine Maastrichtian Jagüel Formation (see Neuquén). The Angostura Colorada Formation and other Upper Cretaceous strata between 41° and 42° south latitude (Lapido et al., 1984) have also been considered correlatives of the upper part of the Neuquén Group (Casamiquela, 1980; see Neuquén).

Southward, between 43° and 44° south latitude and along the Argentine-Chilean international boundary, volcanic and pyroclastic rocks of Early Cretaceous age are intercalated with marine continental shales and sandstones. Up to 1,240-m-thick (?Tithonian) Valanginian-Hauterivian marine shales, limestones, and sandstones interbedded with andesitic volcanic and pyroclastic rocks, the Alto Palena Formation (including the Arroyo Cajón and Cerro Campamento formations) is overlain by up to 710 m of andesites with intercalated continental sandstones and shales, the Cordón de las Tobas or Carrenleufú Formation (see Fuenzalida, 1968; Pesce, 1979a; Thiele et al., 1979; Haller, 1985). The upper part of this volcanoclastic succession bearing plant remains has also been named La Cautiva Formation. K-Ar isotopic data of the Cordón de las Tobas Formation gave values ranging from 113 ± 5 to 64 ± 2 Ma (Pesce, 1979a; Ramos, 1979; Haller and Lapido, 1980; Haller, 1985). The petrology and geochemistry of the Cordón de las Tobas Formation has been studied by Haller (1985). The marine rocks of the Alto Palena Formation could in fact be related to the Lower Cretaceous sedimentary sequence of the Austral Basin (see Austral Basin). This sequence is in turn unconformably covered by 500 m of basalts, agglomerates, tuffs and breccias, acid ignimbrites, conglomerates and sandstones of the Tres Picos Prieto Formation. Absolute age dates for the latter unit give values between 62 ± 3 and 83 ± 10 Ma (Franchi and Page, 1980; Haller et al., 1981).

ANDEAN BASIN

The Andean Basin contains all of the Cretaceous sequence present in west-central Argentina and central and northern Chile.

The outcrops are exposed along the Principal Cordillera of Argentina and Chile and the Coastal Cordillera of Chile (Figs. 6, 10).

The Principal Cordillera extends without interruption from southern Neuquén (about 40° south latitude) to northern Chile (about 19° south latitude) (Fig. 1), along the international boundary of Chile with Argentina and Bolivia (Harrington, 1956c; Aubouin et al., 1973; Gansser, 1973). It consists principally of Mesozoic and Cenozoic rocks, and the Cretaceous is quite well represented along its entire extension at about the 70° west meridian. From about 39° to 32° south latitude, Cretaceous outcrops are well developed in Argentina, but from 32° to about 18° south latitude, they are restricted to Chile (Figs. 1, 6, 10). In the northern part of Chile between 18°15′ and 27°15′ south latitude, the Principal Cordillera is divided into two branches. Thus, between 20° and 26° south latitude, the Domeyko and Claudio Gay (Figs. 6, 10) ranges form independent chains to the west of the Principal Cordillera and are separated from it by tectonic depressions with salars. The Cretaceous is mostly present along the western ranges.

The Coastal Cordillera rises with a north-south trend along the Pacific coast of Chile, from Patagonia to Arica. It consists chiefly of Paleozoic rocks, although on its eastern part, north of 36° south latitude, a very thick volcanic and volcanoclastic sequence with some marine intercalations developed during Hettangian-Barremian time.

Thus, in central and northern Chile, the Cretaceous is mostly represented in a north-south–oriented belt, about 100 to 150 km wide, located between the Principal and Coastal Cordillera and partially interrupted by granitic intrusions and the younger sedimentary cover.

Neuquén

During Tithonian-Hauterivian time, the sea invaded central-west Argentina from the west and northwest (Fig. 17) and formed an embayment, the Neuquén Embayment, with a large eastward expansion. The Cretaceous in that area is therefore quite well known from a number of outcrops in the west (Fig. 6) and from borehole data in the east.

In this region, Groeber (1946) originally divided the Jurassic and Cretaceous rocks into several sedimentary "cycles" (see under Scope and Purpose). A major marine sedimentary "cycle" extended from the Hettangian to the Oxfordian-Kimmeridgian, and was succeeded by Kimmeridgian continental volcanoclastics (Riccardi, 1983a). A second marine sedimentary "cycle" began in the Tithonian and extended well into the Early Cretaceous (Hauterivian-Barremian). In this area the Tithonian is therefore more closely related to the Lower Cretaceous than to the Jurassic succession. On that basis, Groeber (Groeber, 1953; Harrington, 1956c) segregated the Tithonian from the Jurassic and included it in the beginning of the Cretaceous Andean sedimentary "cycle" (Table 17, Fig. 8). In its original definition, the Andean was thought to comprise Tithonian to Coniacian strata, but presently

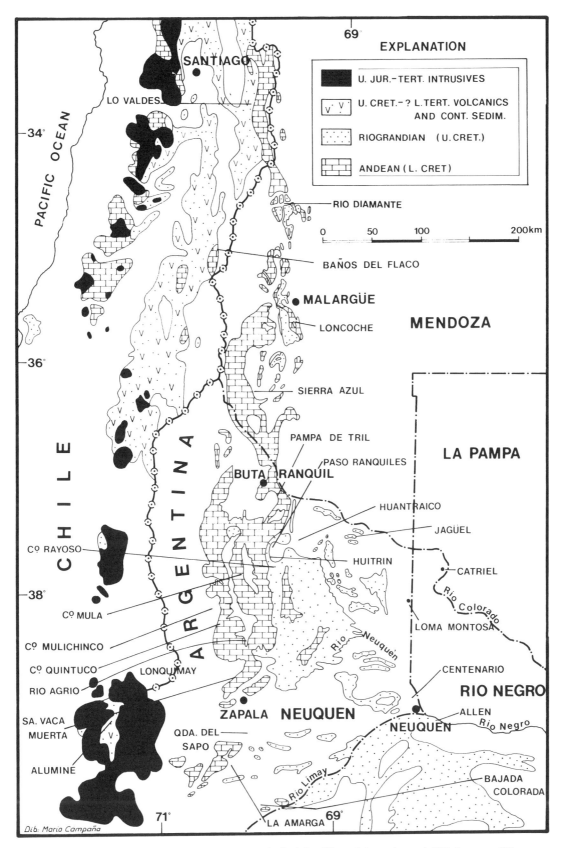

Figure 6. Distribution of major facies within Principal Cordillera of Argentina and Chile between 40° and 33° south latitude.

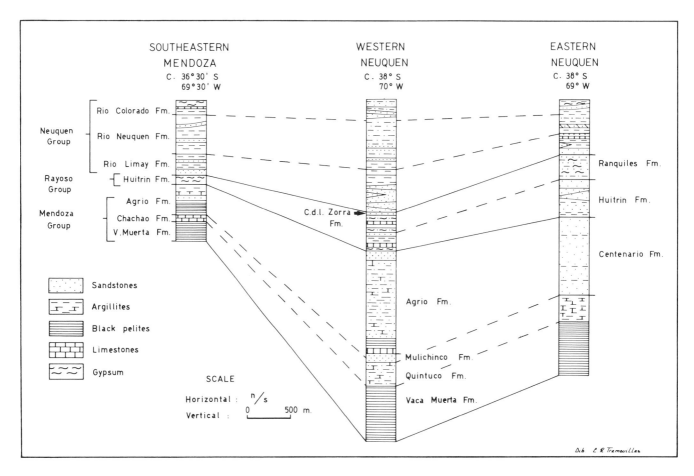

Figure 7. Comparative stratigraphic sections of Tithonian-Campanian in west-central Argentina.

it is considered to range from the Upper Kimmeridgian to the Aptian-Albian (Digregorio, 1978; Digregorio and Uliana, 1980). The Andean strata rest unconformably or paraconformably on Jurassic rocks.

The Andean "cycle" (Table 17) has been divided into two "subcycles" or hemicycles: the Mendocian "subcycle" (Groeber, 1946) and the Rayosian "subcycle" (Uliana et al., 1975a,b). The whole sequence has been described by Groeber (1929, 1946), Gerth (1914), Weaver (1931), Herrero Ducloux (1946), De Ferrariis (1968), Rolleri and Criado Roque (1970), Stipanicic and Rodrigo (1970), Marchese (1971), Digregorio (1972), Leanza H. (1973, 1981c), Leanza H. et al. (1978), Digregorio and Uliana (1975, 1980), Gulisano et al. (1984), Mendiberri (1985), and Mitchum and Uliana (1985).

The Mendocian marine succession is mainly represented by the Mendoza Group (Figs. 8, 9), which rests paraconformably on red sandstones and mudstones of the Kimmeridgian "Tordillo Formation" of western Neuquén and green sandstones and conglomerates of the Catriel and Quebrada del Sapo Formations. These are mostly restricted to the subsurface of eastern and southern Neuquén (Figs. 6, 17). Toward the eastern border of the

basin, however, the basal marine beds of the Mendoza Group transgress over older formations. Thus on the eastern flank of the Azul Range (Fig. 6), where the Tordillo Formation is absent, the Mendoza Group rests on top of older units, like the Oxfordian La Manga Formation, and Dogger, Lias, or Trias rocks to the south. The continental beds of the "Tordillo Formation" have been separated into two sequences separated by an unconformity, a lower one grading downward to the Auquilco Formation, and an upper one transitional to the marine Tithonian Vaca Muerta Formation (see Riccardi, 1983a). As the lower part of this sequence is not represented in the type locality the Tordillo Formation is now included in the Mendoza Group (Legarreta et al., 1981). For some authors (C. A. Gulisano, written communication, 1983), all facies of the Tordillo Formation unconformably overlie the Auquilco Formation.

The Mendoza Group is typically developed in the Andean foothills of Neuquén Province (Fig. 6), where it reaches a maximum thickness of about 2,400 m and consists mostly of rocks of marine origin. It includes a series of stratigraphic units that were recognized and named by Weaver (1931), and can be summarized as follows, base to top (Table 17, Fig. 8): about 200 to 1,700

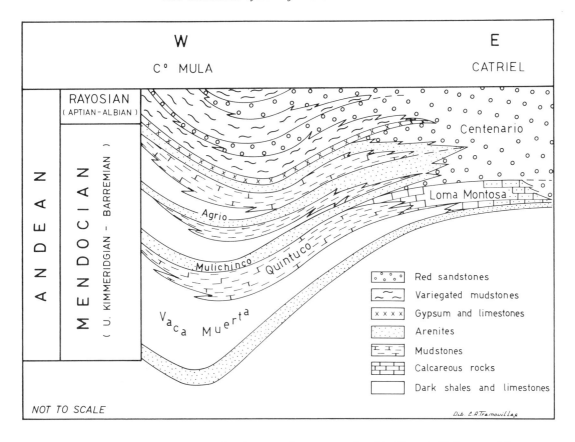

Figure 8. Diagram interpreting relations in Lower Cretaceous rocks of the Neuquén Embayment (after Digregorio and Uliana, 1980).

m of black shales and micritic limestones of the Vaca Muerta Formation; in southern Neuquén (Fig. 9), these grade laterally upward into 400 to 750 m of bioclastic oolitic limestones, argillites, and calcareous sandstones of the Quintuco Formation (including the Picún Leufú Formation of Leanza H., 1973); a unit in turn overlain by 25 to 500 m of red and green conglomerates, sandstones, and argillites of the Mulichinco Formation. The latter is mostly of Valanginian age and reaches its maximum thickness in southern Neuquén (Figs. 7, 9), where it is in part correlative with the Agrio Formation (Dellapé et al., 1979). The Agrio Formation completes the marine sequence and consists of Hauterivian-Barremian limestones, mudstones, marls, and sandstones, with a maximum thickness of approximately 1,600 m. On the basis of seismic stratigraphic analyses, the Lower Tithonian–Valanginian succession has been divided (Gulisano et al., 1984; Mitchum and Uliana, 1985) into 9 to 10 depositional sequences (Fig. 9).

Toward the east and south (Figs. 8, 9; Table 17), the Mendoza Group changes gradually to continental facies. Thus, in the subsurface of eastern Neuquén (Figs. 8, 9) the carbonate platform facies (Carozzi et al., 1982) of the oil-bearing Quintuco

Formation—consisting of dolomites with evaporitic and pelitic-sandy intercalations—is included in the Loma Montosa Formation (Zilli et al., 1979; Rodríguez Schelotto et al., 1981), which, in turn, grades laterally and upward into the conglomeratic sandstones and fangolites of the Centenario Formation (Digregorio, 1972). To the south (Table 17), the Mendoza Group equivalent includes, from the bottom, 46 m of conglomerates of the Pichi Picún Leufú Formation, 16 m of limestones and red beds of the Ortiz Formation, 100 m of red and green sandstones of the Limay Formation, and about 300 m of red argillites, conglomerates, and sandstones of the Bajada Colorada Formation (Dellapé et al., 1979; Rolleri et al., 1984a,b). Locally, some red beds equivalent to the Agrio Formation have been named the La Amarga Formation (Dellapé et al., 1979; Rolleri et al., 1984a,b).

The fossil invertebrates found in the Lower Cretaceous marine formations of west-central Argentina (Tables 5, 9) have been described by Behrendsen (1891–1892), Steuer (1897), Burckhardt (1900a,b, 1903), Haupt (1907), Douvillé (1910), Gerth (1925), Krantz (1928), Weaver (1931), Frenguelli (1944), Leanza (1944, 1945, 1949, 1957), Sokolov (1946), Leanza and Giovine (1949), Giovine (1950, 1952), Indans (1954), Riccardi

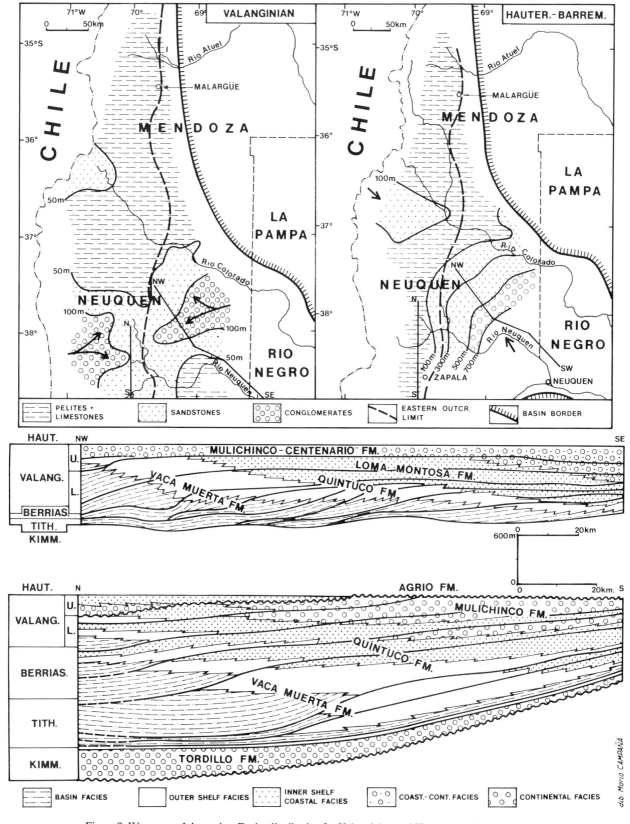

Figure 9. West-central Argentina. Facies distribution for Valanginian and Hauterivian-Barremian (after Uliana et al., 1977), and diagrams interpreting relations for the Tithonian-Valanginian (N–S, after Gulisano et al., 1984; NW–SE, after Mitchum and Uliana, 1985).

et al. (1971), Leanza H. (1972, 1975, 1980, 1981a,b, 1985), Leanza A. and Leanza H. (1973), García and Leanza H. (1976), Musacchio (1978, 1979a,b, 1980, 1981), Leanza H. and Wiedmann (1980), Leanza and Garate Z. (1983, 1986), Manceñido and Damborenea (1984), Musacchio and Abrahamovich (1984), Simeoni (1985), and Simeoni and Musacchio (1986). Fossil vertebrates (Table 12) have been described by Dolgopol de Sáez (1957), Gasparini (1973, 1982), Gasparini and Dellapé (1976), Bocchino (1977), Aramayo (1981), and Gasparini et al. (1982). The microflora (Table 3) has been studied by Volkheimer and Sepúlveda (1976), Volkheimer et al. (1977), Volkheimer and Prámparo (1984), and Quattrocchio and Volkheimer (1985).

The ammonites (Table 13, 14) indicate ages from Tithonian to Hauterivian time, although the presence of crioceratids with Barremian affinities (Giovine, 1950, p. 74; Leanza H., 1981c) and the possible existence of *Silesites* Uhlig, *Holcodiscus* Uhlig and *Desmoceras* Zittel (Windhausen, 1918b, p. 24; Groeber, 1929, p. 35; 1933, p. 19; 1946, p. 186) suggest that, at least in some areas, the marine sediments could be of Barremian and ?Aptian age (Riccardi, 1984a).

The La Amarga Formation has yielded ostracods, charophytes (Musacchio, 1970, 1971, 1978, 1981), pollen (Gamerro, 1975b; Volkheimer, 1978), pterosaurs (Montanelli, 1987) and mammals (Bonaparte, 1986a; Bonaparte and Rougier, 1987), which—together with the stratigraphic relationships—indicate a Barremian-Albian age (Tables 3, 9).

The strata of the Mendoza Group are conformably overlain by those of the Rayoso Group. The Rayoso Group (Huitrinian of Groeber, 1946) has the same distribution as the Mendoza Group, reaching a thickness of about 1,100 m in the Andes (Di Paola and Marchese, 1970). In that region it comprises—from base to top—the Huitrín, the Ranquiles and Cañadón de la Zorra formations (Table 17, Fig. 7).

The Huitrín Formation ("Yeso de Transición" or "Huitriniano", of Groeber, 1929, 1946; Herrero Ducloux, 1946, 1947) is a 200- to 490-m-thick succession of limestones and evaporites (gypsum and anhydrite) that changes upward into detrital alluvial rocks (red sandstones and mudstones). It is overlain by 600 m of variegated mudstones and red sandstones of the Ranquiles Formation bearing charophytes (Table 3), ostracods and foraminifers (Table 9) of Aptian-?Albian age (Musacchio and Palamarczuk, 1976; Musacchio, 1981). This unit, in turn, is overlain by 70 m of variegated mudstones of the Cañadón de la Zorra Formation. Pelites and evaporites are mainly represented in the west; they decrease in thickness toward the south and east, where clastic strata become more important. In the southeastern and southern margins of the Neuquén embayment (Figs. 6, 17), equivalent strata are represented by the upper part of the Centenario Formation and the Bajada Colorada Formation (Fig. 8; Tab. 17).

The Huitrín Formation contains fresh-water mollusks (De Ferrariis, 1968; Digregorio, 1972) and pollen (Volkheimer, 1978; Volkheimer and Salas, 1975, 1976) and is considered to be of Barremian-Albian age (Table 3) (Groeber, 1929, 1946; Weaver, 1931; Digregorio, 1972). To the west, the Huitrín Formation has

intercalations of basic tuffs and basaltic and andesitic flows, which in Chile develop into a thick succession of andesitic flows and tuffs with occasional thin intercalations of terrigenous rocks.

The Rayoso Group is covered paraconformably or unconformably by strata of the third and last sedimentary "cycle" of the Neuquén embayment, the Riograndian "cycle" (Groeber, 1942, 1946). This "cycle" includes a lower continental Neuquenian "subcycle" and an upper marine Malalhueyan "subcycle" (Table 17).

The Neuquenian "subcycle" is represented by red clastic rocks of the Neuquén Group (Groeber, 1929, 1946, 1959) throughout central-east Neuquén, southern Mendoza, and northwest Río Negro provinces (Figs. 6, 19). With a maximum thickness of 1,400 m in western Neuquén, it consists of continental alternating reddish brown and pinkish yellow sandstones and mudstones of fluviatile origin (Cazau and Uliana, 1973).

The Neuquén Group has been studied by Herrero Ducloux (1947), De Ferrariis (1968), and Cazau and Uliana (1973). From base to top (Fig. 7; Table 17), it includes the Río Limay, Río Neuquén, and Río Colorado formations, each with several members. These formations represent three sedimentary sequences of fluvial evolution that begin with conglomerates, continue with sandstones and mudstones, and end with pelites.

The Neuquén Group has been dated as post-Albian and pre-Maastrichtian (?Coniacian-Campanian) on the basis of stratigraphic relationships. It contains charophytes and ostracods (Tables 4, 10; Musacchio, 1973; Uliana and Musacchio, 1979). Dinosaur remains (Table 12), from which the old name "Dinosaurian sandstones" (Ameghino, 1890) originated, are abundant, especially in the upper part (Río Neuquén and Río Colorado formations) of the Neuquén Group (Lydekker, 1893; Estes et al., 1970; Bonaparte, 1978). Most vertebrates are the same as those found farther south in the Chubut Group (Table 12; Bonaparte and Gasparini, 1979).

The Neuquén Group is in turn unconformably overlain by 600 m of strata of the Malargüe Group, which correspond to Groeber's (1946) Malalhueyan sedimentary "subcycle" and are present throughout northern Patagonia at about 38° to 40° south latitude (Fig. 20). The succession has been studied by Windhausen (1918a), Wichmann (1927b), Weaver (1931), Feruglio (1949), Groeber (1959), Leanza (1967b), Bertels (1969), Uliana and Dellapé (1981), Legarreta et al. (1986), and the relationships correctly assessed by Jeletzky (1960). It has been divided into eight depositional sequences (Legarreta et al., 1986).

In the Neuquén embayment, the Malargüe Group (Table 17) begins at the base with variegated marls intercalated with sandstones, limestones, gypsum, and green clays of the Allen Formation. This is lacustrine to shallow marine in origin (Andreis et al., 1974) and bears fresh-water microfossils, bivalves, gastropods (Tables 7, 10), *Ceratodus,* chelonids, and remains of crocodiles, plesiosaurs, and dinosaurs (Table 12) (Doello Jurado, 1922, 1927; Wichmann, 1927a,b; Groeber, 1929, 1933, 1946, 1955, 1959; Leanza, 1964, 1967b; Casamiquela, 1969a,b, 1979, 1980; Bertels, 1969, 1970a, 1972a,b; Camacho, 1968; Leanza and Hünicken, 1970; Angelozzi, 1980; Ballent, 1980; Bonaparte, 1984; Manceñido and Damborenea, 1984; Bonaparte and Novas, 1985). To the west the Allen Formation changes to the red and green sandstones and conglomerates of the Loncoche Formation. The facies variations suggest a paleoslope to the east (Digregorio and Uliana, 1980), although the Loncoche Formation has also been considered of Pacific origin (Zambrano, 1986).

The Allen Formation is overlain by the yellowish to brown-yellowish marine siltstones, claystones, and argillites of the Jagüel Formation (Windhausen, 1914a,b, 1918a,b; Bertels, 1968) and its equivalents, the Coli Toro, Aguada Cecilio, Huantraico and Malargüe formations (Bertels, 1968, 1970a,b; Casamiquela, 1964, 1979, 1980); these strata, which have been included in the Jagüelian Stage, have yielded invertebrate macro- and micro-faunas (Tables 7, 10) and fishes of Early to Middle Maastrichtian age (Leanza, 1964, 1967b; Bertels, 1968, 1970a-c, 1972a,b, 1974, 1975, 1980; Camacho, 1968, 1969; Riccardi, 1975; Weaver, 1927, 1931; Uliana and Musacchio, 1979; Cione and Laffite, 1980; Malumián et al., 1983; Manceñido and Damborenea, 1984).

The Loncoche and Jagüel formations are overlain by the marine regressive beds of the Paleocene Roca Formation (Schiller, 1922; *sensu* Bertels, 1970a,b) with apparent conformity (Casamiquela, 1964, 1969b; Leanza, 1964; Yrigoyen, 1972), although the micropaleontology suggests a paraconformity with a Late Maastrichtian hiatus (Bertels, 1964, 1968, 1969, 1970a,b).

Mendoza

In Mendoza Province, Argentina (Fig. 6), the Cretaceous sequence is similar to that of the Neuquén Province (Gerth, 1914; Lahee, 1927; Fossa Mancini, 1937; Groeber, 1951; Volkheimer, 1970; Yrigoyen, 1972, 1979; Uliana et al., 1977; Legarreta et al., 1981; Legarreta and Kozlowsky, 1984; Ramos, 1985a,b). The Lower Cretaceous sequence (Fig. 7), however, thins to the north-east where it changes to the calcareous shelf facies of the Chachao Formation (Uliana et al., 1977; Mombrú et al., 1979; Carozzi et al., 1981; Legarreta and Kozlowski, 1981, 1984). To the northwest in the Diamante River and Aconcagua areas (Figs. 6, 10), there is an increase in clastic components, e.g., Lindero de Piedra Formation, and the Neuquén subdivisions of the Mendoza Group are not clearly developed (Leanza H., 1981c; Legarreta et al., 1981).

The Mendoza Group (Table 17) is overlain by 50 m of evaporites, limestones, clastics, and red beds, which are correlative to the Rayoso Group of Neuquén and belong to the Huitrín Formation (including part of the Salas Formation of Lahee, 1927). They extend into Chile (Fig. 11), where they are known as the Colimapu Formation. The Upper Cretaceous is represented by continental rocks, the Diamante or Salas Formation (Volkheimer, 1970), which is considered equivalent to the Neuquén Group farther south (Bettini et al., 1979). It is mostly restricted to the western part of Mendoza Province, where it is about 990 to 1,150 m thick.

On top of this continental sequence at the eastern slope of the Andean belt, south of Malargüe (Fig. 6), the Campanian-Maastrichtian marine succession of the Loncoche Formation lies conformably (Bodenbender, 1892; Behrendsen, 1891–1892; Burckhardt, 1900a; Gerth, 1914; Fritzsche, 1919; Weaver, 1931; Groeber, 1933, 1955; Muhlmann, 1937; Fossa Mancini, 1937; Bertels, 1969; Camacho, 1969, 1971; Legarreta et al., 1986). The existence of Maastrichtian-Danian neritic marine sediments in the High Cordillera (Yrigoyen, 1979) suggests a Pacific-Atlantic connection.

Chile

Central Chile. East-Central Chile. A Cretaceous sequence similar to that of west-central Argentina is exposed in the Chilean part of the Principal Cordillera (Figs. 6, 10, 11; Table 17). The stratigraphy has been established by Corvalán (1956), Klohn (1960), González and Vergara (1962), Vergara (1969), Covacevich et al. (1976), Thiele (1980), Charrier (1981), and a paleo-geographical analysis across the Argentine-Chilean boundary at 35° south latitude was presented by Davidson and Vicente (1973). The correlatives of the Cretaceous sequence exposed in the Coastal Cordillera at 30° to 35° south latitude are described in Vergara (1969) and Levi (1970).

Between 36° and 33°15′ south latitude (Figs. 6, 11; Table 17), the Cretaceous has at its base 800 m of conglomerates, sandstones, shales, and limestones of the Leñas Espinoza Formation (Charrier, 1982); this possibly rests conformably on the Kimmeridgian continental Río Damas Formation. The Leñas Espinoza Formation contains Lower-Middle Tithonian marine invertebrates and grades upward into the fossiliferous limestones, marls, and calcareous sandstones of the Baños del Flaco or Lo Valdés Formation, 700 to 1,000 m in thickness and containing Tithonian-Hauterivian marine invertebrates (Corvalán and Pérez, 1958; Biró, 1976, 1980; Hallam et al., 1986) and fish (Schultze, 1981). In the lower part of this unit there are dinosaur tracks (Casamiquela and Fasola, 1968). The Baños del Flaco Formation is equivalent to the San José Formation farther north (Aguirre, 1960; Moscoso et al., 1982).

The Baños del Flaco Formation is conformably overlain by 850 to 3,000 m of continental sandstones, tuffs, conglomerates, red shales, and gypsum of the Colimapu Formation (including the Plan de los Yunques and Cristo Redentor formations; Aguirre, 1960; Jaroš and Zelman, 1969; Moscoso et al., 1982). On the basis of charophytes and stratigraphic relationships this was dated as Late "Neocomian"-Aptian or ?Albian (Martínez and Osorio, 1963), and is thus considered correlative to the Huitrín Formation and the Neuquén Group (Bettini et al., 1979; Godoy, 1981) of the Argentinian Andes, and to the Veta Negra and Las Chilcas Formations of the Coastal Cordillera (Table 17; Fig. 11). The Colimapu Formation is unconformably overlain (Table 17; Fig. 11) by as much as 6,000 m of andesites, basalts, and basaltic breccias intercalated with lacustrine sandstones, siltstones, and conglomerates of the Abanico Formation (including the Río

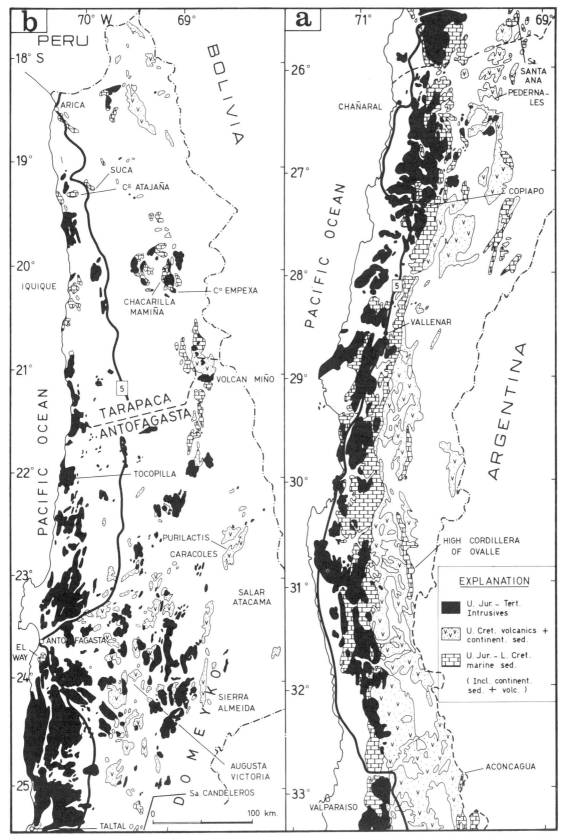

Figure 10. Distribution of major facies of Cretaceous rocks of the Principal and Coastal Cordilleras of Argentina and Chile between 33° and 17° 30′ south latitude.

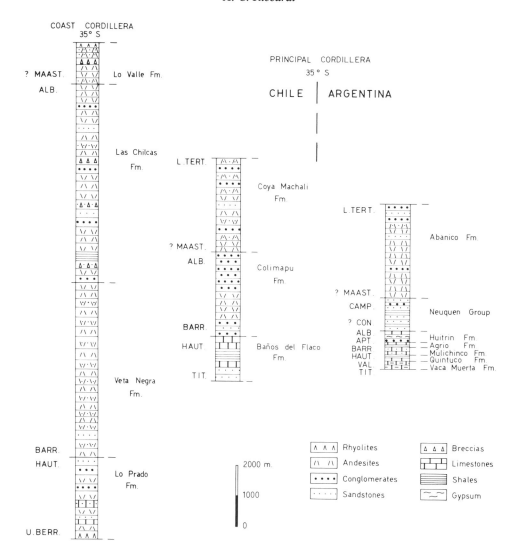

Figure 11. Comparative stratigraphic sections of Cretaceous rocks of the Coastal Cordillera and Principal Cordillera at about 35° south latitude (modified from Aubouin et al., 1973).

Blanco Formation; Gajardo, 1981; Ferraris, 1981). On the basis of radiometric age dates, this unit has been dated as ?Maastrichtian to Early Miocene (Vergara and Drake, 1978, 1979a,b; Drake et al., 1982). This sequence is unconformably overlain by Tertiary volcanics of the Farellones Formation. Similarly, in the Aconcagua area of the Argentine Cordillera (Figs. 1, 10a), the Cretaceous is represented by the Mendoza and Rayoso Groups that rest on the Kimmeridgian Tordillo Formation and are unconformably overlain by the volcanics of the Abanico Group (Yrigoyen, 1972; Ramos, 1985a). The same sequence, but without Tithonian fossils, is known from the High Cordillera of Ovalle between 30°20′ and 31°20′ south latitudes (Fig. 10) (Mpodozis et al., 1973; Mpodozis and Rivano, 1976).

West-Central Chile. The area of west-central Chile between

33° and 34° south latitude was studied by Levi (1960, 1973), Herm (1967), Piraces (1976), and Nasi and Thiele (1982), and between 32° and 33° south latitude by Thomas (1958), Carter (1963), Carter et al. (1961), and Godoy (1982). A strip from the Coastal to the Principal Cordillera at 34°30′ south latitude was described by Vergara (1969).

The Cretaceous sequence here (Table 17) has at its base marine strata intercalated with volcanic and volcanoclastic rocks: the Lo Prado Formation (including the Patagua and Pachacama formations of Thomas, 1958; Aliste et al., 1960; Piraces, 1976). The Lo Prado Formation consists of 3,500 to 6,800 m of acid pyroclastic rocks, mostly ignimbrite and brecciated lava intercalated with fine- to medium-grained sandstones, conglomerates, limestones, and andesitic to dacitic lavas. The Lo Prado Formation, which is synonymous with the La Lajuela Formation (Vergara, 1969), has yielded marine invertebrates of Late Berriasian–Hauterivian age and is conformably overlain by 5,500 to 7,500 m of porphyritic andesitic flows with intercalated continental red sandstones and conglomeratic sandstones of the Veta Negra Formation. Radiometric ages (K-Ar) of plutonic intrusives and andesitic flows indicate 80 and 105 Ma (Vergara and Drake, 1979a,b). The Veta Negra Formation is considered equivalent to the Colimapu Formation (Fig. 11). The Veta Negra Formation is unconformably overlain by 3,500 to 6,000 m of conglomerates, sandstones, shales, and dark brown andesitic flows of the Las Chilcas Formation. The Las Chilcas Formation has yielded plant remains (Torres, 1982; Table 2). Some freshwater bivalves suggest a Turonian age, although absolute ages of plutonic intrusives indicate 110 to 120 Ma (Drake et al., 1976; Vergara and Drake, 1979a,b), and algae remains could indicate an Aptian-Albian age (Godoy, 1981). Marine fossils (Corvalán and Vergara, 1980) suggest correlation to the Upper Cretaceous sediments of Algarrobo (see under Navidad Basin; Fig. 20) and to the Colimapu and ?Abanico formations of the Principal Cordillera (Fig. 11; Table 17).

Bounding or conformably(?) overlying the Las Chilcas Formation (Thomas, 1958; Godoy, 1981; Nasi and Thiele, 1982) are 700 to 3,000 m of green to pinkish continental andesitic, dacitic and rhyolitic flows, tuffs, and breccias intercalated with sandstones and conglomerates of the Lo Valle Formation (Fig. 11; Table 17). This unit has yielded K-Ar ages of 7.18 ± 0.5 and 64.6 ± 0.5 to 77.8 ± 1.0 Ma, indicating a Late Cretaceous–Early Tertiary age (Drake et al., 1976; Vergara and Drake, 1978, 1979a,b; Thiele et al., 1980). The Lo Valle Formation interfingers to the south with the Abanico Formation (Godoy, 1982).

Coquimbo area. In the Principal Cordillera between 29°33′ and 30°15′ south latitude (Fig. 10a), the Cretaceous is restricted to continental and volcanic rocks. According to Thiele (1964) and Dedios (1967, 1978), these rest unconformably on the Jurassic. The sequence (Table 17) begins with gray to reddish conglomerates, sandstones, shales, tuffs, and breccias of the Pucalume Formation that contain ostracods and have a variable total thickness of 150 to 2,000 m. It is dated as Barremian-Albian in age and is regarded as equivalent to the Colimapu Formation farther south. To the south (Rivano, 1980, 1984; Olivares et al., 1985) the Pucalume Formation rests conformably on Neocomian volcanoclastics and on marine calcareous shales and sandstones of the Río Tascadero Formation. The Viñita Formation lies unconformably on the Pucalume Formation and consists of 3,000 to 5,000 m of continental conglomerates, sandstones, andesitic flows, and tuffs. To the west, near Ovalle (Fig. 10a), this unit has yielded dinosaur bones (Casamiquela et al., 1969), suggesting a Maastrichtian age. The Viñita Formation has also been regarded as an equivalent of the Abanico and Coya Machalí formations farther south (Table 17). The Viñita Formation is unconformably overlain by the Lower Tertiary continental volcanics of the Los Elquinos Formation.

Toward the west of the Principal Cordillera, between 29°

and 31° south latitude (Fig. 10a), extensive Cretaceous outcrops have been described by Aguirre and Egert (1965, 1970), Thomas (1967), and Palmer et al. (1980). Here in the southwest, the base of the Cretaceous sequence (Table 17) contains andesites, breccias, and tuffs called the Tamaya Formation. Toward the top of the sequence and to the northeast these volcanic rocks are intercalated with fossiliferous marine limestones and continental red sandstones known as the El Reloj Formation; associated bivalves indicate a Valanginian age. Toward the east-northeast, the volcanic rocks are overlain by the Arqueros Formation. This unit is 2,000 to 5,000 m thick; it consists of three volcanic members including andesites, conglomerates, and breccias, interbedded with two primarily sedimentary members with marine limestones, calcareous sandstones, and shales. This formation has been dated as Hauterivian-Barremian in age on the basis of the ages of some granitic intrusives (125 ± 15 Ma) and by the shallow marine fauna. The whole sequence has been named the Ovalle Group by Thomas (1967) and is considered equivalent to the Lo Prado Formation farther south.

The Arqueros Formation is conformably overlain by 750 to 1,250 m of reddish and purplish continental conglomerates, sandstones, and shales intercalated with andesites of the Quebrada Marquesa Formation. It can be divided into four members, the lowest of which bears shallow-water marine invertebrates. The upper member contains plant remains and seems to be continental in origin. The Quebrada Marquesa Formation has been dated as Aptian–Early Albian on the basis of marine bivalves and is considered equivalent to the Veta Negra Formation (Table 17) on the basis of its fauna and stratigraphic relationships. It reaches 1,900 m in thickness toward the east. According to Palmer et al. (1980), K-Ar analysis yielded dates between 61.3 and 81.7 Ma, although it is possible that these values are too high due to argon loss caused by Cretaceous-Tertiary intrusions.

The Quebrada Marquesa Formation is unconformably overlain (Table 17) by 70 to 80 m of basal conglomerates belonging to the Viñita Formation. This unit attains its maximum thickness of 1,000 to 1,500 m toward the east, where it consists of sandstones and conglomerates; in its upper part, the Viñita Formation contains gray andesitic lavas and reddish-gray tuffs. This unit is largely if not entirely nonmarine and has yielded some poorly preserved fresh-water gastropods and petrified wood (Table 2; Torres and Rallo, 1981). Dinosaurs (Casamiquela et al., 1969) suggest a Maastrichtian age. A K-Ar analysis has given a date of 64.7 Ma (Palmer et al., 1980). The Viñita Formation is unconformably covered by the Tertiary volcanic and volcanoclastic rocks of the Los Elquinos Formation.

Copiapó area. Between 26°45′ and 29° south latitude (Fig. 10a), the succession has been described in a series of papers by Biese (1942, 1958; see also Groeber, 1953), Segerstrom (1959, 1960, 1961, 1962, 1967, 1968), Segerstrom and Parker (1959), Segerstrom et al. (1963), Segerstrom and Ruiz (1962), Segerstrom and Moraga (1964), Corvalán (1974), and Abad (1976, 1980). The Lower Cretaceous marine sequence of the area has been studied by Jurgan (1977a,b); the general geology has been reviewed by Jensen and Vicente (1977), Moscoso (1984), and Cisternas et al. (1985), and the ammonite zonation has been developed by Tavera (1956), Corvalán (1974), and Wiedmann (in Jurgan, 1977a,b). Other marine invertebrates were described by Paulcke (1903).

The complete Lower Cretaceous sequence (Table 17) is present at the latitude of Copiapó (Fig. 10a) toward the south-southeast, where it begins at the base with 680 m of andesitic flows of the Punta del Cobre Formation; its base is not exposed. This unit is disconformably overlain by 60 to 400 m of dark gray to greenish limestones, with a few sandstone and conglomerate intercalations assigned to the Abundancia Formation. This formation has yielded ammonites of Late Valanginian age. It is conformably overlain by 875 m of gray limestones, silty shales, sandstones, and conglomerates of the Nantoco Formation, which bears Early Hauterivian ammonites. Conformably overlying the Nantoco Formation are 150 to 500 m of thin-bedded calcareous siltstones with Late Hauterivian ammonites belonging to the Totoralillo Formation. This marine sequence ends with the Barremian-Aptian? Pabellón Formation, which consists of as much as 2,000 m of thick-bedded gray-green limestones with intercalated sandstones, conglomerates, and breccias, yielding echinoids (Table 5; Paulcke, 1903; Larrain, 1975). The Valanginian-Aptian sequence has been named Chañarcillo Group, although the Punta del Cobre Formation has been excluded by Jurgan (1977a,b) due to its unconformable relationship with the overlying units.

The Chañarcillo Group interfingers to the east-northeast and west (Moraga, 1977) with the volcanic and continental rocks of the Bandurrias Formation (Table 17); the Bandurrias Formation also laterally replaces and partially overlies the Chañarcillo Group toward the northeast. The lower units, the Punta del Cobre and Abundancia Formations, are clearly defined toward the south (27°30' to 27°45' south latitude), where the Bandurrias Formation is present east and west of the marine belt. The entire group cannot be readily divided farther south where marine strata of Late Hauterivian–?Late Barremian age are intercalated with volcanic and detritic facies (Pincheira and Thiele, 1982). The clastic content of the Pabellón Formation increases from south to north, whereas facies changes in the Barremian attest to a regressive phase of sea level.

The Bandurrias Formation consists of 3,000 m of andesitic flows and breccias, cross-bedded red sandstones and conglomerates. It rests in the east on the Upper Jurassic La Ternera Formation, and in the west on the Jurassic La Negra Formation. The lower part has been dated on ammonites as Late Valanginian–Barremian, whereas the upper part is considered post-"Neocomian" and equivalent to the Colimapu Formation farther south (Table 17). Sediments of the Bandurrias Formation are still present as far north as 25°40' south latitude (Fig. 10a) (Davidson and Godoy, 1976; Mercado, 1978; Naranjo, 1981).

Toward the east, at about 29° south latitude, Reutter (1974) and Moscoso (1976) recognized the presence of 1,500 m of andesitic flows, breccias, tuffs, volcanic conglomerates, and sand-

stones. This unit, the Picudo Formation, rests conformably on the Jurassic Lautaro Formation, and has been dated as Upper Jurassic–?Lower Cretaceous (Reutter, 1974).

The Bandurrias Formation and all other older units are unconformably overlain by 4,500 m of continental conglomerates, andesites, tuffs, and breccias of the Cerrillos Formation (including the Pircas Formation of Mercado, 1982). This is older than 105 ± 10 Ma, according to Pb α radiometry of intruded granite (Segerstrom, 1968), and stratigraphic relationships indicate an Aptian–early Late Cretaceous age (Rivera, 1985); Abad (1976), however, considered that the Cerrillos Formation spans Upper Turonian–Maastrichtian time. Where the Bandurrias and/or Chañarcillo Group are absent, south of 27°45' south latitude, the Cerrillos Formation rests unconformably on the Upper Jurassic Lautaro Formation. The Cerrillos Formation is conformably overlain by continental sediments of the Hornitos Formation (Rivera, 1985). The Hornitos Formation bears dinosaur remains (Chong, 1985). According to Rivera (1985) is Aptian–early Late Cretaceous in age and is unconformably covered by 1,500 m of Upper Cretaceous breccias and sandstones with intercalated andesitic flows: the Venado Formation (Sepúlveda and Naranjo, 1982).

North of 27°30' south latitude, the Chañarcillo Group grades laterally into continental sediments, volcanoclastics, and andesitic flows of the Bandurrias, Quebrada Monardes, Quebrada Seca, and Quebrada Paipote formations (Mercado, 1982; Sepúlveda and Naranjo, 1982; Bell and Suárez, 1985). South of the Salar de Pedernales (Fig. 10), basaltic andesites—comparable to those present in the Cerrillos Formation—unconformably overlie Upper Jurassic volcanics and are unconformably covered by Middle Tertiary volcanics (Cisternas and Oviedo, 1979). Between Chañaral and Taltal (Fig. 10a), the Cretaceous is present with up to 3,000 m of andesitic breccias and flows, with intercalated limestones containing Neocomian (?Valanginian) fossils: the Cerros Floridas beds (Naranjo, 1978) or Aeropuerto Formation (Naranjo and Puig, 1984). The base and top of this unit are unknown.

Domeyko Range. The Domeyko Range (Fig. 10b) extends between 22°20' and 26° south latitude and includes an important Cretaceous sequence with both marine and continental facies. After the pioneering work of Harrington (1961), the general stratigraphy of this range has become better known through the work of different authors (see below). To the south, a well-exposed section is present at Quebrada Asientos, west of the Salar de Pedernales (Fig. 10a). According to Harrington (1961), the Cretaceous succession begins at the base with about 50 m of unfossiliferous shales and coarse sandy tuffites that are faulted against Jurassic units. This unit, named the Potrerillos Formation, is overlain by about 405 m of soft green shales of the Pedernales Formation (Table 17), locally grading upward into sandy shales and shaly calcareous sandstones of brownish, purple, and chocolate colors, and grayish-blue massive limestones. In the upper part, the Pedernales Formation contains Tithonian-Valanginian faunas (Reyes and Pérez, 1985). The next youngest unit in this area consists of Lower Cretaceous andesitic flows and tuffs, and, northeast of the Salar de Pedernales, of 965 m of continental reddish-brown argillaceous sandstones and siltstones named the Agua Helada Formation (Table 17; García, 1967).

Marine Cretaceous strata appear to extend northward to Santa Ana and Candeleros Hills (Fig. 10a,b; Chong, 1976, 1977; Naranjo and Covacevich, 1979; Naranjo, 1981; Naranjo and Puig, 1984), where are found 1,000 m of marine continental clastic sequences with andesitic intercalations: the Santa Ana Formation. It yields few bivalves, gastropods, and echinoids in the lower part, and has been assigned a Tithonian-Early Cretaceous age range. These rocks have also yielded the pterosaur *Pterodaustro guinazui* Bonaparte (Casamiquela and Chong, 1980), originally described from San Luis province, Argentina (see under San Luis Basin). Above these marine levels there are lagoonal facies containing pholidophorid fishes and poorly preserved plants. The whole sequence is overlain by red sandstones, conglomerates, and mudstones with silicified wood remains and intercalated andesites. These are, in turn, unconformably overlain by the volcanic rocks of the Augusta Victoria Formation. The Cretaceous sequence is unconformably covered by Tertiary pyroclastic rocks.

Farther north, at Almeida Hill (Fig. 10b; Table 17), the Cretaceous is represented by andesitic flows, tuffs, and sediments of Malm-"Neocomian" age. Andesitic rocks (Late Jurassic?) are overlain by about 600 m of greenish and yellowish tuffs, sandy tuffites, and brecciated tuffs of the Pular Formation. In the middle part the Pular Formation has thin intercalations of black bituminous shales with abundant remains of the phyllopod *Cyzicus*. It has been dated as Tithonian?-Lower Barremian and is overlain by 1,950 m of thick-bedded red conglomerates and sandstones, with interbedded red shales: the Pajonales Formation (?Late Barremian–Early Senonian in age).

In the northern part of the Domeyko Range, the Cretaceous is represented by a thick, sandy conglomeratic group unconformably overlying Upper Paleozoic or Triassic continental sediments intercalated with volcanics (Dingman, 1963; Moraga et al., 1974; Ramírez and Gardeweg, 1982). These Cretaceous beds are well developed in Purilactis (Fig. 10b), where they have been named the Purilactis Formation. This unit is about 1,000 m thick and is unconformably overlain by a younger sequence of sandstones, conglomerates, andesites and breccias of Tertiary age. The Purilactis Formation bears the algae-like structure known as *Pucalithus,* which is characteristic of the Puca or Salta Group of Bolivia and northern Argentina (see under Aimará Basin). Thus, Brüggen (1950) has considered them as contemporaneous units (Reyes, 1972; Schwab, 1984).

West of the Domeyko Range (Fig. 10b), between 23° and 25° south latitude, there are more than 2,000 m of rhyolites, andesites, dacites, and minor continental clastic rocks that García (1967) placed in the Augusta Victoria Formation (Table 17). They appear to consist of at least two units that, farther south, were placed in the Cerrillos and Hornitos formations (see under Copiapó area), and range from Late Cretaceous to Early Tertiary

in age (Ferraris, 1976). The upper part of the Augusta Victoria Formation extends westward to the Tocopilla area, north of Antofagasta (Fig. 10), where K-Ar analysis on rhyolites gave values of 109 ± 2 Ma and 114 ± 3 Ma (Ferraris, 1978) and 78 ± 3 Ma (Boric, 1981).

To the east of the continental facies of the Purilactis Formation, at about 22°15' south latitude and 68°15' west longitude, between the Domeyko and the Andean Cordilleras, a 500-m-thick continental sequence, the Lomas Negras Formation (Marinovic and Lahsen, 1984), consisting of clastic and andesitic rocks, includes 15 m of marine limestones with bivalves, fishes and foraminifers. These marine levels have been considered as an equivalent of the Maastrichtian-Paleocene El Molino and Yacoraite formations of northern Argentina and Bolivia (see under Aimará Basin) (Marinovic and Lahsen, 1984; Salfity et al., 1985).

Quebrada El Way. The marine Lower Cretaceous sediments of Quebrada El Way in the Coastal Cordillera, south of Antofagasta (Fig. 10b), were studied by Brüggen (1950), Harrington (1961), Alarcón and Vergara (1964), García (1967), Fricke and Voges (1968), Jurgan (1974), Ferraris and Di Biase (1978), and Flint et al. (1986a,b). Jurassic andesitic lavas of the La Negra Formation are unconformably covered by 2,000 m of conglomerates and red sandstones, representing an alluvial fan/fan delta complex. The transgressive sequence is completed by 600 to 660 m of fossiliferous calcareous rocks. This succession was previously divided, from below, in the Caleta Coloso and El Way formations (Table 17). Lately the Caleta Coloso Formation has been restricted to the basal alluvial conglomerates, whereas the fan delta complex has been named Lombriz Formation. The overlying fossiliferous beds have been included in the Tableado Formation. The name Way Group has been proposed for the whole sequence.

The El Way Formation has a thickness of 600 to 660 m and contains marine invertebrates that have been described by Leanza and Castellaro (1955), Alarcón and Vergara (1964), Jurgan (1974), and Larrain (1975). Only the Hauterivian-Barremian stages appear to be represented, and the supposed presence of Albian (Leanza and Castellaro, 1955; Harrington, 1961) is suspect (Ferraris and Di Biase, 1978). The sequence consists, in ascending order, of 100 to 150 m of brown and yellow marls and marly to sandy limestones, 200 m of gray-yellow calcareous shale and calcareous sandstones, 250 m of calcareous limestones with intercalated marls, and 60 m of conglomerates with intercalated sandy and marly limestones. The El Way Formation can be correlated to the south with the Pabellón, Totoralillo, and Nantoco formations of the Chañarcillo Group (see under Copiapó area). Immediately to the east, only the Late Cretaceous–Early Tertiary volcanics of the Augusta Victoria Formation are present.

Chacarilla-Mamiña and Volcán Miño areas. The regional geology of the central region of Tarapacá Province in northern Chile has been studied by Galli (1957, 1968), Galli and Dingman (1962), Dingman and Galli (1965), and Thomas (1967).

The Cretaceous that is exposed in the Chacarilla-Mamiña area (Fig. 10b) between 20°45' and 20° south latitude rests unconformably on the Jurassic Chacarilla Formation. The Cretaceous consists of (Table 17) a thick sequence of continental sedimentary rocks and lava flows with poorly stratified reddish gray breccias and conglomerates and gray to grayish violet trachyte flows lying below and above it. The middle part of this sequence, called the Cerro Empexa Formation, includes beds of grayish red fine-grained sandstones and an irregular bed of gypsum. The Cerro Empexa Formation is about 600 m thick and could be of Albian-Turonian age or older, according to the 95 ± 10 Ma Pb α determination of intrusives in the San Juan Morales area (Galli, 1968). It is unconformably overlain by the Late Tertiary–Pleistocene strata of the Altos de Pica Formation.

Toward the southeast (Fig. 10b), at about 21° to 22° south latitude and between 69° and 68°30' west longitude, north, west, and south of Volcán Miño, continental Upper Jurassic (Table 17) strata are unconformably overlain by andesitic and pyroclastic rocks of Tithonian-Hauterivian age, and these are overlain by a rhyolitic sequence of Middle to Late Cretaceous age. These are in turn unconformably overlain by Upper Cretaceous–Lower Tertiary red psammites (Maksaev, 1978, 1979; Vergara, 1978a,b; Skarmeta and Marinovic, 1981). The 1,750 m of andesitic breccias and andesites of the Arca and Macata Formations, at the base of the sequence, are assumed to be of Tithonian-Hauterivian age because they unconformably underlie 1,200 to 2,500 m of rhyolitic and dacitic rocks of the Peña Morada and Colla Huasi formations with K-Ar radiometric age dates of 89.8 ± 2.3 to 114.52 ± 2.82 Ma. The Arca Formation is intruded by diorites with a K-Ar radiometric age date of 68.0 ± 2.5 (Marinovic and Lahsen, 1984). This unit is in turn unconformably overlain by 1,000 m of red conglomerates and sandstones of the Tolar Formation, a possible equivalent of the Purilactis Formation (see under Domeyko Range) of Late Cretaceous–Early Tertiary age (Maksaev, 1978).

Iquique-Arica area. Between Iquique and Arica (Fig. 10b), the Cretaceous is known through the work of Cecioni and García (1960a,b), Salas et al. (1966), Tobar et al. (1968), and Thomas (1970) (Table 17). The basal Jurassic Guantajaya Formation, mostly continental, but also containing Tithonian ammonites (Cecioni, 1961), is unconformably overlain by at least 900 to 1,400 m of continental light colored sandstones, conglomerates, and red siltstones of probable lacustrine origin, named the Atajaña Formation. Along the coastline east of Arica, about 690 m of breccias and andesitic flows of the Sausine Formation are considered to be a lateral equivalent; both formations are included in the Vilacollo Group. The Atajaña Formation thins eastward, and also occurs south and southeast of Iquique. Immediately north of Iquique (Fig. 10b; Table 17), the Atajaña Formation grades upward into 400 m of greenish gray siltstones, calcareous sandstones and conglomerates, and brown laminated shales. Oyster beds in the lower part contain ammonites of probable Berriasian age. This unit, named the Blanco Formation, is unconformably overlain by 2,170 m of porphyritic lava flows

with sheets 3 to 10 m thick; these flows contain a few thin breccias and lenticular red sandstones that seem to wedge toward the west. South of 20° south latitude (Fig. 10b), these rocks (Suca Formation) appear to be replaced by 1,600 m of andesitic breccias of the Punta Barranco Formation that rest unconformably on the Jurassic Caleta Ligate Formation. Where the Blanco Formation is absent, the Suca Formation rests unconformably on the Atajaña Formation. The latter formation is placed in the Upper Cretaceous and is unconformably overlain by Pleistocene rocks.

In the western area of Arica there are about 1,000 m of breccias, tuffs, and andesitic flows interbedded with conglomerates and sandstones assigned to the Lupica Formation (Salas et al., 1966). This unit rests unconformably on ?Precambrian slates and conformably underlies Oligocene rocks. The Lupica Formation is considered Late Cretaceous–early Tertiary in age and is perhaps a correlative of the Suca Formation.

COLORADO BASIN

The central eastern Argentina area (Fig. 17), located between the Somuncurá Massif and southern Brazil, is characterized by two main fault systems trending east-northeast–west-southwest and east-southeast–west-northwest. The basement was thus block-faulted, and differential movements between the blocks gave rise to several basins. Transverse faults occur within the basins and are responsible for secondary basins separated by basement highs. Movement along these faults which began in Late Jurassic time, was related to the opening of the South Atlantic and intrusive and extrusive basic magmatic activity that persisted to Early Cretaceous time. These rocks are extensively developed in the Paraná Basin (see under Paraná Basin) (Zambrano and Urien, 1970).

The southernmost basin in this region is the Colorado Basin, primarily represented offshore and known through borehole data (Figs. 1, 18, 19). South of the Colorado Basin are the small Valdés and Rawson Basins (Fig. 19), where a Cretaceous sequence similar to that of the Colorado Basin is concealed by Cenozoic sediments (Zambrano and Urien, 1970; Urien et al., 1981; Bianchi, 1984). The stratigraphy of the Colorado Basin has been studied by Ewing et al. (1963), Kaaschieter (1963, 1965), Ludwig et al. (1965), Zambrano (1971, 1972, 1975, 1980a), Zambrano and Urien (1970, 1975), Yrigoyen (1975b), Lesta et al. (1979), and Urien et al. (1981).

The Cretaceous sequence in the Colorado Basin (Table 18, in pocket inside back cover) begins at the base with about 1,000 m of predominantly dark brown, purplish, or reddish (partly with greenish decoloration zones) conglomerates, sandstones, siltstones, claystones, and shales of fluvio-lacustrine origin, named the Fortín Formation. The Fortín Formation is distributed in three separate, fault-bound depocenters (Uliana and Biddle, 1987). It rests unconformably on Precambrian or Paleozoic, and is considered Late Jurassic–Early Cretaceous (Lesta et al., 1979) or Cenomanian-Turonian in age (Zambrano, 1980a) on the basis of palynological analysis; it is correlative to the Arata Formation of the Macachín Basin (see under Macachín Basin).

The Fortín Formation is unconformably overlain by the Upper Cretaceous Colorado Formation. This unit consists of 85 to 1,925 m of reddish, brown, green, or variegated claystones and siltstones interbedded with fine- to coarse-grained, cross-bedded sandstones of fluvio-lacustrine to shallow marine origin. The Colorado Formation is restricted to the axial parts of the basin, although it is more widespread than the underlying Fortín Formation (Uliana and Biddle, 1987). A Turonian-Santonian, ?Campanian age for these rocks was based on foraminifers and microplankton (Table 4; Gamerro and Archangelsky, 1981) found in the upper part of the sequence. The Colorado Formation is considered equivalent to the Abramo Formation of the Macachín Basin (see under Macachín Basin).

Conformably (Zambrano, 1974, 1980a) or unconformably (Yrigoyen, 1975b; Lesta et al., 1979) overlying the Colorado Formation are 10 to 204 m of gray, greenish, or dark gray, usually calcareous marine claystones and siltstones with Maastrichtian-Paleocene microfossils (Table 4; Kaaschieter, 1965; Gamerro and Archangelsky, 1981), named the Pedro Luro Formation. This unit is partly covered by pyroclastic and basaltic rocks of the Ranquél Formation, and by continental and marine Tertiary sediments.

MACACHÍN BASIN

Between about 35°30′ and 38° south latitude and 63° to 64° west longitude (Figs. 18, 19), there is a north-northwest–trending basin that may be connected with the Colorado Basin (see under Colorado Basin). This basin is filled with Mesozoic and Cenozoic rocks, with the Cretaceous known from the subsurface (Salso, 1966; Yrigoyen, 1975b; Russo et al., 1979). At the northwest margin of this basin, Precambrian or Lower Paleozoic pelitic and metamorphic rocks are unconformably covered by 120 m of fine- to medium-grained, reddish gray continental sandstones with intercalated dark gray clays; these strata become progressively more clayish upward. This sequence, the Arata Formation, was first considered Permo-Triassic, but later (Yrigoyen, 1975b) included in the Cretaceous. The Arata Formation is in turn unconformably overlain by Plio-Pleistocene sediments. Toward the southeast of the basin the sequence begins at the base with 300 m of red-brown sandy siltstones and light colored, fine-grained continental sandstones and intercalated clays named the Abramo Formation of Cretaceous age. This unit is overlain by 500 m of green siltstones and sandstones with intercalated clays and volcanic glass of the Macachín Formation of Oligocene-Miocene age; the grain size decreases upward. According to Salso (1966), the Abramo and Macachín formations are partially or totally marine in origin.

SALADO BASIN

The Salado Basin (Figs. 18–20) is a 5,000-m-deep graben-like feature (Introcaso and Ramos, 1984) related to northwest-

southeast–trending faults. To the south it is bound by a composite horst extending east-southeast, the Northern Hills of the Buenos Aires Province. To the north the graben is limited by the Martín García High, which separates it from the Santa Lucía and Paraná Basins (Fig. 17). The Salado Basin is probably connected, to the west and south, with the Macachín (see under Macachín Basin) and Colorado (see Colorado Basin) Basins (see Zambrano and Urien, 1970; Bracaccini, 1970, 1980). To the northwest a possible connection with the Paraná Basin exists through the Laboulaye subbasin (see under Chaco–Pampean Plain) (Zambrano, 1974; Bracaccini, 1980).

The Cretaceous sedimentary sequence of the Salado Basin (Table 18) rests unconformably on Precambrian or Paleozoic basement and/or Upper Jurassic–Lower Cretaceous basaltic flows in different areas. At the base it contains about 3,500 m of reddish, brown, and purple cross-bedded continental sandstones and claystones of the Río Salado Formation. The deepest subsurface levels reached by drilling (–2,000 m) contain a microflora no older than Aptian (Zambrano, 1974), and thus the age of the Río Salado Formation is believed to be Cenomanian (Yrigoyen, 1975b). The complete sequence has also been considered correlative to the Late Cretaceous–Paleocene Mariano Boedo Formation of the Chaco-Pampean plain (see under Chaco-Pampean Plain) (Bracaccini, 1980).

The Río Salado Formation is unconformably overlain by 890 m of coarser reddish, greenish, or variegated continental sandstones intercalated with claystones and conglomerates. This unit, the General Belgrano Formation, is considered Late Cretaceous by some authors (Zambrano, 1974; Yrigoyen, 1975b) and correlative to the Abramo and Colorado Formations of the Macachín (see under Macachín Basin) and Colorado (see under Colorado Basin) basins, and perhaps also to the Guichón Formation of the Paraná Basin (see under Paraná Basin) (Zambrano and Urien, 1970). Bracaccini (1980), however, regarded it as lower Tertiary on the basis of regional correlations.

The General Belgrano Formation is conformably overlain (Table 18) by 1,190 m of marine claystones and siltstones with anhydrite and gypsum of the Las Chilcas Formation. This unit is considered to be Maastrichtian-Paleocene and correlatable with the Pedro Luro and Mariano Boedo formations of the Colorado Basin and Chaco-Pampean plain (see under Colorado Basin and Chaco-Pampean Plain) by Zambrano and Urien (1970), Zambrano (1974), and Yrigoyen (1975b), although Bracaccini (1980) placed it in Oligocene–Upper Pliocene time.

SANTA LUCÍA–LAGUNA MERÍN BASINS

In southeastern Uruguay there is a system of faulted basement blocks trending northeast-southwest and extending from Colonia to Puerto Gómez (Figs. 17, 18), which includes at least two main tectonic grabens, the Santa Lucía Graben in the southwest and the Laguna Merín Graben in the northeast. The former trends N60°E-N70°E, and the latter N40°E. The Laguna Merín represents the southern part of the Pelotas Basin (see under

Pelotas-Santos Basins). The Santa Lucía Basin has also been named Canelones Basin (Jones, 1956; Urien and Zambrano, 1973).

These tectonic basins were successively filled by Cretaceous volcanic and sedimentary rocks (Table 18). The volcanic rocks, known as Puerto Gómez Formation (Caorsi and Goñi, 1958, = Canelones and Treinta y Tres flows of Jones, 1956), consist mostly (90%) of spilitic basalts and andesites dated by K-Ar as 130 ± 3 Ma. Other, geographically more restricted, volcanic rocks consist of rhyolites, trachytes, and syenites, and they have been assigned to the Arequita Formation (Bossi, 1966, = Aigüa and Lascano Series of Walther, 1927; Caorsi and Goñi, 1958; including the Valle Chico Formation of Ferrando and Fernández, 1971). These latter rocks overlie the Puerto Gómez Formation, and K-Ar radiometry indicates an age range between 112 and 130 Ma.

In the Santa Lucía Basin these two formations are overlain, and in part interbedded with (Sprechmann et al., 1981), aeolian to lacustrine sandstones intercalated with conglomerates and shales that reach a maximum thickness of 2,400 m and exhibit lateral and vertical changes. On the basis of these lithologic changes, Jones (1956) recognized three different facies—the Migues, Montes, and Tala-facies—that were later included by Bossi (1966) in the Migues Formation. This unit bears Lower Cretaceous (Aptian-Albian) pollen and microfossils (Goñi and Hoffstetter, 1964; Bossi, 1966; Sprechmann et al., 1981). Similar facies are also present in the Laguna Merín Basin. The Migues Formation can be regarded as correlative to the Fortín and Río Salado formations of the Colorado (see under Colorado Basin) and Salado (see under Salado Basin) Basins.

According to Zambrano (1974), the Migues Formation overlies lacustrine bituminous black shales intercalated with gray and green siltstones and fine-grained sandstones of the Castellanos Formation; palynological data indicate that these strata are of Early Cretaceous and ?Late Jurassic age. Above the Migues Formation, Zambrano (1974) also reported the presence of coarse red sandstones and conglomerates that could be equivalent to the Asencio Formation of the Paraná Basin (Sprechmann et al., 1981; see under Paraná Basin).

PELOTAS-SANTOS BASINS

The Brazilian coastline is broken up into a series of Mesozoic-Cenozoic sedimentary basins overlying a Precambrian, or locally a Paleozoic, basement, and extending eastward beneath the Atlantic Ocean (Figs. 18–20).

The basins were formed during Early Cretaceous time by differential movements, the Wealdeanian reactivation (Almeida, 1972), of the Brazilian platform along faults parallel to Late Precambrian alignments. The southernmost Pelotas Basin lies south of Porto Alegre and adjacent to Uruguay (Figs. 18–20). To the north and south the basin is limited by the Florianópolis and Polonio basement highs (Urien et al., 1981). Farther north lies the Santos Basin situated between Florianópolis and Río de Janeiro.

The tectonic and stratigraphic features of one or both basins have been dealt with by Asmus and Ponte (1973), Asmus and Porto (1972), Campos et al. (1974), Almeida (1976), Francisconi and Kowsmann (1976), Ponte and Asmus (1976, 1978), Soares and Landim (1976), Gonçalves et al. (1979), Kumar and Gambôa (1979), Asmus (1981, 1984), Urien et al. (1981), Ojeda (1982), and Williams and Hubbard (1984).

The sequence (Table 18) in this region, from base to top, begins with a section of terrigenous medium- to coarse-grained sandstones, with intercalated shales and subordinate limestones. This unit has an average thickness of 300 to 500 m and exhibits strong facies changes associated with a deltaic-lacustrine complex. On the basis of a diverse nonmarine ostracode fauna found in the northern basins of Brazil, this sequence has been ascribed to the "Neocomian." It seems to be well represented in the Santos Basin, where it has been named Guaratiba Formation. Here it rests on basaltic rocks (121 ± 11 Ma) of the Cananeia Formation, which are probably related to the extensive Serra Geral lava flows of the Paraná Basin (see under Paraná Basin).

The Lower Cretaceous succession in the Santos Basin is overlain by 200 to 800 m of evaporites, euxinic shales, and sandstones; the evaporites are not present in the Pelotas Basin, but north of the Río Grande Rise (Fig. 18) they form a continuous belt, 700 km wide in the south and progressively narrower toward the north. These rocks, called the São Vicente Formation in the Santos Basin, contain nonmarine ostracods, palynomorphs, planktonic foraminifers, mollusks, and fish remains. Stratigraphic relationships (Ojeda, 1982) and palynology indicate an Aptian age. Above this there is a clastic (Florianópolis Formation) and carbonatic (Guarujá Formation) transgressive marine sequence about 400 m thick. The Forianópolis and Guarujá formations range in age from the Early Albian to Cenomanian. Overlying the Guarujá Formation is an Upper Cretaceous regressive marine sequence (1,300 to 5,000 m thick); generally it represents continental margin progradation but includes evidence of several minor transgressive episodes. The sequence passes from nonmarine coastal plain facies (Santos Formation) to inner shelf and deep marine facies (Santa Catarina or Itajai Formation).

The Upper Cretaceous regressive sequence of the Santos Basin was caused by the large volume of sediments shed from the rising Serra do Mar hinge line (Williams and Hubbard, 1984). It differs from a regional picture of continued marine transgression, as represented in the Pelotas Basin and in the other Brazilian marginal basins.

PARANÁ BASIN

The Paraná intracratonic basin covers an area of about 1,600,000 km^2, and extends in a southwest-northeast direction throughout northeastern Argentina, western Uruguay, southeastern Paraguay, and southern Brazil. In the southwest the basin is mostly restricted to the subsurface of the Chaco-Pampean

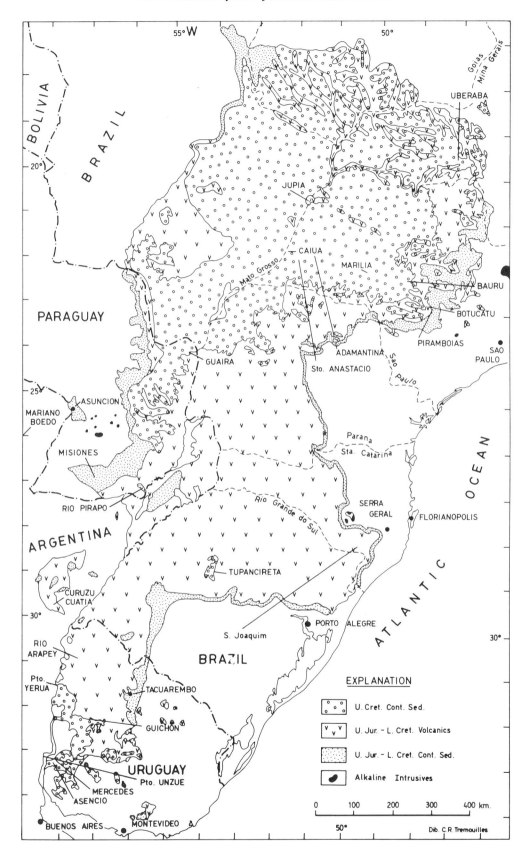

Figure 12. Distribution of major facies of Cretaceous rocks of the Paraná Basin.

Plain (see Chaco-Pampean Plain), whereas in the northeast it is separated from the San Francisco Basin by the Alto Paranaiba Anteclise or Arch (Hasui et al., 1975; Figs. 1, 17). The Paraná Basin is divided by the Ponta Grossa Arch in a northern and southern subbasins. It includes 5,000 m of sediments and basaltic flows of Paleozoic, Mesozoic, and locally, Cenozoic, age, and is floored by Precambrian and Paleozoic rocks.

The general aspects of the stratigraphy of this basin, including the Mesozoic sequence, have been dealt with by many authors: the Brazilian area: Oliveira and Leonardos (1943), Beurlen et al. (1955), Beurlen (1956, 1970), Oliveira (1956), Bischoff (1957), Sanford and Lange (1960), Mendes (1961, 1971), Delaney and Goñi (1963), Barbosa et al. (1964), Bigarella and Salamuni (1967a,b), Mendes and Petri (1971), Bowen (1972), Carraro et al. (1974), Schneider et al. (1974), Figuereido and Bortoluzzi (1975), Millan (1976), Soares (1975, 1981), and Dos Santos et al. (1984); Paraguay: Harrington (1950, 1956a), Hoffstetter and Ahlfeld (1958), Eckel (1959), Putzer (1962), and Banks and Díaz de Vivar (1975); Uruguay: Walther (1919), Lambert (1941), Harrington (1956b), Caorsi and Goñi (1958), Goñi and Hoffstetter (1964), Bossi (1966), and Sprechmann et al. (1981); and Argentina: Frenguelli (1927, 1930a), Padula and Mingramm (1968, 1969), Herbst (1971b, 1980), and Gentili and Rimoldi (1979).

The Cretaceous is represented by continental sediments and volcanics, mostly basaltic flows, that in Brazil have been included in the São Bento Group (White, 1908) and in Uruguay in the "Neo-Gondwana" beds (Falconer, 1937a,b; Caorsi and Goñi, 1958). As originally defined, the São Bento Group included the upper Paleozoic Río do Rasto Formation; it was later enlarged to include the Mesozoic Rosário do Sul, Santa Maria (Triassic), Botucatu, Serra Geral, and Caiuá formations (see Gordon, 1947). Other workers restricted the group to the last three formations (Schneider et al., 1974; Almeida, 1964; Salamuni and Bigarella, 1967; but compare with Figuereido and Bortoluzzi, 1975). Lately the São Bento Group has been considered to include the Pirambóia, Rosário do Sul, Botucatu, and Serra Geral Formations (Soares, 1981) of Middle Triassic to Early Cretaceous age.

The Botucatu Formation (Gonzaga de Campos, 1889) is typically represented at Botucatu Hill, São Paulo State, Brazil (Fig. 12), and covers an area of more than 1,300,000 km^2. The Botucatu Formation extends into western Uruguay and southeastern Paraguay, where it was respectively called the Tacuarembó Formation (Falconer, 1931) and the Misiones Formation (Harrington, 1950; Putzer, 1962). It also extends into northeastern Argentina, where it is known as the Solari Member of the Curuzú Cuatiá Formation (Herbst, 1971b; Gentili and Rimoldi, 1979). In the subsurface of the Chaco-Pampean Plain (see under Chaco-Pampean Plain), it is known as San Cristóbal Formation (Padula and Mingramm, 1969).

The Botucatu Formation consists of about 50 to 320 m of dark red, pink-weathering, fine- to medium-grained and cross-bedded quartzose sandstones (Almeida, 1953; Goñi and Delaney, 1961; Bjornberg and Landim, 1966; Bigarella and Salamuni,

1967b; Bigarella, 1972, 1973a,b; Suguio et al., 1974). It rest unconformably on Precambrian, Paleozoic, and/or Triassic strata. In Bon Retiro do Sul, the Botucatu Formation overlies the Bon Retiro de Sul conglomerates, which in turn rest unconformably on Triassic rocks (Eick et al., 1973; Gamermann et al., 1974).

The Botucatu sandstones were thought to be unconformably overlain by the Serra Geral Formation, although similar sandstones have also been described as being intercalated with the older basaltic lava flows of the Serra Geral. The Botucatu sandstones are mostly aeolian, and, in some places (mostly in the lower part), they are associated with fluviatile or fluvio-lacustrine facies (Pirambóia Formation, or Caturrita and Santana members; Washburne, 1930a; Almeida, 1964; Almeida and Barbosa, 1953; Soares, 1975; Figuereido and Bortoluzzi, 1975). Some authors (e.g., Soares, 1975; Dos Santos et al., 1984) have restricted the Botucatu Formation to the aeolian part of the sequence and named the remaining sandstones the Pirambóia Formation. Schneider et al. (1974) even considered that the Pirambóia Formation unconformably underlies the Botucatu Formation (see also Millan, 1976), within which they also recognized fluviatile sediments. Radiometric age analysis (Rb/Sr) indicated 197 ± 5 Ma for the upper part of the Pirambóia Formation (see Soares, 1981). Soares (1981) has also regarded the fluvial sandstones of the northern and southern areas of the Paraná Basin as distinct units: the Pirambóia and Rosário de Sul formations (including the Río Pardo, Santa Maria, and Caturrita members). Thus, the Pirambóia Formation could be a northern equivalent of the Rosário do Sul Formation, which is present south of the Ponta Grossa Arch (Fig. 17). Therefore, the Botucatu Formation is generally restricted to aeolian strata, although it also includes fluviatile and lacustrine facies. Despite the fact that the lower contact of the Botucatu Formation is not clearly defined and seems to be discontinuous, all of these units are now included, together with the Serra Geral Formation, within the São Bento Group.

Paleowind patterns for the Botucatu Formation trend from west or southwest in the south and from the north-northeast in the north (Bigarella and Salamuni, 1961, 1964); the prevalent winter winds seem to have been from the southwest, with less important northwestern summer winds (Bossi et al., 1978). The only fossils described so far from the "Botucatu sandstones" (Tables 11, 12) are *Conchostraca,* fresh-water ostracods (Almeida, 1950; Mezzalira, 1966; Salamuni and Bigarella, 1967; Souza et al., 1971) and ganoid fishes (Walther, 1932) from the Pirambóia facies. Vertebrate ichnofossils (Leonardi, 1980, 1981) and fossil wood (Souza et al., 1971) occur in the upper aeolian levels. Mones (1980) has also described a Pholydosauridae from the Tacuarembó Formation. Fossil vertebrates indicate Middle to Late Triassic age for the Pirambóia and Rosário do Sul formations, whereas stratigraphic relationships suggest a Jurassic age for the Botucatu Formation (Dos Santos et al., 1984).

The Serra Geral Formation (White, 1908) extends for about 1,200,000 km^2 throughout the Paraná Basin (Fig. 12) and represents one of the most extensive continuous lava fields of the world. The individual tholeiitic basaltic flows, without known pyroclastics (Mendes and Frakes, 1964), vary in thickness from 2 to 100 m, with most values ranging between 30 and 40 m. Average thickness is about 650 m, locally reaching about 1,529 to 1,800 m (see Beurlen, 1970; Bellieni et al., 1983).

The Serra Geral lavas range from labradorite andesites to olivine basalts; they have high iron and alkalies content and low magnesium and calcium content (Baker, 1923; Walther, 1927; Frenguelli, 1927; Leinz, 1949; Teruggi, 1955; Milton and Eckel, in Eckel, 1959; Schneider, 1964, 1970; Freitas, 1964a; Rüegg and Dutra, 1965, 1970; Rüegg and Vandoros, 1965; Leinz et al., 1966, 1968; Melfi, 1967; Cordani and Vandoros, 1967; Issler, 1968, 1969a, b, 1970; Leterrier et al., 1972; Halpern et al., 1974; Sartori et al., 1975; Szubert, 1979; Bellieni et al., 1983). Geochemical trends suggest an eastward migration of volcanism (Bellieni et al., 1983), and that the whole province can be subdivided into at least three different subprovinces (Rüegg, 1970, 1976; Rüegg and Amaral, 1976). Vertical variation in facies has also been shown to exist; at the base, lava flows consist of tholeiitic basalts, whereas the middle and upper lava flows exhibit intermediate to acid compositions (Cordani et al., 1980; Dos Santos et al., 1984). According to Leinz et al. (1968), these volcanic rocks belong to two magmatic cycles—an older intrusive one and a younger intrusive and extrusive one.

The Serra Geral Formation interfingers with the Botucatu Formation; otherwise, it rests unconformably on the Triassic Rosário do Sul Formation. It is unconformably overlain by the Caiuá, Bauru (see below), or Cachoeirinha (Tertiary) formations in different areas. The underlying and intercalated sandstones of the Botucatu Formation and the overlying sandstones of the Caiuá Formation are lithologically alike and are believed to be of the same origin. Some authors (Cordani and Vandoros, 1967; Mendes and Petri, 1971) have included all basalts and sandstones in this region in a single stratigraphic unit, the Botucatu Formation (but see below).

The Serra Geral lavas seem to be related to dikes, sills, and other intrusive bodies of basic diabasic and alkalic igneous rocks in Brazil, Paraguay, and Uruguay (Herz, 1977). In Río Grande do Sul, Brazil, they have been named the Caneleiras Formation (Ribeiro, 1971). They were studied by Marini et al. (1967) and Fúlfaro and Suguio (1967) in the Paraná and São Paulo States of Brazil, where they form dike swarms with a southeast-northwest orientation. Their relationships with the Botucatu sandstones have been dealt with by Giudiccini and Campos (1968). Their chronology in eastern Paraguay was discussed by Comte and Hasui (1971).

The Serra Geral Formation and the related Botucatu sandstone were thought to be Triassic and/or Jurassic (see Harrington, 1950; Eckel, 1959; Putzer, 1962; Goñi and Hoffstetter, 1964). However, K-Ar and Rb-Sr radiometric age analysis showed that the Serra Geral Formation ranges between 100.5 and 147.7 Ma in age, with most values clustering around 120 to 130 Ma (Creer et al., 1965; Amaral et al., 1966, 1967; McDougall and Rüegg, 1966; Bossi, 1966; Melfi, 1967; Vandoros et al.,

1966; Cortelezzi and Cazeneuve, 1967; Hasui and Cordani, 1968; Stipanicic and Linares, 1969; Comte and Hasui, 1971; Sartori et al., 1975; Freitas, 1976; Cordani et al., 1980). The lower basaltic rocks are 130 to 150 Ma, whereas the intermediate and acid lava flows are, respectively, 123 to 130 and 118 to 125 Ma (Dos Santos et al., 1984). This indicates an Early Cretaceous age, although some flows and intrusions, as well as part of the Botucatu Formation, may be Late Jurassic in age (Amaral et al., 1966; Cordani and Vandoros, 1967). Similar rocks with younger ages, i.e., 51 to 82 Ma, have also been recognized to the north and east, where they exhibit a similar northeast trend (see Amaral et al., 1967). These alkaline rocks seem to belong both to an Early Cretaceous (127 to 133 Ma) magmatic phase, and to a Late Cretaceous–Early Tertiary (51 to 82 Ma) phase as developed along the eastern margin of the Paraná Basin in the northern coast of São Paulo State, west of Minas Gerais and southern Goias (Amaral et al., 1967; Hasui and Cordani, 1968; Minioli, 1971; Freitas, 1976).

The Serra Geral basalts have also been named the "Arapey lavas" at a borehole location in northwest Uruguay (Fig. 12). In northeast Argentina, these basalts are included as a member in the San Cristóbal Formation (Padula and Mingramm, 1968), or as the Posadas member of the Curuzú Cuatiá Formation (Gentili and Rimoldi, 1979). Similar basalts are also present in other basins of Uruguay (see Santa Lucía-Laguna Merín Basins), southern (see Pelotas–Santos Basins), and northeastern Brazil; this suggests that Early Cretaceous volcanic activity affected the entire eastern part of the Uruguayan and Brazilian platform (Minioli et al., 1971). The Serra Geral Formation is unconformably overlain by about 250 to 270 m of fine- to medium-grained, cross-bedded red quartzose sandstones of the Caiuá Formation (Washburne, 1930a,b; Bosio and Landim, 1971; Mezzalira, 1964; Freitas, 1973) with fluviatile and aeolian origin. Prevalent winds were probably from the southwest (Petri, 1983).

The Caiuá Formation is present along the Paraná river and its larger tributaries, north of the Ponta Grossa Arch between Jupia, São Paulo State, Brazil (Figs. 12, 18) and in Guaira (Paraná State, Brazil), south-southeast of Mato Grosso, west of São Paulo State, Brazil, and in southeastern Paraguay. The arkosic sandstones were at first believed to be a separate post-Serra Geral unit with unclear relationships to the Bauru Formation (Moraes Rego, 1935; Maack, 1947; Scorza, 1952; Freitas, 1955; Mezzalira, 1964; Landim and Fúlfaro, 1972). They were later incorporated into the Botucatu Formation (Almeida, 1953, 1964; Salamuni and Bigarella, 1967; Cordani and Vandoros, 1967; Mendes and Petri, 1971; Freitas, 1973). Although some authors (Landim and Fúlfaro, 1972) have regarded the Caiuá sandstones as Cenozoic in age, according to Mezzalira and Arruda (1965), Freitas (1973), Mezzalira (1974), Millan (1976), Schneider et al. (1974), and Dos Santos et al. (1984), they lie between the Serra Geral and Bauru Formations and are therefore placed in the Albian-Turonian. These sandstones are thought to have been deposited on the eroded surface of the Serra Geral volcanics (Soares and Landim, 1976; Schneider et al., 1974). The light-colored sandstones of the Río Pirapó Formation (Harrington, 1950), which in southern Paraguay overlie the Serra Geral Formation, could be equivalents of the Caiuá Formation.

Reddish sediments, mostly fluviatile and lacustrine, resting on the Serra Geral, Botucatu, Caiuá, or Middle to Upper Paleozoic formations west of São Paulo, in northeast Minas Gerais, south of Goias, and southeast of Mato Grosso (Fig. 12) have been placed in the Bauru Formation (Gonzaga de Campos, 1905). This unit, extensively reviewed by Mezzalira (1974), Arid (1977), Soares et al. (1980), and Pires (1982), includes in its lower part 230 m of reddish brown fine-grained fluviatile sandstones; the upper part of the formation consists of 100 to 200 m of coarse-grained and conglomeratic calcareous sandstones. Recently (Soares et al., 1980; Soares, 1981), the lower part of the Bauru Formation has been included in the Santo Anastácio and the Adamantina formations (the latter including the São Jose do Río Preto, Araçatuba, Taciba, and Ubirajara facies); the upper part has been named the Marília Formation. In Minas Gerais the Marília Formation grades laterally into tuffaceous sandstones of the Uberaba facies or Formation (Mendes and Petri, 1971; Grossi et al., 1971; Soares, 1981). These are related to alkaline and ultrabasic volcanic activity that occurred between 55 and 85 Ma.

Soares et al. (1980) and Soares (1981) have elevated the Bauru Formation to the rank of group, including, from the base, the Caiuá, Santo Anastácio, Adamantina, and Marília formations (Table 18). There is a disconformity between the Caiuá and Santo Anastácio formations. According to Pires (1982), the stratigraphic relationships and rank of this unit have yet to be established for the whole basin.

The stratigraphic, sedimentary, and tectonic features of the Bauru Formation were discussed by Freitas (1955, 1964b), Björnberg et al. (1970), Mezzalira (1974), Suguio et al. (1975), Barcha and Arid (1977), Soares et al. (1980), and Soares (1981), all of whom considered it to be typically fluviatile in origin. Soares and Landim (1976) and Soares (1981) have related the different facies to the tectonic evolution of the area.

The Bauru Group has yielded bivalves, gastropods, fishes, crocodiles, chelonids, dinosaurids, and plants (Ihering R., 1909; Ihering H., 1913; Huene, 1929a, 1931, 1934a,b; Roxo, 1936; Price, 1945, 1950a,b, 1951, 1953, 1954, 1955; Staesche, 1937, 1944; Petri, 1955; Mezzalira, 1966, 1974; Suárez, 1969; Arid and Vizotto, 1966, 1971, 1975; Estes and Price, 1973; Báez, 1985) (Tables 11, 12), indicating a Late Cretaceous age. K-Ar analysis in intercalated alkaline flows gave ages of 52 to 61 Ma (Coutinho et al., 1982).

According to Soares et al. (1980) and Soares (1981), the saurids found in the upper part of the Adamantina Formation and lower part of the Marília Formation indicate a Santonian-Maastrichtian age. The same authors have also suggested an Aptian-Albian age for the Caiuá Formation, and an Albian-Cenomanian age for the Santo Anastácio Formation. In east-central Mato Grosso, the Bauru Group is unconformably overlain by the Tertiary Cachoeirinha Formation. Rocks with similar lithology and stratigraphic position (Cretaceous-Tertiary) have also been recorded to the south in Río Grande do Sul where they have been named the Tupanciretã Formation (Menegotto et al., 1968) and are considered to be equivalent to some of the Chaco-Pampean deposits (Coulon et al., 1973).

Partially equivalent rocks cropping out in western Uruguay (Fig. 12; Table 18) have been divided, from base to top, into the Guichón sandstones (Lambert, 1940; Bossi et al., 1963), the Mercedes sandstones and conglomerates (Serra, 1945), and the Asencio sandstones (Caorsi and Goñi, 1958, = Sorianense Stage of Kraglievich, 1932). The Guichón sandstones are about 100 m thick and have yielded vertebrates described by Rusconi (1933) and Huene (1934a,b). This unit consists of aeolian, fine-grained, reddish, shaly, massive sandstones with some occasional cross-bedding. It rests unconformably on the Lower Cretaceous Serra Geral Formation and grades upward into 30 to 71 m of unfossiliferous sandstones and conglomerates with intercalated limestones of lighter colors of the Mercedes Formation (includes the Chileno Conglomerates of Lambert, 1939). On this unit the Asencio Sandstone rests conformably (Caorsi and Goñi, 1958); it consists of about 40 m of fine-grained sandstones from which Huene (1929a) and Mones (1980) described several titanosaurids (Table 12). Insect remains have also been described (Table 11; Roselli, 1939; Frenguelli, 1946; Francis, 1975; Langguth, 1978). The Asencio sandstones are disconformably covered by Tertiary and Quaternary deposits. Their lateritized equivalents were named the Palacio sandstone (Walther, 1919; Palacense Stage of Kraglievich, 1932).

Similar sandstones in northeast Argentina (Fig. 12; Table 18) are included in the Mariano Boedo Formation (see Padula and Mingramm, 1968; Amos and Rocha Campos, 1970) and the Puerto Yeruá plus Pay Ubre, and Puerto Unzué formations (Herbst, 1971b, 1980; Gentili and Rimoldi, 1979; Santa Cruz, 1981).

CHACO-PAMPEAN PLAIN

The Chaco-Pampean Plain (Figs. 17–20) is a depressed physiographic feature interposed between the epicratonic Parana Basin and the Pampean Range (Fig. 20). It corresponds to the Chaco-Paraná Basin, which is connected to the Paraná Basin. Toward the northwest the Chaco-Paraná Basin is separated from the Aimará Basin by a structural high, the Hayes Arch (Fig. 19), while toward the northeast it is partially separated from the Paraná Basin by the Asunción Arch (Fig. 17). Toward the south it seems to be connected to the Salado and Macachín basins (Figs. 18, 19) through two small depressions, the Laboulaye and Rosario "basins" (Zambrano, 1974; Yrigoyen, 1975b). The Chaco-Paraná Basin, as is the case with all similar basins, was produced during Late Jurassic–Early Cretaceous basement fracturing and block faulting. In central Argentina a positive feature, the Pampean Range, provided most of the detrital sediments for the Chaco-Paraná Basin.

The Chaco-Pampean Plain is covered by a thin veneer of Quaternary deposits. Mesozoic strata present in the region rest unconformably on Paleozoic or Precambrian rocks, and are mostly known through borehole data. Outcrops are almost entirely restricted to eastern Argentina, near the boundary with Uruguay and Brazil, in part of the Paraná Basin (Fig. 12). The subsurface Mesozoic of the Chaco-Pampean Plain consists of units similar to those of the Paraná Basin (see under Paraná Basin). The Serra Geral basalts are therefore quite well represented in northeastern Argentina, where they extend in a southwest-northeast direction and are intercalated with sandstones of the Tacuarembó Formation. Toward the west and north (Fig. 17), however, the basalts are replaced by the Tacuarembó sandstones, and as much as 800 to 1,600 m of basalts and sandstones are included in the San Cristóbal Formation (Table 18; Padula and Mingramm, 1968). The San Cristóbal Formation rests on Upper Triassic sediments or older Paleozoic rocks and is unconformably overlain by light brown sandy argillites with intercalated shales containing calcareous nodules and gypsum. These sediments, which attain a maximum thickness of 350 m, are named the Mariano Boedo Formation and have been assigned a Late Cretaceous–Paleocene age (Padula, 1972).

To the northwest, the Mariano Boedo Formation is equivalent to the upper part of the Salta Group; to the south it correlates with the General Belgrano Formation of the Salado Basin (see under Salado Basin); and to the east, it correlates with the Bauru Group and the Mercedes, Guichón and Asencio Formations of the Paraná Basin (Table 18; see under Paraná Basin).

CENTRAL ARGENTINA

Pampean Range

The Pampean Range constitutes a system of isolated mountains rising abruptly from the surrounding plains of central Argentina. The range is formed by uplifted and downfaulted blocks of Precambrian rocks. The fault block tectonics that characterize the Pampean Range of central Argentina produced restricted longitudinal basins that were filled by continental sediments of different ages. All of these rocks, including Paleozoic and Mesozoic continental strata and volcanics, rest unconformably on metamorphic-plutonic basement and form erosive relics, covered unconformably by Cenozoic rocks.

Sedimentary rocks from the Pampean Range and surrounding areas (discussed below) were previously grouped under the Córdoba Basin by Weaver (1942) or Intra-Pampean Basin by González and Aceñolaza (1972). Along its northern margin, the Pampean Range includes Cretaceous strata closely related to those forming part of the Aimará Basin (see under Aimará Basin); they are discussed in the section dealing with that area.

Along its eastern margin, the Pampean Range includes Cretaceous sediments and volcanics that disappear to the east below the thick cover of sedimentary rocks extending to the Paraná River in the Chaco-Pampean Plain (see under Chaco-Pampean Plain). In western Santiago del Estero and near the northeastern border of the Pampean Range, Cretaceous rocks are present in the Guasayán Hill (Figs. 1, 17; Table 18) where 250 m of continental red sandstones and conglomerates of the Los Cerrillos Formation rest on Paleozoic tuffs and Precambrian metamorphic rocks. They appear to be equivalent to olivine basalts, of the Ichagón Formation that intrude metamorphic basement. These units are considered equivalents of the Sierra de los Cóndores Group farther south (Lucero, 1979).

The Sierra de los Cóndores Group (Fig. 17; Gordillo and Lencinas, 1967, 1970, 1972, 1979) is the proximal equivalent of the San Cristóbal Formation of the Chaco-Pampean Plain (Table 18). It consists of as much as 400 m of red sandstones and conglomerates, in some places interbedded with basalts, calc-alkalic trachytes, and trachybasalts that are contemporaneous with the Serra Geral eruptives of the Paraná Basin (see under Paraná Basin). The Sierra de los Cóndores Group rests unconformably on plutonic-metamorphic basement rocks. It varies in thickness from a few meters to around 450 m and has been divided into two sedimentary and two volcanic formations, in chronological order: the Embalse Río Tercero sandstones and conglomerates, the Cerro Colorado trachybasalts, the Cerro Libertad conglomerates, and the Rumipalla trachybasalts. The isotopic ages of the volcanic rocks (K-Ar) range from 112 ± 6 to 129 ± 3 Ma (Stipanicic and Linares, 1969; Valencio, 1972; Linares and Valencio, 1974, 1975; Vilas, 1976; Mendía, 1978; Cortelezzi et al., 1981). Basaltic rocks with K-Ar radiometric ages of 76 ± 5 Ma and 80 ± 5 Ma have also been reported farther south, east of Mercedes, in San Luis Province (Fig. 17; Santa Cruz, 1980).

Along the northwestern margin of the Pampean Range and the eastern part of the Famatina System, between $28°2'$ and $28°52'$ south latitude (Figs. 1, 17), are about 1,500 m of brown to purple continental conglomerates, sandstones, and tuffs of the Crestón Formation that rest unconformably on upper Paleozoic continental sediments and that are, in turn, unconformably overlain by Cenozoic rocks of continental origin. The Crestón Formation has been dated as Triassic by some authors (Bodenbender, 1922; Frenguelli, 1948; Turner, 1964) and as Upper Cretaceous by others (Bodenbender, 1924; De Alba, 1970).

San Juan–La Rioja provinces

At the boundary between La Rioja and San Juan provinces (Figs. 1, 17) there is a small basin (Ischigualasto) with important terrestrial upper Paleozoic–Triassic strata, as well as sedimentary rocks and basaltic sills of probable Cretaceous age. They seem to be related to similar rocks of that age present in central San Juan Province, northern and southern Mendoza Province, and northwest San Luis Province (Stipanicic, 1967; Stipanicic and Bonaparte, 1972, 1979; Stipanicic and Linares, 1969, 1975; Regairaz, 1970; Yrigoyen, 1975a; Días and Massabié, 1974; Cingolani et al., 1981).

The Triassic continental rocks are unconformably covered by as much as 500 m of conglomerates and reddish fine-grained sandstones and siltstones of the Cerro Rajado Formation (Stipanicic and Bonaparte, 1972), which appear to be equivalent to the La Cruz Formation of San Luis Province. The Cerro Rajado Formation is in turn disconformably? overlain by about 500 m of brown and green conglomerates and pelites of the Quebrada del Médano Formation (Parker, 1974), considered to be correlative to the Lagarcito Formation of the San Luis Basin (Yrigoyen, 1975a). It has also been considered as Maastrichtian-Paleocene (Zambrano, 1986) or Tertiary (Bossi, 1977; Malizzia and Limeres, 1984) in age. The Quebrada del Médano Formation is unconformably covered by lower Tertiary sedimentary rocks.

San Luis Basin

In San Luis Province (Flores, 1969, 1979; Flores and Criado Roque, 1972; Criado Roque et al., 1981) there is a relatively small basin (Figs. 1, 17) with Cretaceous continental strata (Table 18) that rest unconformably on lower Paleozoic or Precambrian basement rocks. The Cretaceous consists of 400 m of red conglomerates—the Los Riscos Formation—which grade laterally into 100 to 250 m of fluviatile sandy conglomerates, mudstones, and sandstones of the El Jume Formation and 18 to 60 m of lacustrine gray greenish and reddish siltstones of the La Cantera Formation. The La Cantera Formation is conformably overlain by 118 m of fine- to medium-grained red sandstones of the El Toscal Formation; this unit grades laterally upward into 130 to 400 m of red conglomerates called the La Cruz Formation. In marginal areas, the whole sequence—named the Gigante Group—becomes conglomeratic. The La Cruz Formation is interbedded with basalts that gave K-Ar radiometric ages of 106 ± 5 to 161 ± 3 Ma (Yrigoyen, 1975a; Llambías and Brogioni, 1981). The microflora found in the La Cantera Formation indicates an Early Cretaceous age (Flores, 1969; Yrigoyen, 1975a; Bonaparte, 1981). Pterosaurs (Table 12) of Late Jurassic–Early Cretaceous age have been found in the La Cruz Formation (Bonaparte and Sánchez, 1975), and the La Cantera and El Toscal formations have yielded other vertebrates and invertebrates (see Bonaparte, 1978, 1981). The basaltic rocks in the upper part of this group are considered equivalent to the trachybasalts of the Los Cóndores Group (see under Pampean Range).

The Gigante Group is unconformably overlain by 250 m of reddish sandstones with intercalated conglomerates bearing some vertebrates (Table 12) (Bonaparte, 1970, 1971; Sánchez, 1973; Bocchino, 1973, 1974). The sandstones were included in the Lagarcito Formation and placed in the Upper Cretaceous, although the vertebrates suggest Lower Cretaceous. The whole sequence is unconformably covered by continental Neogene rocks.

Cacheuta-Alvear Basin

A basin that is oriented north-northwest by south-southeast crosses Mendoza Province, in western Argentina (Fig. 17). At the latitude of the Diamante River, a structural high divides it into two subbasins, the Cacheuta Subbasin to the north and the Al-vear Subbasin to the south. The substratum is formed by Paleozoic marine sediments, and the subbasins are filled mostly by Triassic and Tertiary continental strata.

In the Cacheuta Subbasin (Table 18), the Cretaceous seems to be restricted to subsurface units, known from east-southeast of Mendoza City. The Punta de las Bardas Formation is about 6,500 km^2 in area and 50 to 180 m thick, consisting of olivine basalts intercalated with sandy siltstones (Rolleri and Criado Roque, 1968, 1970). Isotopic age determinations based on Rb-Sr analysis indicate ages of 124 ± 5 and 134 ± 5 Ma (Stipanicic, 1967; Stipanicic and Linares, 1969). This unit rests unconformably on 100 m of red conglomerates and sandy siltstones and sandstones included in the Barrancas Formation and dated as Triassic (Rolleri and Fernández Garrasino, 1979) or Early Cretaceous (Yrigoyen, 1975a). The basalts of the Punta de las Bardas Formation are unconformably covered by Tertiary continental conglomerates. Dolerites 105 ± 10 Ma in age (K-Ar) have also been reported in this area (Días and Massabié, 1974).

In the Alvear Subbasin the Cretaceous is also represented by basalts included in the Punta de las Bardas Formation (Criado Roque, 1979). The radiometric age determinations (?K-Ar) reported by Criado Roque (1979), however, gave values ranging from 107 to 291 Ma, with two principal clusters of data: the major are of Middle Jurassic age and the minor are of Early Permian age. Because of the position of these basalts above Triassic continental sediments, the unit has also been considered to be of Jurassic age.

In the Alvear Subbasin a sequence of argillites and red siltstones intercalated with coarse to conglomeratic sandstones, gypsum, and anhydrite has been named the Pozo Chimango Formation. Yrigoyen (1975a) places this unit below the Punta de las Bardas Formation, whereas Criado Roque (1979) considers it to lie above this unit.

AIMARÁ BASIN

Throughout northern Argentina and west-central Bolivia (Fig. 13), the Cretaceous consists of continental rocks with some marine intercalations. These sedimentary rocks were deposited in a series of basins and subbasins that are separated by structural highs. They are widely distributed (Figs. 13, 14), from west to east, throughout the Puna, Eastern Cordillera, and Subandean structural units; in the Chiquitanas Hills of Bolivia; and from the latitude of Tucumán city northward to the Lake Titicaca area. From here they extend into Perú. Deposits of this basin also extend eastward beneath the surface of the Beni-Chaco Plains of Bolivia and western Paraguay (Fig. 14).

The Cretaceous rocks of the Altiplano and the eastern Cordillera of Bolivia, and all of those present in the Eastern and Subandean Cordillera of northern Argentina, belong to a large western basin, sometimes called the Andean Basin. Those Cretaceous rocks present in eastern Bolivia were deposited in a large eastern basin, often called the Subandean Basin (Fig. 14). The entire area embracing both the Western and Eastern Basins, how-

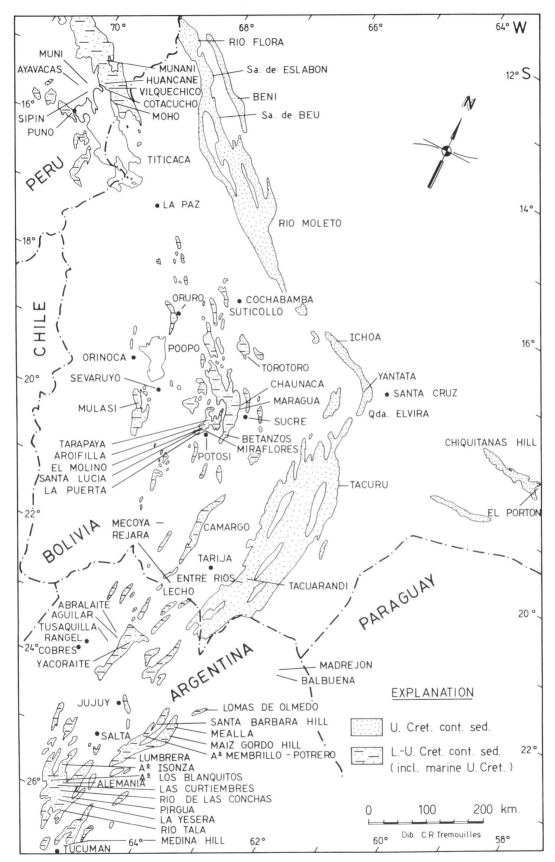

Figure 13. Distribution of major facies of Cretaceous of northern Argentina and Bolivia.

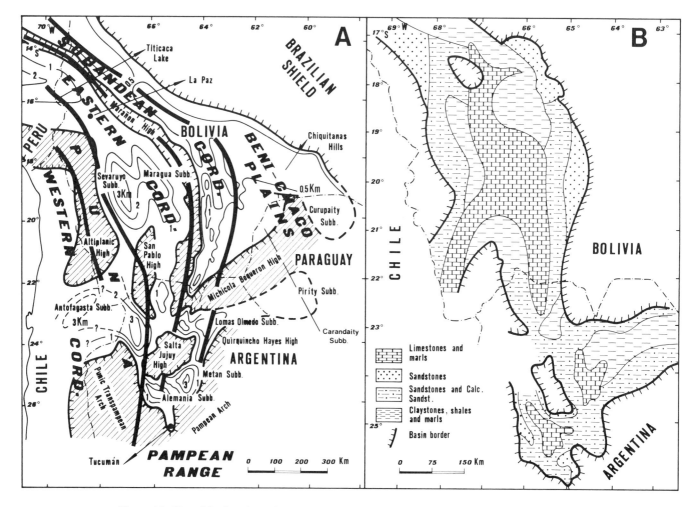

Figure 14. Aimará Basin. (A) Main subbasins and structural units (modified from Reyes and Salfity, 1973; Banks and Díaz de Vivar, 1975; Reyes, 1974; Salfity, 1980; Schwab, 1984). (B) Facies distribution of Maastrichtian strata in the western part of the Aimará Basin (after Moreno, 1970; Cherroni, 1977).

ever, has also been designated as the Subandean Basin by Aubouin et al. (1973) to stress its separation from the Andean Basin of northern Chile and southern Perú (see Andean Basin). More restricted names, such as the Salta or Northern Basin, have also been applied to these structures (Weaver, 1942; Malumián and Báez, 1978; Aceñolaza and Toselli, 1981). In order to eliminate nomenclatural confusion, the entire Cretaceous basin of northern Argentina and Bolivia has been called by Riccardi (1987) the Aimará Basin, following Groeber (1953).

Within the Aimará Basin (Fig. 14), both main subbasins were separated by a north-south–oriented positive feature known as the Aiquile-Marañón High. This uplift, called in Perú the Marañón High by Benavides Cáceres (1956), disappears at 7° south latitude where both basins merge. Toward the south, at the international boundary between Bolivia and Argentina, the Marañón High merges with a northeast-southwest–trending uplift,

the Michicola-Boquerón High, and continues to the south as the Salta-Jujuy High, to disappear in northern Argentina before reaching the southern border of the basin. Thus, the Western Subbasin extends eastward around the Salta-Jujuy High into an area that, today, belongs to the Subandean structural belt, but which is separated from the Eastern subbasin of Bolivia by the Michicola-Boquerón High. The Eastern and Subandean Cordilleras (Fig. 14) grade southward into the Pampean Range, which is mainly formed by Precambrian rocks and plunges northward at the latitude of Tucumán.

The Western Subbasin was bordered to the west by the Altiplanic and Punic-Transpampean (or Calchaquí) Highs and to the east by the Marañón High. In southern Bolivia it appears to connect to the southwest with northern Chile, and to the southeast with northern Argentina. Toward the south it was limited by the Pampean Range, whereas it is separated from the Chaco-

Paraná Basin to the east-southeast, and within Argentina and Paraguay (Pirity Subbasin) by a structural high, i.e., the Quirquincho-Hayes High.

The Eastern Subbasin is bordered to the west by the Marañón High, to the northeast by the Brazilian shield, and to the southeast by the Michicola-Boquerón High (see Reyes, 1974; Banks and Díaz de Vivar, 1975).

Northern Argentina

In northern Argentina the Mesozoic age is represented by a sedimentary succession developing from continental to lacustrine facies with some minor marine intercalations (Figs. 17-20; Table 18). These rocks were originally named the Petrolífera Formation and later the Salta Formation or Group (Brackebusch, 1883, 1891; Bonarelli, 1921; Turner, 1959). For general descriptions of these strata, see Palmer (1914), Lohmann (1970), Mingramm and Russo (1970), Moreno (1970), Turner (1970), Méndez et al. (1979), Mingramm et al. (1979), Turner and Méndez (1979), Turner and Mon (1979), Turner et al. (1979), Salfity (1979, 1980), and Aceñolaza and Toselli (1981). These beds are distributed (Figs. 13, 14) from west to east throughout the Puna, Eastern Cordillera, and Subandean geologic belts, and from the latitude of Tucumán City (Calchaquíes and San Javier Hills) and Santiago del Estero (27°30′ to 28° south latitude) in the south to the Bolivian-Argentinian boundary in the north. In Bolivia they were named the Puca Group by Steinmann (1906) (see under Bolivia).

From the northeast to southwest, the Salta Group overlies, with angular unconformity, successively older Paleozoic to Precambrian rocks. The Salta Group is unconformably overlain by Tertiary (post-Middle Eocene) sandstones and conglomerates. The Salta Group was first divided (Bonarelli, 1914, 1921, 1927) into (from base to top): lower sandstones, dolomitic limestones, and variegated marls. Its assumed age was said to vary from Carboniferous to Early Tertiary; Triassic and/or Cretaceous were the most common designations (Groeber, 1953; Harrington, 1956c). At present the group is divided, in ascending order, into three stratigraphic units: the Pirgua, Balbuena, and Santa Bárbara subgroups (Table 18), whose ranks vary according to different authors. On the margins of the basin, strong facies changes occur, especially in a west-east direction; these units are not so clearly defined as in the more centrally located areas. Therefore, some authors have recorded the absence of certain of these units and changes in others (Danieli and Porto, 1968; Mon, 1971; Mon and Urdaneta, 1972; Porto et al., 1982), whereas others preferred to use a different stratigraphic nomenclature (Bossi and Wampler, 1969).

The Pirgua Subgroup (Vilela, 1951; Boso et al., 1984) consists of as much as 1,000 m of fluviatile and alluvial red conglomerates, which are overlain by 1,500 to 2,000 m of red mudstones, argillites, and shales, and 1,640 m of flood stream and sheet flood medium- to coarse-grained red sandstones. Accordingly, this unit has been divided (Table 18) into three formations by Reyes and

Salfity (1973; Reyes, 1972); in ascending order, they are the La Yesera, Las Curtiembres, and Los Blanquitos formations. The La Yesera Formation has intercalations of pyroclastics that increase in abundance toward the east-southeast, where they are distinguished as the El Cadillal Formation (Bossi, 1969; Mon and Urdaneta, 1976). Basaltic and rhyolitic flows in the El Cadillal Formation, such as the Alto de las Salinas Complex and the Isonza basalt, gave radiometric ages (based on K-Ar) ranging from 96 to 128 ± 5 Ma (Bossi and Wampler, 1969; Valencio et al., 1976).

The Las Conchas basalt (Valencio et al., 1976), containing flows, dikes, and sills, is also present in the Las Curtiembres Formation. Radiometric dating based on K-Ar indicates various flows ranging in age from 76.4 ± 3.5 to 78 ± 5 Ma (Valencio et al., 1976, 1977; Reyes et al., 1976). Vulcanites are also intercalated in the Los Blanquitos Formation (Reyes et al., 1976). Thus, two main volcanic events appear to be present within the Pirgua Subgroup, the first in the La Yesera Formation and the second in the Las Curtiembres Formation.

The continental strata of the Pirgua Subgroup were deposited in various subbasins separated by structural highs. Its thickness and facies distribution patterns are therefore quite variable. The Pirgua Subgroup has yielded (Table 12) anurans in the Las Curtiembres Formation (Reig, 1959; Ibañez, 1960; Parodi Bustos, 1962; Parodi and Kraglievich, 1960; Parodi et al., 1960; Báez, 1982), and dinosaurs in the upper part of Los Blanquitos Formation (see Bonaparte and Bossi, 1967; Powell, 1979). It has a maximum thickness of about 1,600 m, thinning from north to south where the basin is limited by the northern extension of the Pampean Range. Near Tucumán City (Fig. 13), the Pirgua Subgroup includes andesites and olivine basalts; in the southern Medina Hill, north of Tucumán City, equivalent beds are included in the El Cadillal Formation (Bossi and Wampler, 1969; Powell and Palma, 1981; Porto et al., 1982). The Pirgua Subgroup is supposed to be Valanginian-Campanian in age, although its upper part could be slightly younger, according to the recorded dinosaurs (Powell, 1979). The La Yesera Formation spans the Valanginian-Cenomanian Stages, the Las Curtiembre Formation includes Turonian-Coniacian, and the Los Blanquitos Formation is Santonian-Campanian in age.

The Pirgua Subgroup is unconformably overlain (Fig. 15; Table 18) by a stratigraphic succession, the Balbuena Subgroup, consisting of the following layers, in order of oldest to youngest: about 300 m of fluvio-lacustrine and aeolian yellowish to white, massive, fine- to medium-grained calcareous sandstones of the Lecho Formation (Turner, 1959; Salfity, 1980); around 800 m of oil-bearing, shallow, marine light brown to yellow grayish, oolitic limestones with interbedded marls, siltstones, and sandstones of the Yacoraite Formation (Turner, 1959; Gómez Omil et al., 1984); and as much as 900 m of lacustrine dark shales and mudstones of the Olmedo Formation (Moreno, 1970). Lately, the Olmedo Formation has been included in the overlying Santa Bárbara Subgroup (see below) on the basis of its supposed disconformable relationship with the Yacoraite Formation and interfin-

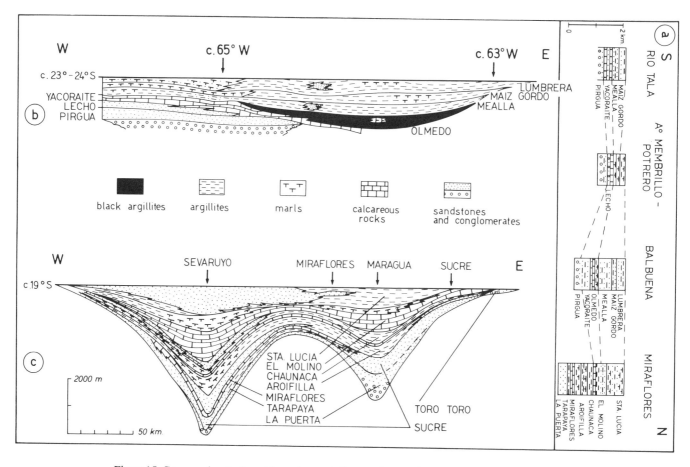

Figure 15. Comparative stratigraphic sections and diagrams interpreting relations in Cretaceous rocks of Bolivia and northern Argentina (modified after Russo and Rodrigo, 1965; Moreno, 1970; Mingramm et al., 1979).

gering with the Mealla Formation (Boll and Hernández, 1985, 1986).

The Balbuena Subgroup, which reflects brackish to shallow marine depositional environments, thins southward (Fig. 15) where the Olmedo Formation changes facies to green clayish sandstones of the Tunal Formation (Turner et al., 1979). The other two formations become indistinguishable in this area. Sandy facies found in the southern part of the basin, which are totally or partially equivalent to the Lecho or Yacoraite Formation, have been separated as the Quitilipi and Pala Pala formations (Salfity and Marquillas, 1981). To the west (Fig. 15), the Yacoraite Formation overlaps the Pirgua and Lecho formations, and is the only unit of this subgroup present in the eastern part of the Puna geologic province (Fig. 14). Near the Chilean boundary, however, at about 24° south latitude and 67° west longitude, the Pirgua Subgroup and the Lecho? and Yacoraite formations reach as much as 4,600 m in thickness. This could indicate the existence of another large basin that extends from the Puna to San Pedro de Atacama in Chile (Fig. 14); this reconstruction is supported by the similarity between the Pirgua-Yacoraite and Purilactis-Lomas

Negras formations (Lencinas and Salfity, 1973; Schwab, 1973, 1984; Salfity et al., 1985; see under Domeyko Range). It is possible, however, that these sedimentary rocks could also include Tertiary deposits (J. Salfity, written communication, 1981; Alonso et al., 1984). A connection between the Aimará Basin and the Pacific Ocean through Chile has been suggested by Bonaparte et al. (1977) and Bonaparte and Powell (1980).

The Lecho Formation contains dinosaur and bird remains (Bonaparte and Powell, 1980; Walker, 1981). The Yacoraite Formation has yielded gastropods, bivalves, ostracods, foraminifers, charophytes, palynomorphs, algae-like structures (*Pucalithus* and *Gymnosolen saltensis*; Frenguelli, 1937), fishes, crocodiles, and dinosaurs (Tables 11, 12; Bonarelli, 1921, 1927; Fritzsche, 1924; Cossmann, 1925; Frenguelli, 1945; Aceñolaza, 1968; Leanza, 1969a; Benedetto and Sánchez, 1971, 1972; Musacchio, 1972a; Méndez and Viviers, 1973; Powell, 1979; Alonso, 1980; Gasparini and Buffetaut, 1980; Moroni, 1982; Stach and Angelozzi, 1984; Cione and Pereira, 1985; Cione et al., 1985; Alonso and Marquillas, 1986). The Balbuena Subgroup is equivalent to the El Molino Formation of Bolivia (see below) and has been placed in the Maastrichtian, although K-Ar radiometric ages of tuffs from the middle part of this unit indicate an age of 60 ± 2 Ma (Fernández, 1976), and palynological analysis of the Olmedo Formation suggests a Paleocene age (Bianucci et al., 1981; Moroni, 1982). Evidence based on foraminifers needs to be more thoroughly examined (Malumián and Báez, 1978).

Conformably overlying the Balbuena Subgroup are 1,000 to 2,000 m of continental marly sandstones, marls, and clays of the Santa Bárbara Subgroup; this has been subdivided (Table 18; Fig. 15), from base to top, into the Mealla, Maíz Gordo, and Lumbrera formations (Moreno, 1970). Marginal facies consisting of red sandstones and pelites have been included in the El Madrejón Formation (Cazau et al., 1976). The Santa Bárbara Subgroup has yielded palynomorphs (Quattrocchio, 1978, 1980; Volkheimer et al., 1984), insects (Cockerell, 1936; Murature de Sureda and Alonso, 1980), fishes (Fernández et al., 1973; Fernández, 1976), and reptiles and mammals (Pascual and Odreman, 1973; Carbajal et al., 1977; Pascual et al., 1979; Pascual, 1980), which suggests an age that is Middle–Late Paleocene to Early Eocene.

At the southern margin of the basin, in the Calchaquíes and Medina hills (Fig. 13), correlatives of the Santa Bárbara Subgroup have been included in the Yacomisqui Formation (Galván and Ruiz Huidobro, 1965; Porto and Danieli, 1979; Porto et al., 1982; Torres, 1985), in the lower Río Loro Formation and overlying Río Salí Formation of the Tucumán Group (Bossi, 1969), and in the Los Pocitos, Cañada Ancha, and Agua San Cristóbal formations of Danieli and Porto (1968).

Bolivia

Altiplano and Eastern Cordillera. The Cretaceous in this region (Fig. 14) is represented by the Puca Group (meaning red in the Quechua language; Steinmann, 1906; Ahlfeld and Braniša, 1960), or the Pilcomayo Group (Braniša, 1968), or the Potosí Group (Rivas and Carrasco, 1968), which extends from southern Perú to northern Argentina.

The Puca Group rests unconformably on Paleozoic rocks, and, due to its transgressive character, onlaps with successively younger units from the center toward the periphery of the basin (Fig. 15). The basin is elongated and narrow with an irregular surface and a number of subbasins, so that the Puca Group exhibits striking facies changes over short distances. Its remnants are preserved in small thrusts and synclines. A general account can be found in Ahlfeld (1946, 1956, 1970), Ahlfeld and Braniša (1960), Sonnenberg (1963), Radelli (1964), Fricke et al. (1965), Russo and Rodrigo (1965), Schlatter and Nederlof (1966), Lohmann (1970), YPFB (1972), Maeda and Urdininea (1976), Cherroni (1977), Pareja et al. (1978), Oblitas et al. (1978), and Sempere (1986).

One of the most complete successions of the Puca Group is that of the Miraflores Syncline (Figs. 13-15), near Potosí City, which was first visited by d'Orbigny (1842). The Puca Group here has been divided into seven formations (Schlagintweit, 1941; Lohmann and Braniša, 1962; Russo and Rodrigo, 1965); these

are (Table 18), from base to top, the La Puerta, Tarapaya, Miraflores, Aroifilla, Chaunaca, El Molino, and Santa Lucía formations. The group consists of more than 2,000 m of mostly continental sediments, with marine intercalations, the Miraflores and El Molino formations. It reaches a maximum thickness of 5,600 m in Sevaruyo, 120 km west of Potosí (Fig. 15).

The La Puerta Formation was removed from the Puca Group by Cherroni (1977), and the Tarapaya, Miraflores, and "lower part of the Aroifilla Formation" were named "Lower Puca" Group, whereas the "upper part of the Aroifilla Formation" and the Chaunaca, El Molino, and Santa Lucía formations were included in the "Upper Puca" Group. More recently (Sempere, 1986), the Puca Group was divided into three subgroups: the Sucre (Kimmeridgian to Barremian, including the La Puerta Formation), the Potosí (Aptian-Campanian, including the Tarapaya, Miraflores, Aroifilla, and Chaunaca formations), and the Camargo (Maastrichtian-Oligocene, including in its lower part the El Molino Formation) subgroups.

The Puca Group begins with 343 to 1,000 m of yellowish and whitish, cross-bedded, medium- to coarse-grained, unfossiliferous sandstones, the La Puerta Formation (or "lower sandstones" of Bonarelli, 1914, 1921, 1927) topped by the Maragua basalt. This is unconformably covered by 67 to 1,480 m of violet marls, siltstones, and clays with some red sandstones and tuffaceous material of the Tarapaya Formation (Reyes and Salfity, 1973). Toward the west, in the southeastern Altiplan area, 500 m of sandstones are equivalent to the La Puerta and Tarapaya formations and are included in the Lacaya Formation. Basal red conglomerates underlying the La Puerta Formation at the western margin of the basin, east of Poopo Lake (Fig. 13), were named the Condo Formation (Vargas, 1970). The Tarapaya Formation, in turn, is overlain with local unconformity, by 22 to 100 m of gray, dense, thick-bedded limestone, containing variegated clays and marls near the base and increasing in silt content in the upper part; toward the margins of the basin this part of the sequence laterally grades into sandstones and conglomeratic sandstones. These rocks, placed in the Miraflores Formation in central Bolivia, the Anta Formation in the Altiplan, and the Ayavacas Formation in southern Peru (Ahlfeld, 1959), contain echinoids, mollusks, decapods, and ostracods (Table 11; Fritzsche, 1924; Berry, 1922, 1932, 1939; Lohmann and Braniša, 1962; Braniša et al., 1966; Secretan, 1972). The ammonite *Neolobites* sp. (Braniša, 1968; Plate 15, Figs. 1-3 herein) indicates a Cenomanian age, although generic affinities are uncertain, according to Reyes (1972, p. 115; Braniša et al., 1966, p. 308-309). The presence of *Mortoniceras* in the Ayavacas Formation of southern Perú suggests an Albian age.

The Miraflores Formation is conformably overlain by 459 m of calcareous marls with some clays in the upper part; this has been named the Aroifilla Formation, and is divided by an unconformity and the Betanzos basalt into a lower and an upper part. In the southeastern Altiplan, both parts are included in the Orinoca Formation.

Conformably overlying the Aroifilla Formation are 198 to

540 m of variegated bituminous and gypsiferous clays and marls of the Chaunaca Formation (= Mulasi Formation of the Altiplan), which contain ostreids, ostracods, characeans, bivalves, and microflora (Braniša et al., 1966; Pérez Leytón, 1987). The Chaunaca Formation is disconformably overlain by 392 to 1,750 m of white oolitic limestones with some intercalated variegated marls and shales, the El Molino Formation. These facies contain characeans, ostracods, mollusks, fishes, reptiles, the algae-like structure *Pucalithus* (Table 11, 12; d'Orbigny, 1842; Steinmann, in Fritzsche, 1924; Pilsbry, 1939; Bonarelli, 1921; Ahlfeld and Braniša, 1960; Schaeffer, 1963; Braniša et al., 1964, 1969; Wenz, 1969; Broin, 1973; Cappetta, 1975; Gayet, 1982a,c), and microflora (Pérez Leytón, 1987).

The equivalents of the El Molino Formation are the Vilquechico Formation of the Lake Titicaca area (Fig. 13) to the north, and the Balbuena Subgroup of northern Argentina to the south. In the Altiplan it is represented by as much as 650 m of sandstones of the Coroma and Pahua formations. Possible equivalents in northern Chile (see under Domeyko Range) are strata found in the Purilactis Hills (Harrington, 1961; Cherroni, 1977).

The uppermost Cretaceous lies within 207 to 1,200 m reddish and greenish, sandy, and gypsiferous marls and clays yielding some ostracods and characeans of Maastrichtian-Early Tertiary age, the Santa Lucía Formation. This represents the final stage of evaporite formation in the El Molino regression. In the Altiplan, 300 m of equivalent sediments are included in the Candelaria Formation.

Toward the margins of the basin (Fig. 15; Table 18), Lower Cretaceous rocks are absent, and the sequence begins with a red sandy facies, the Sucre and Toro Toro formations. The Sucre Formation consists of as much as 700 m of poorly sorted, dark violet, medium- to fine-grained sandstones and conglomeratic layers. It is a near-shore lateral equivalent of the Tarapaya, Miraflores, and lower part of the Aroifilla formations. Near Sucre (Fig. 15), the Sucre Formation overlies the Lower Paleozoic and underlies the Chaunaca and El Molino formations. The status of the Sucre Formation, however, needs revision (Cherroni, 1977).

The Toro Toro Formation includes 290 to 1,210 m of sandstones, which are medium to fine grained, well sorted, bedded, pink, white, and red, and locally slightly calcareous, and which contain conglomeratic levels and basaltic flows at the base. The Toro Toro Formation commonly contains trace fossils and represents the initial transgressive phase of the El Molino Formation. It is a near-shore equivalent of the Upper Aroifilla and Chaunaca formations (Fig. 15). Similar sedimentary rocks in the Titicaca area (Fig. 13; Table 18) were placed in the Cotacucho Formation (see below), and in the eastern border of the basin in the Chaupiuno Formation (Ponce de León, 1966).

Basaltic flows are interbedded in the Toro Toro, Aroifilla, La Puerta, and Sucre Formations (Russo and Rodrigo, 1965; Barth, 1973; Avila-Salinas, 1986). The Betanzos basalt, intercalated in the Aroifilla Formation, gave a K-Ar age of 82.5 Ma, suggesting a Coniacian-Santonian boundary age (Evernden et al., 1966; Reyes and Salfity, 1973).

The Puca Group varies greatly in thickness, reaching 3,300 m in the Sevaruyo Subbasin (Fig. 15), 2,200 m at Miraflores, and only 780 m in the Camargo–Las Carreras syncline (Fig. 13).

In the Lake Titicaca area (Fig. 13; Table 18), the Cretaceous sequence has been divided by Newell (1949) into the following layers, in order from base to top: Huancané, Lower Moho, Ayavacas (= Miraflores), Upper Moho, Cotacucho (= Toro Toro = Upper Aroifilla + Chaunaca), Vilquechico (= El Molino) and Muñani (= Santa Lucía) formations. Additional information was provided by Audebaud et al. (1973) and Laubacher (1978), and Newell's (1949) stratigraphy was modified by Ellison (1985) and Palacios and Ellison (1986). The Cretaceous succession begins with up to 225 m of marine limestones, calcareous sandstones and conglomerates—the Sipin Formation—which to the northwest change to sandy facies—the Angostura Formation—at most 100 m in thickness. These units are conformably covered to the southeast by about 300 m of marine limestones—the Ayavacas Formation—and to the northeast by 967 m of red sandstones and shales—the Muni Formation. The Muni Formation is conformably covered by 500 m of coarse, brown, continental sandstones of the Huancané Formation, which in turn are overlain by 800 m of estuarine mudstones and sandstones, the Moho Formation. The Moho Formation is conformably covered by 800 m of continental red sandstones and gypsum of the Cotacucho Formation. The Vilquechico Formation, consisting of 680 m of marine gray shales and bearing Campanian-Maastrichtian charophytes, rests on the Cotacucho Formation. The Cretaceous succession is covered by the Tertiary Muñani Formation.

Subandean Belt and Chiquitanas Hills. In the southeastern part of this belt (Figs. 1, 13), the Cretaceous sequence (Fig. 16, Table 18) begins with a 140-m-thick basaltic flow, the Entre Ríos Basalt (see Saavedra, 1970), which has yielded an age of 83 Ma (Pareja et al., 1978). Above the lava are 600 to 1,200 m of gray, red, and yellow cross-bedded sandstones and sandy shales, the Tacurú Formation or Group (Mather, 1922; Pareja et al., 1978). Some clays and shales recur at irregular intervals throughout its various members.

Where this sequence thickens northward (Fig. 16), it can be divided into two or three different units. The name Tacurú Group has here been used to include the entire Cretaceous sequence of the Subandean Belt of central and southern Bolivia. Thus, Reyes (1974) has extended the northern subdivision of this group to the southern sequences, whereas other authors (Pareja et al., 1978) have maintained independent sets of stratigraphic names for the different areas.

In the central area of the Subandean Belt of Bolivia (Figs. 13, 16), the sequence reaches approximately 500 m in thickness and consists of 350 m of yellow calcareous sandstones and 180 m of cross-bedded red sandstones. These compose the Peña Blanca (= Castellanos) and Elvira members of Reyes (1974) or the Ichoa Formation of Pareja et al. (1978; Sanjines-Saucedo, 1982), which are followed conformably by 90 m of yellow sandstones these authors named the Izozog Member and the Yantata Formation, respectively. These two formations or three members—the latter

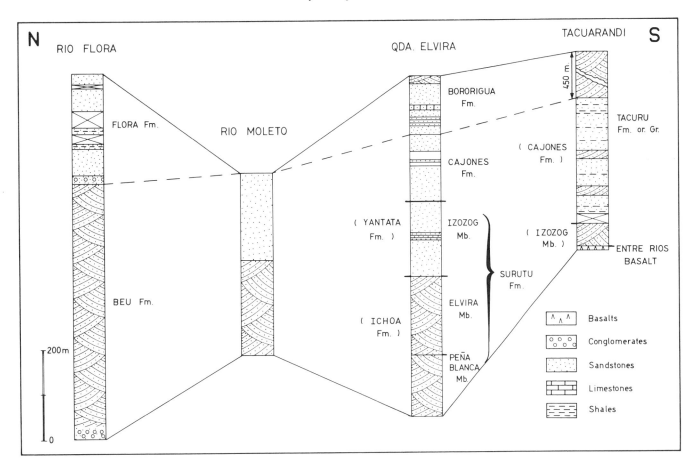

Figure 16. Comparative stratigraphic sections of Cretaceous rocks in the Subandean Range of Bolivia (modified from Reyes, 1974).

included in the Surutú Formation (Lamb and Truitt, 1963)—are conformably overlain by 120 to 200 m of gray, whitish limestones and calcareous sandstones with siliceous and calcareous concretions in the Cajones Formation (Heald and Mather, 1922). According to Reyes (1974), this unit is (?)unconformably overlain by 75 m of reddish, medium-grained argillaceous and cross-bedded sandstones, the Bororigua Formation. This unit, together with the Surutú and Cajones Formations, comprises the Tacurú Group. The Tacurú Group is unconformably covered by Tertiary sediments and toward the south is present only with its upper part, beginning with Izozog Member of the Surutú Formation.

Pareja et al. (1978) included the entire upper part of the sequence (including the Bororigua Formation) of central Bolivia in the Cajones Formation, whereas toward the south the entire sequence is included in the Tacurú Formation and considered equivalent to the Ichoa and most of the Yantata formations farther north.

In the northern part of the Subandean Belt of Bolivia (Figs. 13, 16; Perry, 1963; Martínez et al., 1971; YPFB, 1972) the Cretaceous sequence overlies Paleozoic and consists of 600 m of red, cross-bedded sandstones of the Beu Formation, discussed below, which are apparently equivalent to the Yantata plus Ichoa formations or to the Elvira plus the Izozog member of the Surutú Formation. The Beu Formation is unconformably overlain by 82 to 600 m of reddish purple conglomerate containing chert and quartzite pebbles and cobbles. This is overlain by variegated, fine- to coarse-grained sandstones interbedded with variegated clayey to sandy siltstones and sandy shales containing *Pucalithus,* gastropods, ostracods, fishes, and charophytes. This unit of the sequence, named the Flora Formation by Perry in 1963 (including the Eslabón Formation of Reyes, 1960), thins southward. On the basis of its fossil content and lithologic similarities, the Flora Formation is considered to have been deposited in a marine near-shore or brackish environment, and to belong in Campanian-Maastrichtian time. The Flora Formation is unconformably overlain by yellowish to brownish quartzose sandstones, the Bala Formation.

This eastern basin continues beneath the subsurface of the Beni-Chaco plains (Fig. 14), which are located east of 63° west longitude, between the Subandean Ranges and the Brazilian

Shield. In the subsurface of the Paraguayan Chaco, 3,000 m of red beds have been reported (Carandaity Subbasin of Banks and Díaz de Vivar, 1975), consisting of sandstones, siltstones, and conglomerates. These overlie Devonian rocks. These strata, however, have been variously ascribed to the Paleozoic (Eckel, 1959), Triassic (Gerth, 1955; Putzer, 1962). Cretaceous (Bentz, in Putzer, 1962), and Tertiary (Rassmuss, 1957). Similar rocks are also present in the Pirity Subbasin, which extends from northern Argentina into western Paraguay and is separated from the Carandaity Subbasin by the Boquerón Arch (Fig. 14). Possible equivalents of these strata have been found with the Misiones sandstones (see under Paraná Basin) or the Tacurú Formation of Bolivia.

In the Chiquitanas Hills of southeastern Bolivia (Fig. 13; Table 18), the Cretaceous is represented by 450 m of whitish, yellowish, and pinkish calcareous sandstones, the El Portón Formation, which are considered equivalent to the Ichoa and Yantata Formations and to the Tacurú Formation of the Subandean Hills. This unit rests unconformably on Devonian rocks and is covered by conglomerates of probable Tertiary age.

PACIFIC BASINS

Along the Coastal Cordillera of Chile south of 33° south latitude, there are several small basins containing Upper Cretaceous sediments and extending off-shore into the Pacific Ocean (Fig. 20). They are detailed in the following paragraphs.

Navidad Basin

The Navidad sedimentary basin is 130 km southwest of Santiago City, Chile. First described by Darwin (1846), it has a north-northwest strike. General accounts of the stratigraphy have been given by Cecioni (1978, 1980).

The Navidad Basin includes 500 to 1,000 m of Cretaceous and Tertiary sediments, which have been included in the Navidad Group. The Cretaceous (Table 18) is represented by as much as 150 m of coquina, conglomerates, calcareous sandstones, and some coal-bearing sandstones—the Punta Topocalma Formation. This unit contains large silicified tree trunks and Campanian invertebrates of Antarctic affinities. According to Pérez and Reyes (1980), it is a correlative of the Quiriquina Formation (see below). Similar fossils are also present north of Topocalma in the fossiliferous conglomerate of Algarrobo (Levi and Aguirre, 1960, 1962). The Punta Topocalma Formation is overlain by the Eocene Río Topocalma Formation.

Arauco Basin

At about 36°30′ to 37°45′ south latitude, in Quiriquina Island and the shores of Arauco Bay (Fig. 20), there are 15 to 160 m of marine conglomerates, coquina, and glauconitic sandstones, the Quiriquina Formation. This bears an abundant Campanian-Maastrichtian marine fauna. This unit rests unconformably on Paleozoic basement and is unconformably overlain by Eocene continental sediments (Biró, 1982c; Frutos et al., 1982). This Cretaceous-Tertiary succession has been named the Arauco Group (Hoffstetter et al., 1957).

Eastward, the Quiriquina Formation thins and becomes more shallow water in aspect (see Maeda et al., 1972). Offshore, 40 km northwest of Talcahuano, it is about 1,000 to 1,200 m thick (Wenzel et al., 1975; Biró, 1979).

The Quiriquina Formation contains gastropods, bivalves, ammonites, coleoids, decapods (Table 7), fishes, reptiles, birds (Table 12), and microflora (Table 4) (d'Orbigny, 1848; Möricke, 1895; Steinmann, 1895; Wilckens, 1904; Oliver Schneider, 1923; Lambrecht, 1929; Wetzel, 1930, 1960; Broili, 1930; Casamiquela, 1969b; Hünicken and Covacevich, 1975; Takahashi, 1978, 1979; Biró, 1982a,b; Gasparini and Biró, 1986; Stinnesbeck, 1986).

PALEONTOLOGY

FLORA

The Cretaceous flora of southern South America has been studied in detail in Argentina (Tables 1-4). Archangelsky (1978) gave an account of the Cretaceous megaflora from Neuquén, Argentina, and information about the Argentina flora was included in general reviews prepared by Menéndez (1969), Archangelsky (1970), Volkheimer and Baldis (1975), Baldoni (1980, 1981), and Romero and Arguijo (1981).

The Lower Cretaceous megaflora (Tables 1, 2) includes a large proportion of taxa that were also present in the Jurassic. The main differences lie in the more modern aspect of filicals and conifers. Most genera are pandemic, although in Patagonia some endemic taxa also occur. The articulates, including *Equisetites*, are more abundant in Early Cretaceous time than in Late Cretaceous. The Gleicheniacea diversified in the Early Cretaceous, but declined afterward. *Nathorstia* reached the Albian, and *Hausmannia* reached the Lower Cretaceous as the only representative of the Dipteridacea. The Corystopermacea are absent, with the possible exception of *Pachypteris*, which is present in the Lower Cretaceous. The Osmundaceae also declined at the end of the Cretaceous. The Cycadales are also present with taxa such as *Nilssonia*.

The Lower Cretaceous of Patagonia is rich in Bennettitales, e.g., *Dictyozamites, Williamsonia, Ptilophyllum, Otozamites, Zamites*, and *Cycadolepis*. The Ginkgoales, represented by *Ginkgoites, Allicospermum, Karkenia*, and *Baiera*, are rare but persisted into the Tertiary.

The conifers are also well represented in the Lower Cretaceous, especially the Podocarpacea, e.g., *Podocarpus, Tomaxellia, Trisacocladus, Apterocladus*, which reach the present time. The Araucariacea are also well documented, and the Toxodiaceae are present in the Lower Cretaceous with *Athrotaxis*.

The beginning of the angiosperms has not yet been properly established in the area, although *Nothofagus* became dominant toward the end of the Cretaceous, together with such plants as the Podocarpacea, Myrtacea, Sterculiacea, and Proteaceae. At the same time the Pteridosperms, Bennettitales, and Caytoniales declined or became extinct.

Most Upper Cretaceous plants include genera now known from tropical or subtropical regions, and there are some taxa indicating relationships with Australia and New Zealand. The Lower Cretaceous microflora (Tables 3, 4) is quite well known. In central-west Argentina it has been studied by Volkheimer (e.g., Volkheimer, 1978; Volkheimer and Quattrocchio, 1981; Quattrocchio and Volkheimer, 1985), who has published a stratigraphic survey of all taxa recorded in that area. The southern Patagonian microflora is primarily known through the work of Archangelsky (1963a,d, 1964, 1965, 1967a,b, 1977b). The composition of the microflora is in agreement with what is known about the megaflora.

INVERTEBRATES

Most studies of Lower Cretaceous invertebrates have been made on Argentine material, particularly from central-west Argentina, and from southern Patagonia. A list, including references, of the described and figured taxa found in that area has been published by Camacho and Riccardi (1978, Table 5). These faunas can be considered representative for most of the Tithonian-Cretaceous of the Andean Basin of Argentina and Chile.

One of the oldest studies on the Cretaceous invertebrates of southern South America was published by Forbes (1846), who studied material collected by Darwin in southern Patagonia. In the Andean Basin the first important paper was by Behrendsen (1891–92), who studied Lower Cretaceous material collected by Bodenbender (1892) in southern Mendoza and northern Neuquén provinces. Subsequent general studies of the Cretaceous invertebrates were published by Philippi (1899), Burckhardt (1900a,b, 1903), and Weaver (1931), the last publication being an outstanding contribution to our knowledge of the entire Mesozoic of the Argentinian Principal Cordillera.

The Cretaceous invertebrates of southern Patagonia (Table 6) have been studied by such authors as Stanton (1901), Wilckens (1905, 1907a, 1921), Paulcke (1907), Favre (1908), Bonarelli and Nágera (1921), Richter (1925), Feruglio (1936-1937, 1938), and Piatnitzky (1938); see Riccardi and Rolleri (1980) for a more complete listing.

Most published studies deal chiefly with the ammonites and bivalves, although other groups are sometimes included. In general, it can be said that the Lower Cretaceous invertebrates are mostly known from the Andean Basin, whereas Upper Cretaceous material comes almost exclusively from the Austral Basin.

Much work has been done on the Tithonian and Lower Cretaceous fauna. Papers, mostly on ammonites, have been published by Behrendsen (1891–92), Steuer (1897), Stanton (1901), Haupt (1907), Favre (1908), Douvillé (1910), Gerth (1925),

Krantz (1928), Feruglio (1936–37), Leanza A. (1945, 1968), Leanza and Giovine (1949), Giovine (1950, 1952), Indans (1954), Corvalán (1959), Cloos (1961, 1962), Riccardi et al. (1971), Biró (1980), Leanza H. (1980, 1981a), and Leanza and Wiedmann (1980). The more important fossils guide in Chile was published by Corvalán and Pérez (1958), and the ammonite zonal sequence has been dealt with by Leanza (1947), Leanza and Hugo (1978), Wiedmann (1980), Leanza (1981a,c), and Riccardi (1984a,b).

The Tithonian-Berriasian ammonite fauna (Table 13, 14) of the Andean Province is somewhat differentiated from those of other parts of the world, but its affinities lie with the Himalayan and Mediterranean Provinces (see Enay, 1972), as well as with northern South America. The Andean Province extended along the Pacific margin of South America, including ammonite faunas of the Andean and Austral basins, and is characterized by *Argentiniceras, Andesites, Cuyaniceras, Choicensisphinctes, Frenguelliceras, Groebericeras, Parodontoceras, Pseudinvoluticeras, Substeueroceras, Wichmanniceras,* and *Windhauseniceras*. During Valanginian-Hauterivian time, the southern limit of the Andean Province became established at about 42° south latitude. Characteristic genera are *Acantholissonia, Lissonia, Pseudofavrella, Holcoptychites,* and *Weavericeras*. The presence of Boreal genera, i.e., *Polyptychites, Virgatites, Simbirskites* (Douvillé, 1910; Leanza A., 1957) has yet to be demonstrated (Uhlig, 1911; Windhausen, 1918b; Rawson, 1971; Thieuloy, 1977; Leanza and Wiedmann, 1980). The Austral Basin was characterized in the Hauterivian-Barremian by a low-diversity ammonite fauna with endemic (*Favrella, Hatchericeras*), boreal (*Protaconeceras, Aegocrioceras*) or South African and Caucasian (*Colchidites*) genera. Faunal diversity became higher during Aptian-Albian time, with clear similarities to southern Africa, Madagascar, and Australia.

The Upper Cretaceous ammonites from Patagonia (Table 16) have clear Indo-Pacific affinities, as indicated by the dominance of the Kossmaticeratidae (Macellari, 1985a).

The trigonids (Tables 5, 6) from the Andean and Austral Basins (see Fuenzalida, 1964; Pérez and Reyes, 1978, 1980, 1983a,b; Reyes and Pérez, 1978, 1979, 1985; Pérez et al., 1981; Reyes et al., 1981; Camacho and Olivero, 1985; Leanza, H., 1985; Leanza H. and Garate, 1986) exhibit affinities with those from the Uitenhage Formation of South Africa (see Uhlig, 1911; Hallam, 1967; Reyment and Tait, 1972). Late Cretaceous bivalves (Table 7) from Patagonia have closer affinities with those from Australia, New Zealand, and New Caledonia (Freneix, 1981).

Lower and Upper Cretaceous foraminifers have been studied in southern Patagonia (Table 8), where they have been used as the basis for a local sequence of stages (Natland et al., 1974). They show low diversity, and are dominated by benthic species. In spite of some similarities to those from the Boreal, Australoasiatic, and Pacific regions, they seem to be endemic and unreliable for correlation (Malumián, 1979; Bertels, 1979, 1987).

Lower Cretaceous foraminifers and ostracods are also known from the Andean Basin (Table 9). Most foraminiferal

TABLE 1. Early Cretaceous Fossil Plants Described from Southern South America

Age columns (both halves): BERRIASIAN | VALANGINIAN | HAUTERIVIAN | BARREMIAN | APTIAN | ALBIAN

TAXON (left half)

- Microtyriacites baqueroensis Martinez
- Trichopeltis reptans Martínez
- Brefeldiellites argentina Martínez
- Marchantites hallei Lundblad
- Equisetites sp.
- Todites williamsoni (Bgt.)
- Gleichenites argentinica Berry
- G. feruglioi Herbst
- G. medinensis Ruiz
- G. sanmartinii Halle
- G. vegagrandis Herbst
- Coniopteris baldonii Ruiz
- C. hymenophylloides (Bgt.)
- Asplenites lanceolatus Halle
- Nathorstia pectinata (Goeppert)
- Hausmannia papilio Feruglio
- H. patagonica Feruglio
- Sphenopteris baqueroensis Archangelsky
- S. cf. fittoni Seward
- S. goeppertii Seward
- S. cf. naktongensis Yabe
- S. patagonica Halle
- S. psilotoides (Stokes & Webb)
- S. sueroi Archangelsky
- Cladophlebis antarctica Nathorst
- C. haiburnensis rectimarginata Herbst
- C. patagonica Frenguelli
- C. tripinnata Archangelsky
- Scleropteris (?) sp.
- Taeniopteris argentina Feruglio
- T. dissecta Baldoni
- T. patagonica Feruglio
- T. sp.
- Pachypteris elegans Archangelsky
- Mesosingeria coriacea Archangelsky
- M. herbstii Archangelsky
- M. mucronata Archangelsky
- M. (?) obtusa Archangelsky
- M. striata Archangelsky
- Mesosingeria sp.
- Nilssonia clarkii Berry
- Pseudoctenis crassa Archangelsky & Baldoni
- P. dentata Archangelsky & Baldoni
- P. ensiformis Halle
- Ticoa harrisii Archangelsky
- T. lamellata Archangelsky
- T. magallanica Archangelsky
- T. magnipinnulata Archangelsky
- Ruflorinia pilifera Archangelsky
- R. sierra Archangelsky
- R. (?) thoriana Archangelsky
- Mesodescolea plicata Archangelsky
- Almargemia incrassata Archangelsky
- Sueria rectinervis Menéndez

TAXON (right half)

- Williamsonia bulbiformis Menéndez
- W. umbonata Menéndez
- Dictyozamites areolatus Archangelsky & Baldoni
- D. crassinervis Menéndez
- D. latifolius Menéndez
- D. minusculus Menéndez
- Otozamites grandis (Oldham)
- O. parviauriculata (Menéndez)
- O. sanctaecrucis Feruglio
- O. waltonii Archangelsky & Baldoni
- Pterophyllum sp.
- P. trichomatosum Archangelsky & Baldoni
- Ptilophyllum acutifolium Morris
- P. antarcticum (Halle)
- P. ghiense Baldoni
- P. hislopi (Oldham)
- P. longipinnatum Menéndez
- Zamites decurrens Menéndez
- Z. aff. gigas (Lindley & Hutton)
- Z. grandis (Menéndez)
- Cycadolepis baqueroensis Baldoni
- C. coriacea Menéndez
- C. involuta Menéndez
- C. cf. jenckensiana (Tate)
- C. lanceolata Menéndez
- C. menendezii Baldoni
- C. oblonga Menéndez
- C. petriellai Baldoni
- Ginkgoites skottsbergi Lundblad
- G. ticoensis Archangelsky
- G. tigrensis Archangelsky
- Karkenia incurva Archangelsky
- Allicospermum patagonicum Archangelsky
- Podozamites sp.
- Tomaxellia biforme Archangelsky
- T. degiustoi Archangelsky
- Araucaria grandifolia Feruglio
- Araucarites baqueroensis Archangelsky
- A. chilensis Baldoni
- A. minimus Archangelsky
- Podocarpus dubius Archangelsky
- Trisacocladus tigrensis Archangelsky
- Apterocladus lanceolatus Archangelsky
- Elatocladus heterophylla Halle
- E. sp.
- Athrotaxis ungeri (Halle)
- Brachyphyllum brettii Archangelsky
- B. feistmantellii (Halle)
- B. irregulare Archangelsky
- B. mirandai Archangelsky
- B. mucronatum Archangelsky
- B. tigrense Traverso

(After: Halle, 1913; Frenguelli, 1935, 1953a, b; Archangelsky, 1963a, b, c, d, 1965, 1967a, b, 1977 a; Menéndez, 1965 b, 1966 a, 1973; Traverso, 1966, 1968; Martinez, 1968; Herbst, 1971 a; Archangelsky and Baldoni, 1972 a, b; Baldoni, 1975, 1977 a,b, 1979, 1980, 1981; Baldoni and De Vera, 1981; Baldoni and Ramos, 1981; Ruiz, 1984; Longobucco et al., 1985).

TABLE 2. Late Cretaceous Fossil Plants Described from Southern South America

TAXON \ AGE	CEN.	TUR.	CON.	SANT.	CAMP.	MAAST.
Gleichenites piatnitzkyi Berry			– –	– –		
G. sp.						│
Hymenophyllum priscum Menéndez						│
Thyrsopteris antiqua Menéndez						│
Adiantum patagonicum Berry			– –	– –		
Dennstaedtia patagonica Berry			– –	– –		
Dryopteris problematicus Berry	│	–	– –	– –		
Asplenium dicksonianum Heer	│					│
Nilssonia sp.	│	–				
Agathis sp.						│
Araucarites patagonica Kurtz						│
Araucarioxylon pichasquensis Torres & Rallo						│
Podocarpus inopinatus Florin						│
cf. *Acmopyle? antarctica* Florín						│
Protophyllocladus australis Berry			– –	– –		
Pseudoaraucaria valentini (Kurtz)						│
Sequoia brevifolia Heer						│
Liriodendron meeki Heer						│
Menispermites obtusiloba Lesquereux						│
M. piatnitzkyi Berry	│	–	– –	– –		
Rollinia? patagonica Berry			– –	– –		
Sassafras acutilobum Lesquereux						│
S. cretaceum Newberry						│
S. mudgei Lesquereux						│
S. subintegrifolium Lesquereux						│
Cinnamomum heeri Lesquereux						│
Perseophyllum hauthalianum Kurtz						│
Protophyllum cf. *rugosum* Lesquereux						│
Laurophyllum chalianum Berry			– –	– –		
L. hickenii Menéndez	│	–				
L. kurtzi Berry	│	–	– –	– –		
L. proteaefolium Berry	│	–	– –	– –		
Laurelia amarillana Berry			– –	– –		
Peumus clarki Berry			– –	– –		
Sterculia patagonica (Spegazzini)	│	–				
S. platanoides (Engelhardt)	│	–				
S. sehuensis Berry			– –	– –		
S. washburni Berry			– –	– –		
Paranymphaea proteaefolia Berry	│	–	– –	– –		
Populus acerifolia Newberry						│
P. cf. *microphylla* Newberry						│
P. nebrascensis Newberry						│
Salix proteaefolia Lesquereux						│
Populites lancastriensis Lesquereux						│
Hydrangea (?) *incerta* Berry			– –	– –		
Liquidambar integrifolium Lesquereux						│
Platanus obtusiloba Lesquereux						│
P. primaeva grandidentata Lesquereux						│
Cissites affinis Lesquereux	│	–				
C. parvifolius argentina Menéndez	│	–				
Schinopsis? dubia Frenguelli			– –	– –		
Carpolithus sp.			– –	– –	– –	– –
Quercus primordialis Lesquereux						│
Araliophyllum cachetamanense Menéndez	│	–				
Myrcia acutifolia Frenguelli	│	–	–			
M. santacruzensis Berry	│	–	–			
Bignonites chalianus Berry	│	–	–			
Ruprechtia (?) *castilloensis* Menéndez	│	–				
Nothofagoxylon pichasquensis Torres & Rallo						│
Myrtoxylon pichasquensis Torres & Rallo						│
Palmoxylon chilensis Torres	?	?	?			
Elaeocarpoxylon pichasquensis Torres & Rallo						│
Apocinophyllum chalianum Berry			– –	– –		

(After: Kurtz, 1902, Berry, 1906, 1916, 1918, 1924, 1937; Frenguelli, 1930b, 1953a; Cecioni, 1957b: Menendez, 1959, 1966b, 1972a, b; Hünicken, 1967, 1971a, b; Romero and Arguijo, 1981; Torres and Rallo, 1981; Torres, 1982).

TABLE 3. Early Cretaceous Microflora Described from Southern South America

TAXON \ AGE	Be.	Va.	Ha.	Ba.	Ap.	Al.
Atopochara trivolvis aff. *triquetra* Grambast						
Diectochara andica Musacchio						
Flabellochara aff. *harrisi* (Peck)						
Gobichara sp.						
Mesochara sp.						
Porochara sp.						
Stellatochara aff. *mundula* (Peck)						
Triclypella aff. *calcitrapa* Grambast						
Biretisporites spp.						
Calamospora mesozoica Couper						
Cibotiumspora cf. *jurienensis* (Balme)						
Concavisporites sp.						
Concavissimisporites sp.						
Cyathidites australis Couper						
Deltoidospora spp.						
D. australis (Couper)						
D. minor (Balme)						
Dictyophyllidites spp.						
D. cf. *mortoni* de Jersey, Playford & Dettmann						
Lygodiumsporites sp.						
Stereisporites sp.						
Todisporites cinctus (Maljawkina)						
T. major Couper						
T. minor Couper						
Triletes gamerroi Baldoni & Taylor						
T. sp.						
Anapiculatisporites cf. *dawsonensis* Reiser & Williams						
A. sp.						
Apiculatisporites cf. *taroomensis* Reiser & Williams						
Baculatisporites cf. *comaumensis* (Cookson)						
Ceratosporites cf. *helidonensis* De Jersey						
Converrucosisporites spp.						
Granulatisporites sp.						
Hamulatisporites sp.						
Horriditriletes sp.						
Leptolepidites cf. *crassibalteus* Filatoff						
L. cf. *macroverrucosus* Schulz						
L. cf. *major* Couper						
L. cf. *proxigranulatus* (Brenner)						
L. sp.						
Osmundacidites sp.						
O. wellmannii Couper						
Pilosisporites sp.						
Rotverrucosisporites sp.						
Rugulatisporites neuquensis Volkheimer						
Uvaesporites cf. *minimus* Volkheimer						
U. sp.						
Verrucosisporites sp.						
Appendicisporites spp.						
A. cf. *stylosus* (Thiegart)						
Cicatricosisporites annulatus Archangelsky & Gamerro						
C. australiensis (Cookson)						
C. baqueroensis Archangelsky & Gamerro						
C. cf. *ethmos* (Delcourt & Sprumont)						
C. hughesii Dettman						
C. ticoensis Archangelsky & Gamerro						
C. spp.						
Favososporites feruglioi (Archangelsky)						
F. spinulosus Gamerro						
Foveosporites sp.						
Klukisporites labiatus (Volkheimer)						
K. scaberis (Cookson & Dettmann)						
K. tuberosus (Döring)						
K. variegatus Couper						
Lycopodiumsporites austroclavatidites (Cookson)						
Reticulatisporites sp.						
Staplinisporites cf. *caminus* (Balme)						
Tripartina sp.						
Ischyosporites volkheimeri Filatoff						
I. sp.						
Matonisporites tenuilabratus Baldoni & Archangelsky						
Trilites cf. *tuberculiformis* Cookson						
Trilobosporites apiverrucatus Couper						
T. canadensis Pocock						
T. purverulentus (Verbitskaya)						
T. trioreticulosus Cookson & Dettmann						
T. spp.						
Antulsporites baculatus (Archangelsky & Gamerro)						
Asterisporites chlonovae (Döring)						
Camarazonosporites microalveolatus Archangelsky & Gamerro						

TAXON \ AGE
Contignisporites cooksonii (Balme)
C. fornicatus Dettmann
C. sp.
Cyatheacidites tectifera Archangelsky & Gamerro
Densoisporites cf. *circumundulatus* (Brenner)
D. corrugatus Archangelsky & Gamerro
D. velatus (Weyland & Krieger)
Duplexisporites spp.
Foraminisporis dailyi (Cookson & Dettmann)
F. microgranulatus Archangelsky
F. variornatus Archangelsky
F. wonthaggiensis (Cookson & Dettmann)
Interulobites distannulatus Archangelsky
I. intraverrucatus (Brenner)
I. pseudoreticulatus Archangelsky
I. cf. *triangularis* (Brenner)
I. variabilis Volkheimer & Quattrocchio
I. sp.
Muricingulisporis sp.
M. annulatus Archangelsky & Gamerro
Polycingulatisporites reduncus (Bolkhovitina)
P. trabeculatus Archangelsky
P. sp.
Polypodiaceoisporites elegans Archangelsky & Gamerro
Sestrosporites pseudoalveolatus (Couper)
S. sp.
Taurocusporites cf. *segmentatus* Stover
T. spp.
Gleicheniidites argentinus Volkheimer
G. aff. *cercinidites* (Cookson)
G. sp.
Coptospora striata Dettmann
Aequitriradites baculatus (Döring)
A. spinulosus (Cookson & Dettmann)
A. verrucosus (Cookson & Dettmann)
Henrisporites elegans Gamerro
H. heteracanthus Gamerro
Minerisporites cheblii Gamerro
M. holobrochatus Gamerro
M. labiosus Baldoni & Taylor
Rouseisporites segmentatus Pocock
Hughesisporites patagonicus Archangelsky
Paxillitriletes magellanica Baldoni & Taylor
P. menendezii Baldoni & Taylor
Laevigatisporites belfordii Burger
Punctatosporites scabratus (Couper)
P. sp.
Schizaeoisporites spp.
Verrucatosporites sp.
Alisporites cf. *similis* (Balme)
A. sp.
Callialasporites dampieri (Balme)
C. cf. *microvelatus* Schulz
C. segmentatus (Balme)
C. trilobatus (Balme)
C. turbatus (Balme)
Dacrycarpites cf. *australiensis* Cookson & Pike
Microcachrydites antarcticus Cookson
M. sp.
Phrixipollenites sp.
Platysaccus sp.
Podocarpidites cf. *ellipticus* Cookson
P. verrucosus Volkheimer
P. sp.
Trisaccites microsaccatus (Couper)
T. cf. *variabilis* (Dev.)
Vitreisporites pallidus (Reissinger)
Araucariacites australis Cookson
A. fissus Reiser & Williams
A. pergranulatus Volkheimer
Balmeiopsis cf. *limbatus* (Balme)
Inapertisporites sp.
Inaperturopollenites turbatus Balme
I. spp.
Pilasporites cf. *allenii* Batten
Asteropollis asteroides Hedlund & Norris
Clavatipollenites hughesii
C. sp.
Cycadopites adjectus (de Jersey)
C. cf. *deterius* (Balme)

TABLE 3. (Continued)

AGE	Be.	Va.	Ha.	Ba.	Ap.	Al.	TAXON	AGE	Be.	Va.	Ha.	Ba.	Ap.
							C. cf. *granulatus* (de Jersey)						
							C. nitidus Balme						
							C. punctatus Volkheimer						
							C. spp.						
							Equisetosporites spp.						
							E. caichiguensis Volkheimer & Quattrocchio						
							Eucommiidites cf. *minor* Groot & Penny						
							Liliacidites sp.						
							Monosulcites minimus Cookson						
							M. subgranulosus Couper						
							M. sp.						
							Retitricolpites sp.						
							Steevesisporites sp.						
							Stephanocolpites mastandreai Volkheimer & Salas						
							Tricolpites cf. *sagax* Norris						
							Classopollis cf. *classoides* (Pflug)						
							C. intrareticulatus Volkheimer						
							C. itunensis Pocock						
							C. simplex (Danzé-Corsin & Laveine)						
							C. torosus (Reisinger)						
							C. sp.						
							Exesipollenites tumulus Balme						
							Gliscopollis sp.						
							Perinopollenites sp.						
							Acanthaulax sp.						
							Achomosphaera sp.						
							Baltisphaeridium spp.						
							Celyphus rallus Batten						
							Comasphaeridium sp.						
							Cribroperidinium spp.						
							C. orthoceras (Eisenack)						
							cf. *Ctenidodinium* sp.						
							Ctenidodinium tenellum Deflandre						
							Cyclusphaera crassa Archangelsky						
							C. patagonica Archangelsky						
							C. psilata Volkheimer & Sepulveda						
							C. (?) *radiata* Archangelsky						
							C. sp.						
							Cymatiosphaera sp.						
							Deflandrea fueguiensis Menendez						
							Endoceratium sp.						
							Filisphaeridium sp.						
							Gonyaulacysta globosa Brideaux						
							G. cf. *jurassica* (Deflandre)						
							G. spp.						
							Hyalosphaera balsensis Volkheimer & Salas						
							H. sp.						
							Hystrichosphaera furcata (Ehrenberg)						
							Hystrichosphaeridium spp.						
							H. cf. *tubiferum* (Ehrenberg)						
							Hystrichosphaerina neuquina Quattrocchio & Volkheimer						
							H. sp.						
							Leiosphaeridia dellapei Volkheimer, Caccavari & Sepulveda						
							L. cf. *hyalina* (Deflandre)						
							L. cf. *staplinii* Pocock						
							L. menendezii Volkheimer, Caccavari & Sepulveda						
							L. spp.						
							Lithodinia sp.						
							Meiourogonyaulax sp.						
							Micrhystridium lymensis gliscum Wall						
							Microdinium sp.						
							Microfasta evansii Morgan						
							Millioudodinium sp.						
							M. ambiguum (Deflandre)						
							Muderongia cf. *staurota* Sarjeant						
							Navifusa sp.						
							Oligosphaeridium pulcherrimum Deflandre & Cookson						
							O. sp.						
							Pareodinia cf. *ceratophora* Deflandre						
							Prolixosphaeridium sp.						
							Pterospermella cf. *goslarensis* (Mädler)						
							Pterosphaeridia volkheimeri Quattrocchio						
							P. sp.						
							Rhaetogonyaulax sp.						
							Schizosporis reticulatus Cookson & Dettmann						
							Scriniodinium sp.						
							Sentusidinium sp.						
							Spiniferites ramosus ramosus (Davey & Williams)						
							S. sp.						
							Systematophora sp.						

(After: Archangelsky, 1964, 1977b; Archangelsky and Gamerro, 1964, 1966a, b, c, 1967; Archangelsky and Seiler, 1980; Archangelsky et al., 1981, 1983, 1984; Gamerro, 1965a, b, 1975a, b, 1977; Quattrocchio and Volkheimer, 1985; Seiler, 1979; Seiler and Moroni, 1984; Volkheimer, 1978; Volkheimer and Pramparo, 1984; Volkheimer and Salas, 1975, 1976; Volkheimer and Sepulveda, 1976; Volkheimer et al., 1977; Musacchio, 1981; Volkheimer and Quattrocchio, 1981; Baldoni and Archangelsky, 1983; Baldoni and Taylor, 1985).

TABLE 4. Late Cretaceous Microflora Described from Southern South America

AGE / TAXON	CENOMANIAN	TURONIAN	CONIACIAN	SANTONIAN	CAMPANIAN	MAASTRICHT
Gobichara groeberi Musacchio						
Gobichara tenuis (Musacchio)				─	─	─
G. walpurgica (Musacchio)				─	─	─
Grambastichara sp.						─
Mesochara ameghinoi Musacchio				─	─	─
Nothochara apiculata Musacchio				─	─	─
Peckisphaera portezueloensis Musacchio				─	─	─
Platychara aff. *caudata* Grambast						─
P. cruciana (Horn af Rantzien)						─
Porochara cf. *gildemeisteri* Koch & Blisenbach						─
Tolypella grambasti Musacchio						─
Cyathidites australis Couper						─
C. cf. *minor* Couper						─
Deltoidospora quiriquinaensis Takahashi						─
Deltoidospora spp.						─
Leiotriletes spp.						─
Henrisporites musacchioi Gamerro			─			
Quiriquinaspora chilensis Takahashi						─
Triletes cf. *parvallatus* Krutzsch						─
Laevigatosporites dehiscens Takahashi						─
L. ovatus Wilson & Webster						─
Polypodiidites inangahuensis Couper						─
Punctatosporites sp.						─
Dacrydiumites florinii Cookson & Pike						─
Parvisaccites cf. *radiatus* Couper						─
Phyllocladidites mawsoni Cookson						─
Piceaepollenites sp.						─
Pinuspollenites concepcionensis Takahashi						─
P. sp.						─
Podocarpidites exiguus Harris						─
P. marwickii Couper						─
P. microreticuloidata Cookson						─
P. multesimus (Bolkhovitina)						─
P. spp.						─
Trisaccites microsaccatus (Couper)						─
Tsugaepollenites mesozoicus Couper						─
Inaperturopollenites laevigatus Takahashi						─
Clavatipollenites hughesii Couper						─
Liliacidites crassibaculatus Freile						─
L. variegatus Couper						─
Monocolpopollenites sp.						─
Monosulcites cf. *granulatus* Couper						─
M. pseudospinosus Freile						─
Retitricolporites sp.						─
Rhoipites sp.						─
Tricolpites microreticulatus (Takahashi)						─
T. retielegans Freile						─
T. spp.						─
Tricolpopollenites fallax (Potonié)						─
Tricolporites spp.						─
Tricolporopollenites minor Takahashi						─
T. microporifer Takahashi						─
T. punctulosus Takahashi						─
T. sp.						─
Anacolosidites sp.						─
Beaupreaidites sp.						─
Myrtaceidites cf. *mesonesus* Cookson & Pike						─
Nothofagidites acromegacanthus Menendez & Caccavari						─
N. anisoechinatus Menéndez & Caccavari						─
N. dorotensis Romero						─
N. fortispinulosus Menéndez & Caccavari						─
N. fueguensis Menéndez & Caccavari						─
N. kaitangatus Te Punga						─
N. paucispinosus Menéndez & Caccavari						─
N. rocaensis Romero						─
N. saraensis Menendez & Caccavari						─
N. suggatei Couper						─
N. cf. *visserensis* Romero						─

AGE / TAXON	CENOMANIAN	TURONIAN	CONIACIAN	SANTONIAN	CAMPANIAN	MAASTRICHT
N. waipawensis Couper						─
Proteacidites minor Takahashi						─
Psilatricolpites patagonicus Freile						─
P. pulcherrimum Freile						─
Triatriapollenites maedae Takahashi						─
Triorites cf. *minor* Couper						─
T. minusculus Mc Intyre						─
Triporopollenites shimensis Takahashi						─
T. festatus Takahashi						─
Amphorosphaeridium fenestratum Davey						─
Areoligera sp.						─
Areosphaeridium argentinensis Pothe de Baldis	─					
Baltisphaeridium sp.						─
Callaiosphaeridium asymmetricum (Defl. & Court.)	─					
Chatangiella cf. *vnigri* Vozzhennikova	─					
Chlamydophorella nyei Cookson & Eisenack	─					
Cleistosphaeridium cf. *armatum* (Defl.)	─					
Cordosphaeridium inodes (Klumpp)	─					
Coronifera oceanica Cooks. & Eisen.	─					
Cyclonephelium cf. *compactum* Deflandre	─					
C. cf. *distinctum* Defl. & Cooks.	─					
Deflandrea dartmooria Cooks. & Eisen.						─
D. diebelii Alberti						─
D. cretacea Cookson					─	─
D. fueguensis Menendez						─
D. quiriquinaensis Takahashi						─
D. striata Drugg.						─
? Diconodinium sp.						─
Eisenackia ornata Cookson & Eisenack						─
Florentinia cf. *deanei* (Davey & Williams)	─					
F. laciniata (Davey & Verdier)	─					
Heslertonia heslertonensis (Neale & Sarjeant)	─					
Heterosphaeridium heteracanthum (Defl. & Cooks.)	─					
Homotryblium cf. *tenuispinosum* Davey	─					
Hystrichodinium cf. *isodiametricum* (Cooks. & Eisen.)	─					
H. pulchrum Deflandre	─					
H. voigtii (Alberti)		─				
Hystrichosphaeridium echinatum Menendez						─
H. tubiferum (Ehrenberg)	─					
Hystrichosphaeropsis ovum Defl.						─
Isabelidinium ? acuminatum (Cooks. & Eisen.)	─					
I. cretaceum (Cookson)						─
I. pellucidum (Defl. & Cooks.)						─
Lanternosphaeridium doubingeri Troncoso						─
Leiosphaeridia sp.						─
Lejeunia tricuspis (Wetzel)						─
L. magnifica (Stanley)						─
Michrystridium deflandrei Valensi						─
M. piliferum Deflandre						─
Nelsoniella aceras Cookson & Eisenack						─
Nematosphaeropsis sp.						─
Odontochitina operculata (Wetzel)						─
O. aff. *porifera* Cooks.						─
O. striatoperforata Cooks. & Eisen.						─
Oligosphaeridium cf. *antophorum* (Cooks. & Eisen.)	─					
Operculodinium centrocarpum (Defl. & Cooks.)						─
Palaeocystodinium australianum (Cooks.)						─
Palaeohystrochophora infusorioides Delfl.						─
Palambages morulosa Wetzel						─
Saeptodinium ? sp.						─
Spiniferites ex gr. *crassipellis* (Defl. & Cooks.)						─
S. ramosus Loeblich & Loeblich						─
Tanyosphaeridium variecalamum Davey & Williams						─
Tenuahystricella sp.						─
Trichodinium chilensis Troncoso & Doubinger					─	─
Trythyrodinium fragile Davey					─	─
T. vermiculatum (Cooks. & Eisen.)	─					
Xenascus ceratioides (Deflandre)	─					

(After: Menéndez 1965; Freile, 1972; Musacchio, 1973; Romero, 1973; Archangelsky and Romero, 1974; Menéndez and Caccavari, 1975; Takahashi, 1978, 1979; Uliana and Musacchio, 1979, Troncoso and Doubinger, 1980; Gamerro and Archangelsky, 1981; Pothe de Baldis, 1987).

TABLE 5. Early Cretaceous Invertebrates (Exclusive of Ammonites) Described from Chubut and Andean Basins

Age columns: BERRIASIAN | VALANGINIAN | HAUTERIVIAN | BARREMIAN | APTIAN | ALBIAN

PORIFERA
Paleospongilla chubutensis Ott
Rhaphidonema meandraceum Fritzsche

CNIDARIA
Actinastrea hexamera Fritzsche
A. minima From.
"Cyathophora" steinmanni Fritzsche
Isastrea cf. *eturbensis* From.
Placocoenia neuquensis Gerth
Columnastrea antigua Gerth

BRACHIOPODA
"Terebratula" collinaria d'Orb.
"T". tamarindus Sow.
Lingula truncata Sow.

BIVALVIA
"Solemya" neocomiensis (Haupt)
Grammatodon mendozanus (Weaver)
G. securis (Leymerie)
Cucullaea gabrielis Leymerie
C. lotenoensis Weaver
C. cf. *salevensis* Loriol
Megacucullaea? neuquensis (Weaver)
Modiolus cf. *ligeriensis* (d'Orbigny)
M. subsimplex (d'Orbigny)
Pinna robinaldina d'Orbigny
Gervillaria alaeformis (Sowerby)
G. militaris (Burckhardt)
Gervillia anceps Deshayes
Inoceramus backlundi Sokolov
"Inoceramus" curacoensis Weaver
Inoceramus sp.
Retroceramus? retrorsus (Keyserling)
Isognomon aff. *americanus* (Forbes)
I. lotenoensis (Weaver)
I. nanus (Behrendsen)
I. quintucoensis (Weaver)
I. ricordeanus (d'Orbigny)
Chlamys (Chlamys) cf. *puzosiana* (Matheron)
C. (C.) robinaldina (d'Orbigny)
C. vacaensis Weaver
C. wayensis Leanza & Castellaro
Probinnites leymerii (Deshayes)
Eopecten covuncoensis (Weaver)
"Plicatula" vacaensis Weaver
Anditrigonia eximia(Philippi)
A. lamberti Levy
A. picunensis (Weaver)
A. subnodosa Levy
Antutrigonia opistolophophora (Lambert)
A. groeberi (Weaver)
Apiotrigonia (A.) progonos (Paulcke)
Buchotrigonia steinmanni (Lisson)
Mediterraneotrigonia hondeana (Lea)
Myophorella (M.) clavellata (Parkinson)
M. coihuicoensis (Weaver)
M. (Haidaia) volkheimeri Leanza & Garate
M. (Promyophorella) garatei Leanza
Pterotrigonia (P.) aliformis (Parkinson)
P. delafossei (Bayle & Coquand)
P. nepos (Paulcke)
P.(Scabrotrigonia) crenulata peruana (Paulcke)
P. (S.) transatlantica (Behrendsen)
Quoiecchia sigeli Leanza & Garate
Rutitrigonia agrioensis (Weaver)
R. longa Agassiz
R. l. undulatostriata (Paulcke)
Steinmannella (S.) erycina (Philippi)
S. (S.) e. haupti (Lambert)
S. (S.) neuquensis (Burckahrdt)
S. (S.) splendida (Leanza)
S. (S.) steinmanni (Philippi)
S. (S.) transitoria (Steinmann)

S. (S.) t. curacoensis (Weaver)
S. (Macrotrigonia) vacaensis (Weaver)
Trigonia angustecostata Behrendsen
T. (T.) carinata Agassiz
T. (T.) c. aliexpandita Leanza & Garate
T. wiedmanni Leanza & Garate
Vaugonia (V.) chunumayensis (Jaworski)
Anophistodon? obesum (Philippi)
Aulacopleurum? trapezoideum (Philippi)
"Lucina" lotenoensis Weaver
L. leufuensis Weaver
L. neuquensis Haupt
L. subporrecta Leanza & Castellaro
Eriphyla agrioensis Weaver
E. argentina Burckhardt
E. a. picunleufuensis Weaver
E. lotenoensis Weaver
Solecurtus neuquensis Weaver
Protocardia (Protocardia) aff. *multistriata* (Shumard)
P. cf. *peregrinorsa* (d'Orb.)
Cyprina (?) argentina Behrendsen
Ptychomia koeneni Behrendsen
Sphaera (?) koeneni (Behrendsen)
Aphrodina quintucoensis (Weaver)
A.? sp.
Thracia aequilatera Behrendsen
T. ? sp.
Myoconcha transatlantica Burckhardt
Astarte notica Leanza & Castellaro
Disparilia? sp.
Corbula bodenbenderi Behrendsen
C. inflata Behrendsen
C. nana Behrendsen
Panopea dupiniana d'Orbigny
P. neocomiensis (Leymerie)
Pholadomya agrioensis Weaver
P. gigantea (Sowerby)
P. sanctaecrucis Pictet & Campiche
Homomya antofagastensis Leanza & Castellaro
H. subandina Leanza & Castellaro
Pleuromya harringtoni Leanza & Castellaro
Plectomya cf. *silenensis* (Richards)
Deltoideum lotenoensis (Weaver)
D. roemeri (Quenstedt)
Aetostreon latissimum (Lamarck)
Ceratostreon minos (Coquand)

GASTROPODA
Protohemichenopus acutus (Behrendsen)
P. neuquensis Camacho
Pleurotomaria gerthi Weaver
Tylostoma jaworskii Weaver
Turritella aff. *lineolata* Roemer
Cerithium cf. *heeri* Pictet & Campiche
Capulus argentinus Haupt
Natica cf. *bulimoides* (Deshayes)
Cinulia andina (Haupt)
Homalopoma sp.
Lissochilus sp.
"Turritella" sp.
Polinices ? bodenbenderi (Behrendsen)
Euspira sp.
Cinulia ? sp.

ARTHROPODA
Meyeria rapax Harbort

ANNELIDA
Parsimonia antiquata (Sowerby)
Sarcinella occidentalis (Leanza & Castellaro)
Rotularia (R.) phillipsii (Roemer)

ECHINOIDS
Pygaster gerthi Weaver
Phymosoma mollense (Paulcke)
Holectypus ? numiserialis Gabb
Clypeopygus robinaldinus d'Orbigny
Heteraster chilensis (Philippi)
Hemiaster wayensis Larrain

(After: Behrendsen 1891-92; Philippi 1899; Burckhardt 1900 a, b, 1903; Paulcke 1903; Haupt 1907; Weaver 1931; Frenguelli 1944 Lambert 1944; Sokolov 1946; Leanza and Castellaro 1955; Corvalan and Perez 1958; Ott and Volkheimer 1972; Volkheimer and Ott 1973; Larrain 1975, 1985; Reyes and Perez 1978, 1979; Leanza 1981b, 1985; Perez et al. 1981; Reyes et al. 1981; Perez and Reyes 1983 a, b; Leanza and Garate 1983, 1986; Manceñido and Damborenea 1984; Aguirre Urreta 1985b).

TABLE 6. Early Cretaceous Invertebrates (Exclusive of Ammonites) Described from Austral Basin

TAXON \ AGE	BERRIASIAN	VALANGINIAN	HAUTERIVIAN	BARREMIAN	APTIAN	ALBIAN
BRACHIOPODA						
Kingena (?) andina Feruglio	—					
BIVALVIA						
Nucula pueyrrydonensis Stanton			—			
Nuculana (?) corbuliformis (Stanton)				—		
Isoarca (?) eximia Feruglio	—					
Barbatia erix Leanza	—					
Megacucullaea cf. kraussi (Tate)		—				
Mytilus (?) argentinus Stanton			—			
Pinna patagoniensis Favre			—			
Pinna sp.			—			
Gervillia hatcheri Stanton	—					
Inoceramus anomiaeformis Feruglio			—			
I. aff. cripsi Sowerby			—			
Maccoyella bonarellii (Leanza)					—	
Oxytoma tardensis Stanton			—			
Entolium argentinum (Stanton)			—			
Camptonectes pueyrrydonensis Stanton			—			
Chlamys (Aequipecten) octoplicoides (Hertlein)	—	–	–	–	—	
Pecten cascadensis Leanza	—					
P. quemadensis Feruglio	—					
P. aff. orbicularis Sowerby						—
Aucellina radiatostriata Bonarelli	—					—
Lima sp.	—					
Trigonia heterosculpta Stanton			—			
T. palaeopatagonica Piatnitzky						—
T. subventricosa Stanton			—			
Iotrigonia stowi aisenensis Reyes		—				
Megatrigonia conocardiiformis (Krauss)	—					—
M. rogersi (Kitchin)	– —					
Pterotrigonia (Rinetrigonia) feruglioi (Piatnitzky)						—
Steinmanella (S.) herzogi (Hausmann)		—				
S. (S.) holubi (Kitchin)		—				
S. (S.) cf. transitoria (Steinmann)		—				
S. (Macrotrigonia) katterfeldensis Camacho & Olivero			—			
S. (M.) maxima Camacho & Olivero		—				
S. (M.) posadensis Camacho & Olivero		—				

TAXON \ AGE	BERRIASIAN	VALANGINIAN	HAUTERIVIAN	BARREMIAN	APTIAN	ALBIAN
S. (M.) vacaensis (Weaver)			—			
?Lucina antartandica Leanza		—				
Astarte fossamancinii Leanza	—					
A. peralta Stanton			—			
A. postsulcata Stanton			—			
Opis gortanii Feruglio			—			
Protocardia (Tendagurium) sp.			—			
Mactra (?) sp.			—			
Tellina sp.			—			
Solecurtus (?) limatus Stanton	—					
"Cyprina" feruglioi Leanza			—			
Tapes (?) patagonica Stanton			—			
Corbula crassatelloides Stanton			—			
Martesia argentinensis Stanton			—			
Pleuromya latisulcata Stanton			—			
Gryphaea cf. usta Feruglio	—					
G. aff. vesiculosa Sowerby			—			
Exogyra aff. ariana Stol.			—			
Aetostreon sp.			—			
Ostrea bononiaformis Leanza	—					
GASTROPODA						
Pleurotomaria tardensis Stanton			—			
Vanikoro (?) sp.			—			
Aporrhais patagonica Stanton			—			
Euspira constricta (Stanton)			—			
E. pueyrrydonensis (Stanton)			—			
Tornatellaea patagonica Stanton			—			
Cinulia australis Stanton			—			
ARTHROPODA						
Palaeastacus terraereginae (Eth.)				—		—
P. sp.				—		
Eryma cf. sulcata Harbort			—			
Enoploclytia sp.				—		
Hoploparia longimana Sow.				—		
Hoploparia sp.				—		
Protocallianassa patagonica Aguirre Urreta				—		
ANNELIDA						
Rotularia callosa (Stol.)	—	–	–	–	–	—

(After: Stanton 1901; Favre 1908; Bonarelli and Nágera 1921; Richter 1925; Feruglio 1936 - 7; Piatnitzky 1938; Camacho 1949; Leanza 1967 c, 1968; Levy 1967 a, b; Fuenzalida 1968; Reyes 1970; Waterhouse and Riccardi 1970; Cecioni and Charrier 1974; Riccardi 1977; Macellari 1979; Aguirre Urreta and Ramos 1981 b; Aguirre Urreta 1983, 1985 b; Olivero 1983; Camacho and Olivero 1985).

TABLE 7. Late Cretaceous Invertebrates (Exclusive of Ammonites) Described from Patagonia and Chile

AGE / TAXON	CENOMANIAN	TURONIAN	CONIACIAN	SANTONIAN	CAMPANIAN	MAASTRICH
BIVALVIA						
Malletia gracilis Wilckens						
Neilo pencana (Phil.)						
cf. *Cucullaea antarctica* Wilckens						
C. argentina Feruglio						
Glycymeris cuneiformis Hupé						
G. feruglioi (Celeste)						
Modiolus sp.						
Pinna morenoi Wilckens						
Inoceramus andinus Wilckens						
I. lateris García & Camacho						
I. menchaquilensis Camacho						
I. steinmanni Wilckens						
Chlamys modestus (Camacho)						
Pecten bagualensis Wilckens						
Chlamys chilensis (d'Orb.)						
"*Pecten*" *delicatulus* Phil.						
P. mabuidaensis Weaver						
P. modestus Camacho						
P. sp.						
Aucellina cf. *parva* Stol.						
Anomia parva Stol.						
A. solitaria Wilckens						
Paranomia? sp.						
Lima? patagonica Wilckens						
Limatula angusta Camacho						
Diplodon amptheatri Manceñido & Damborenea						
D. bodenbenderi Doello Jurado						
D. pehuenchensis Doello Jurado						
"*Mactra*" *tumida* Phil.						
Mulinoides araucana (d'Orbigny)						
M. colossea (Phil.)						
Ceroniola australis Gabb						
Tellina largillierti d'Orbigny						
Donax (Notodonax) annaeugeniae Feruglio						
Solecurtus sp.						
Neocorbicula dinosauriorum (Doello Jurado)						
N. pehuenchensis (Doello Jurado)						
"*Venus*" *cyprinoides* Feruglio						
"*Meretrix*" *alta* (Phil.)						
M. pacifica (Möricke)						
M. rothi Wilckens						
"*Cytherea*" *auca* d'Orbigny						
C. australis Feruglio						
C. orbignyana Gabb						
C. steinmanni Phil.						
Pectunculus feruglioi Celeste						
Corbula chilensis Wetzel						
C. sehuena Ihering						
Panopaea inferior Wilckens						
P. simplex Hupé						
P. thomasi Ihering						
Martesia cazadoriana Wilckens						
Teredo cf. *rugaardensis* Grönwall & Harder						
Pholadomya leanzai García & Camacho						
Patagocardia peterseni Doello Jurado						
Buchotrigonia (B.) topocalmensis Perez & Reyes						
Iotrigonia byriformis (Wilckens)						
Linotrigonia (Oistotrigonia) antarctica (Wilckens)						
L. (O.) pygoscelium (Wilckens)						
Megatrigonia aff. *conocardiiformis* (Piatnitzky)						
Pacitrigonia alamensis Levy						
P. banetiana (d'Orb.)						
P. patagonica (Feruglio)						
P. regina (Wilckens)						
Pterotrigonia cazadoriana (Wilck.)						
P. pseudocaudata (Hector)						
P. (Rinetrigonia) feruglioi (Piatnitzky)						
P. (R.) nullorum Levy						
P. (R.) windhauseniana (Wilckens)						
P. sp.						
Trigonia palaeopatagonidica Piatnitzky						
Lucina grangei d'Orbigny						
L. sp.						
Venericardia feruglioi Petersen						
Astarte venatorum Wilckens						
Cardium acuticostatum d'Orbigny						
Protocardia shebuenensis Feruglio						
Labillia ferrieri Phil.						
L. luisa Wilckens						
L. veneriformis (Hupé)						
Mactra gabbi Philippi						
M. bualpensis Philippi						
Veniella pampaensis (Leanza & Hünicken)						
Gryphaea mendozana Ihering						
G. miradorensis Petersen						
G. pyrotheriorum Ihering						
G. regionis Camacho						
Exogyra guaranitica (Ihering)						
E. mendozana Ihering						
Gryphaeostrea callophylla (Ihering)						
Ostrea cf. *arkotensis* Stol.						
O. clarae Ihering						
O. hippopodium Nilsson						
O. lingua Ihering						
O. rionegrensis Ihering						
O. wilckensi Ihering						
GASTROPODA						
Aphrodina auca (Gabb)						
"*Trochus*" *ovallei* Phil.						
T. quiriquinae Phil.						
Solariella unio Phil.						
Turritella cf. *delmar* Gardner						
T. donarium García & Camacho						
T. malaspina? Ihering						
T. soaresana Hartt.						
T. spp.						
Paleoanculosa amegbiniana (Doello Jurado)						
P. patagonica Parodiz						
Melania macrochilinoides Doello Jurado						
M. pehuenchensis (Doello Jurado)						
Lioplacodes feruglioi Parodiz						
L. wichmanni (Doello Jurado)						
Potamolithus? windhauseni (Parodiz)						
Potamides patagonensis Ihering						
Aporrhais sp.						
Perissoptera sp.						
Struthiolariopsis ferrieri Phil.						
Pugnellus aff. *uncatus* (Forbes)						
"*Natica*" *cerreria* Wilckens						
N. sehuena Piatnitzky						
N. sp.						
Gyrodes euryomphala Phil.						
G. cf. *tenella* Stol.						
Euspira singularis (Möricke)						
E. sp.						
Ampullospira dubia Petersen						
Ampullella australis (d'Orbigny)						
Charona sp.						
Galeropsis laevis Phil.						
Cominella praecursor Wilckens						
C. patagonica Feruglio						
Fusinus chilinus (d'Orbigny)						
F. difficilis (d'Orb.)						
F. metzdorfi (Phil.)						
Pyropsis hombroniana d'Orb.						
Tudicula rugosa Phil.						
Cancellaria cf. *nitidula* Holzapfel						
Struthioptera pastorei Camacho						
Drillia cf. *acutinoda* Phil.						
"*Acteonina*" sp.						
Acteon australis Feruglio						
A. meridionalis Feruglio						
Eoactaeon linteus (Conrad)						
Trochactaeon? patagonica (Feruglio)						
Cinulia pauper Wilckens						
Eriptycha chilensis d'Orbigny						
Scaphander chilensis d'Orbigny						
"*Atys*" *subglobosa* Phil.						
Retusa scutata Wilckens						
"*Scalaria*" *araucana* Phil.						
S. chilensis d'Orbigny						
S. quiriquinae Möricke						
S. steinmanni Möricke						
S. sp.						
"*Physa*" *doeringi* Doello Jurado						
"*P.*" *wichmanni* Parodiz						
Planorbis sp.						
SCAPHOPODA						
Dentalium cazadorianum Wilckens						
D. chilense d'Orbigny						
D. sp.						
ARTHROPODA						
Glyphea oculata Woods						
Hoploparia antarctica Wilckens						
Callianassa burckhardti Boehm						
C. aff. *darchiaci* Milne Edw.						
ECHINOIDS						
Holaster feruglioi Melinossi						
H. mortenseni Bernasconi						
Cardiaster patagonicus Steinmann						
Schizaster deletus Wilckens						
Micraster? sp.						

(After: White 1890; Behrendsen 1891-2; Möricke 1895; Ihering 1899, 1907; Wilckens 1904, 1907 a, 1921; Steinmann and Wilckens 1908; Doello Jurado 1922, 1927; Wichmann 1927 a, b; Heinz 1928; Weaver 1931; Feruglio 1935, 1936 a, b, 1936 - 1937; Melinossi 1935; Piatnitzky 1938; Petersen 1946; Bernasconi 1954; García and Camacho 1965; Camacho 1967 b, 1968, 1969, 1971; Levy 1967 a, b, 1985; Parodiz 1969; Leanza and Hünicken 1970; Manceñido and Damborenea 1984; Aguirre Urreta, 1985 b).

TABLE 8. Early Cretaceous Microfauna Described from Austral Basin

TAXA	STAGE	ESPERANZIAN (TITH.-HAUT.)	PRATIAN (BARREMIAN)	TENERIFIAN (APTIAN-ALBIAN)	PENINSULIAN (ALBIAN-CENOM.)	LAZIAN (CENOM.-SANT.)	RIESCOIAN (SANT.-MAASTR.)
FORAMINIFERA							
Bathysiphon sp.							
B. cf. *vitta* Nauss							
Saccammina cf. *lathrami* Tappan							
Ammodiscus cretaceus (Reuss)							
Glomospira corona Cushman & Jarvis							
Glomospirella sp.							
Reophax cf. *troyeri* Tappan							
Psamminopelta minima (Cushmann & Renz)							
Haplophragmoides sp.							
H. cf. *gigas minor* Nauss							
H. cf. *bulloides* (Beissel)							
Cyclammina cancellata Brady							
Ammobaculites sp.							
A. barrowensis Tappan							
Spiroplectammina grzybowskii Frizzell							
S. gutierrezi Cañon & Ernst							
S. cf. *laevis* Roemer							
S. spp.							
Textularia sp.							
Trochammina ex gr. *bulloides* (Parker & Jones)							
T. cf. *bulloidiformis* (Grzybowski)							
Gaudryina complanata Reuss							
G. healyi Finlay							
G. juliana Malumian & Masiuk							
G. cf. *laevigata* Franke							
Spiroplectinata annectens (Parker & Jones)							
Pseudospiroplectinata ona Malumian & Masiuk							
Tritaxia porteri Cañon & Ernst							
T. gaultina gaultina (Morozova)							
T. g. australis Malumian & Masiuk							
Uvigerinammina cf. *jankoi* Majzon							
Gaudryinopsis ex gr. *taylleuri* Tappan							
Dorothia bulleta (Carsey)							
D. mordojovichi Cañon & Ernst							
D. oxycona (Reuss)							
Marssonella kummi Zedler							
M. subtrochus Bartenstein							
Nodosaria fontannesi (Berthelin)							
N. obscura Reuss							
N. lorneiana (d'Orbigny)							
Astacolus gibber gibber (Espitalie & Sigal)							
A. g. cf. *barremianus* (Michael)							
A. microdictyotos Espitalie & Sigal							
A. cf. *mundus* (Cushman)							
A. mutilatus Espitalie & Sigal							
A. skyringensis Todd & Kniker							
A. tricarinella (Reuss)							
Citharina geisendoerferi (Franke)							
C. sparsicostata (Reuss)							
Dentalina sp.							
D. linearis (Roemer)							
D. nana Reuss							
Frondicularia cf. *archiaciana* d'Orb.							
F. disjuncta Belford							
F. mucronata Reuss							
F. simplicissima Ten Dam							
Lagena hauteriviana cylindracea Bartenstein & Brand							
Lenticulina ambanjabensis Espitalie & Sigal							
L. biexcavata (Myatliuk)							
L. crepidularia (Roemer)							
L. gaultina (Berthelin)							
L. aff. *haesitans* Espitalie & Sigal							
L. heiermanni Bettenstaedt							
L. cf. *malakialinensis* (Espitalie & Sigal)							
L. nodosa (Reuss)							
L. n. malumiani Aubert & Bartenstein							
L. pulchella (Reuss)							
L. reyesi Cañon & Ernst							
Marginulina ex gr. *bronni* (Roemer)							
M. bullata Reuss							
M. robusta robusta (Reuss)							
M. spp.							
Marginulinopsis bettenstaedti (Bartenstein & Brand)							
M. sp.							
Palmula cf. *asiatica* (Fursenko)							
P. cf. *malakialinensis* (Espitalie & Sigal)							
Planularia crepidularis Roemer							
Pseudonodosaria humilis (Roemer)							
P. manifesta (Reuss)							
Saracenaria compacta Espitalie & Sigal							
S. pravosiavlevi Fursenko & Polenova							
S. triangularis (d'Orb.)							
S. tsaramandrosoensis Espitalie & Sigal							
Vaginulina sp.							
V. kochii Roemer							
Lingulina bettenstaedti (Zedler)							
L. nodosaria Reuss							
Globulina prisca Reuss							

TAXA	STAGE	ESPERANZIAN (TITH.-HAUT.)	PRATIAN (BARREMIAN)	TENERIFIAN (APTIAN-ALBIAN)	PENINSULIAN (ALBIAN-CENOM.)	LAZIAN (CENOM.-SANT.)	RIESCOIAN (SANT.-MAASTR.)
Pseudopolymorphina martinezi (Cañon & Ernst)							
Webbinella subhemisphaerica Franke							
Bullopora sp.							
B. tuberculata (Sollas)							
Ramulina spp.							
R. aculeata Wright							
Tristix acutangulus (Reuss)							
T. insignis (Reuss)							
Pyramidina minima (Brotzen)							
Sphaeroidina bulloides d'Orb.							
Bolivina incrassata Reuss							
B. i. gigantea Wicher							
Bolivinoides dracodorreeni Finlay							
Stilostomella aspera (Reuss)							
Praeglobobulimina kickapooensis (Cole)							
P. pupoides (d'Orb.)							
Discorbis minima Vieaux							
Epistominella texana (Cushman)							
Valvulineria fueguina Malumian & Masiuk							
Notoconorbina leanzai Malumian & Masiuk							
Heterohelix globulosa (Ehrenberg)							
H. moremani (Cushman)							
H. punctulata (Cushman)							
H. reussi (Cushman)							
H. striata (Ehrenberg)							
H. ultima tumida (Cushman)							
Globigerinelloides asperus (Ehrenberg)							
G. bentonensis Morrow							
G. caseyi (Bolli, Loeblich & Tappan)							
G. gyroidinaeformis (Moullade)							
G. multispina (Lalicker)							
G. sp.							
Hedbergella delrioensis (Carsey)							
H. hohndelensis Olsson							
H. planispira (Tappan)							
H. portsdownensis (Williams-Mitchell)							
Globotruncana chileana Cañon & Ernst							
G. cretacea (d'Orb.)							
G. coronata Bolli							
G. lapparenti Brotzen cf. *tricarinata* (Quereau)							
G. marginata (Reuss)							
Whiteinella sp.							
W. baltica Douglas & Rankin							
"*Rugoglobigerina*" *bulbosa* Belford							
R. pilula Belford							
R. plana Belford							
R. rugosa (Plummer)							
Hastigerina escheri escheri (Kaufmann)							
Globigerina washitensis (Carsey)							
Archaeoglobigerina wenzeli (Cañon & Ernst)							
Planulina popenoei (Trujillo)							
Notoplanulina australis Malumian & Masiuk							
N. rakauroana (Finlay)							
Cibicides cf. *djaffaensis* Sigal							
Caucasella hoterivica (Subbotina)							
Allomorphina conica Cushman & Todd							
A. paleocenica Cushman							
Quadrimorphina allomorphinoides (Reuss)							
Nonionella austiniana Cushman							
Pullenia bulloides (d'Orb.)							
P. cretacea Cushmann							
P. natlandi Cañon & Ernst							
Alabamina australis australis Belford							
Globorotalites sp.							
Lingulogavelinella globosa (Brotzen)							
L. magallanica Malumian & Masiuk							
Gyroidinoides globosa (Hagenow)							
G. nodus (Belford)							
G. spp.							
Anomalina rubiginosa Cushman							
Anomalinoides sp.							
Gavelinella eriksdalensis (Brotzen)							
G. murchisonensis (Belford)							
G. sp.							
G. (*Berthelina*) ex gr. *berthelini* (Keller)							
Stensioina infrafossa (Finlay)							
Ceratolomarckma? sp.							
Conorboides sp.							
Epistomina caracola caracola (Roemer)							
Hoeglundina elegans (d'Orb.)							
"*Pseudosigmoilina*" cf. *antiqua* Franke							
OSTRACODA							
Novocythere santacruziana Rossi de Garcia							
Cytherella algoaensis Brenner & Oertli							
Cytheropteron? sp.							
RADIOLARIA							
Spumellina gen. et. sp. indet.							

(After: Malumian 1968; Malumian et al. 1971, 1972; Flores et al. 1973; Malumian and Masiuk 1973, 1975, 1976 a, b, 1978; Natland et al. 1974; Malumian and Proserpio 1979; Rossi de García, 1979; Malumian and Nañez 1983; Kielbowicz et al. 1984).

TABLE 9. Early Cretaceous Microfauna Described from Chubut and Andean Basins

TAXON / AGE	BERRIASIAN	VALANGINIAN	HAUTERIVIAN	BARREMIAN	APTIAN	ALBIAN
FORAMINIFERA						
Reophax aff. *neominutissima* Bartenstein			—			
Ammobaculites subcretaceous Cushman & Alexander			—			
Trochammina depressa Lozzo			—			
Marssonella sp.			—			
Nodosaria bettenstaedti Musacchio		—	—			
N. aff. *fontannessi flexocostata* Khan			—			
Astacolus cf. *ambajabensis* Espitalié & Sigal			—			
A. calliopsis (Reuss)			—			
A. explicatus Espitalié & Sigal			—			
A. mutilatus Espitalié & Sigal			—			
A. schloembachi (Reuss)			—			
Citharina cristellarioides (Reuss)			—			
C. orthonota (Reuss)			—			
Dentalina cf. *communis* d'Orbigny			—			
D. linearis Roemer			—			
D. terquemi d'Orb.			—			
Frondicularia sp.			—			
Lagena sulcata (Walker & Jakob)			—			
Lenticulina cf. *circumcidanea* (Berthelin)			—			
L. collignoni Espitalie & Sigal			—			
L. nodosa (Reuss)			—			
L. aff. *subtilis* (Wisniovsky)			—			
L. sp.			—	—		
Marginulina pyramidalis (Koch)			—			
M. sp.			—			
Marginulinopsis bettenstaedti (Bart. & Brand)			—			
M. piculeufuensis Musacchio		—	—			
M. p. tenuistriata Simeoni			—			
Neoflabellina ? sp.			—			
Palmula sp.			—			
Planularia crepidularis (Roemer)			—			
P. cf. *madagascariensis* Espitalié & Sigal			—			
Saracenaria tsaramandrosoensis Espitalié & Sigal			—			
S. valanginiana Bartenstein & Brand			—			
S. ? sp.			—	—		
Vaginulina kummi Eichenberg			—			
V. cf. *riedeli* Bartenstein & Brand			—			
Vaginulinopsis sp.			—			
Lingulina nodosaria (Reuss)			—			
L. simplicissima (Ten Dam)			—			
Polymorphina oblonga Masiuk & Viña			—			
P. sp.			—			
Globulina prisca ? Reuss			—			
Eoguttulina anglica Cushman & Ozawa			—			
E. cf. *liassica* (Strickland)			—			
Guttulina spp.			—			
Pseudopolymorphina cf. *roanokoensis* Tappan			—			
Pyrulina cylindroides (Roemer)			—			
Spirofrondicularia frondicularioides (Chapman)			—			
Tristix acutangulus (Reuss)			—			
T. sp.			—			
Patellina subcretacea Cushman & Alexander			—			
Hedbergella infracretacea (Glaessner)			—			
Conorboides ? sp.			—			
Epistomina australis Masiuk & Viña			—			
E. caracolla (Roemer)			—			
E. cf. *ornata* (Roemer)			—			
Reinholdella hofkeri (Bart. & Brand)			—			

TAXON / AGE	BERRIASIAN	VALANGINIAN	HAUTERIVIAN	BARREMIAN	APTIAN	ALBIAN
OSTRACODA						
"Mantelliana" ulianai Musacchio & Palamarczuk						—
Cypridea amerikana Musacchio				—		
C. australis Musacchio			—			
C. craigi Musacchio			—			
C. cymerata Musacchio			—			
C. aff. *diminuta* Vanderpool			—			
C. elongata Musacchio			—			
C. feruglioi Musacchio			—			
C. ludica Musacchio				— -		
C. modesta Musacchio			—			
C. ranquiliensis Musacchio & Palamarczuk			—			
C. r. papilata Musacchio & Palamarczuk			—			
C. subcuadrata Musacchio			—			
C. volubilis Musacchio			—			
C. v. elipsoidea Musacchio			—			
C. sp.		—			— -	
Rayosoana quilimalensis Musacchio			—			
R. sp.			—			
Clinocypris ? sp. A			—			
Darwinula aff. *oblonga* (Roemer)			—			
D. spp.			—			
Leanzacythere leanzai García & Leanza		—				
L. trahuncuraensis García & Leanza		—				
Monoceratina sp.			—			
Looneyellopsis chinamuertensis (Musacchio)			—			
Wolburgiopsis oguitorum (Musacchio & Palamarczuk)			—			
W. plastica (Musacchio)			—			
"Wolburgia" sarunata Musacchio & Palamarczuk			—			
Rostrocytheridea ? covuncoensis Musacchio			—			
Cushmanidea ? sp.			—			
"Paranotacythere" maruchoensis Musacchio			—			
Paranotacythere sp.			—			
Paracytheridea ? spp.				—		
Procytherura kroemmelbeini Musacchio			—			
P. aff. *maculata* Brenner & Oertli			—			
Bisulcocypris barrancalensis Musacchio			—			
Huillicythere grambasti Musacchio			—			
"Gomphocythere" dorsoacuminata Musacchio			—			
"G." herreriensis Musacchio			—			
G. neuquensis Musacchio				— -		
Theriosynoecum ? piculeufuensis Musacchio			—			
Progonocythere reticulata Dingle			—			
Amphicytherura (Sondagella) lestai Musacchio		—				
A. (Sondagella) theloides Dingle			—			
Cytherelloidea agrioensis Musacchio & Abrahamovich			—			
C. cf. *agyroides* Dingle			—			
C. amosi (Musacchio)			—			
C. andica Musacchio			—			
C. argentina Musacchio & Abrahamovich			—			
C. frenguellii Musacchio & Abrahamovich			—			
C. pseudoagyroides Musacchio & Abrahamovich			—			
C. volkheimeri Musacchio & Abrahamovich			—			

(After: Musacchio 1970, 1971, 1972 b, 1978, 1979 a, b, 1980, 1981; Musacchio and Chebli 1975; García and Leanza 1976; Musacchio and Palamarczuk 1976; Musacchio and Abrahamovich 1984; Simeoni 1985; Masiuk and Viña 1986; Simeoni and Musacchio 1986).

TABLE 10. Late Cretaceous Microfauna and Calcareous Nannoplankton Described from Northern Patagonia

TAXON	CENOMANIAN	TURONIAN	CONIACIAN	SANTONIAN	CAMPANIAN	MAASTRICHTIAN
FORAMINIFERA						
Spiroplectammina laevis (Roemer)						—
Gaudryina boltovskoyi Bertels						—
G. rudita Sandidge						—
Tritaxia pyramidata Reuss						—
Dorothia bulleta (Carsey)						—
Nodosaria marcki (Reuss)						—
N. aff. *torsicostata* Ten Dam						—
Dentalina vertebralis Batsch						—
Lagena acuticostata Reuss						—
L. amorpha paucicostata Franke						—
L. hexagona (Williamson)						—
Lagenoglandulina neuquensis (Bertels)						—
Lenticulina navarroensis (Plummer)						—
Globulina inaequalis Reuss						—
Guttulina lactea (Walker & Jacob)						—
G. problema (d'Orb.)						—
Ramulina globulifera Brady						—
Buliminella pseudoelegantissima Bertels						—
B. aff. *pulchra* (Terquem)						—
Neobulimina argentinensis Bertels						—
N. aspera (Cushman & Parker)						—
N. canadensis Cushman & Wickenden						—
Pyramidina paleocenica (Brotzen)						—
P. prolixa (Cushman & Parker)						—
P. rugosa (Brotzen)						—
Bolivina decurrens (Ehrenberg)						—
B. incrassata Reuss						—
B. sp.						—
Tappanina? sp.						—
Hiltermanella kochi (Bertels)						—
Stilostomella spinosa (Hofker)						—
Praeglobobulimina jaguelensis Bertels						—
Uvigerina elongata Cole						—
Pseudouvigerina aff. *cimbrica* (Troelsen)						—
"Discorbis" correcta Carsey						—
Guembelitria cretacea Cushman						—
Guembelitriella? sp.						—
Heterohelix globulosa (Ehrenberg)						—
H. navarroensis Loeblich						—
H. planata (Cushman)						—
H. striata (Ehrenberg)						—
Planoglobulina acervulinoides (Egger)						—
Pseudotextularia elegans (Rzehak)						—
Globigerinelloides multispina (Lalicker)						—
Loeblichella coarctata (Bolli)						—
Archaeoglobigerina blowi Pessagno						—
A. cretacea (d'Orb.)						—
Globotruncana cf. *gansseri* Bolli						—
Rugotruncana subpennyi (Gandolfi)						—
Rugoglobigerina cf. *bulbosa* Belford						—
R. macrocephala Brönnimann						—
R. rugosa (Plummer)						—
R. sp.						—
Eponides lunata Brotzen						—
Cibicidina reinholdi Ten Dam						—
Cassidella tegulata (Reuss)						—
Coryphostoma plaitum (Carsey)						—
Nonionella cretacea Cushman						—
Pullenia sp.						—
Alabamina kaasschieteri Bertels						—
Anomalinoides pinguis (Jennings)						—
A. sp.						—
Gavelinella camachoi (Bertels)						—
G. jagueliana Bertels						—
G. ? *neuquense* Bertels						—
OSTRACODA						
Bythocypris jaguelensis Bertels						—
Bythocypris? sp.						—
Cypridopsis (Pionocypris)? pasocordobensis Musacchio					---	---
Eucandona huantraicoensis Bertels					---	---
E. sp.					---	---
Ilyocypris alleniensis Angelozzi						—
I. triebeli Bertels						—
I. wichmanni Musacchio						—
I. zampalenis Angelozzi						—
I. (Neuquenocypris) calfucurensis Musacchio					---	---
I. (N.) pecki Musacchio						—
Cypridea sp.					---	---
Paracypris aff. *triangularis*						—
P. spp.						—
Brasacypris? morigerata Musacchio						—
?Argilloecia sp.						—
Darwinula aff. *kwangoensis* Grekoff					---	---
Alatacythere? rocana Bertels						—
?Pterygocythereis sp.						—

TAXON	CENOMANIAN	TURONIAN	CONIACIAN	SANTONIAN	CAMPANIAN	MAASTRICHTIAN
Jonesia? sp.						—
Wolburgiopsis neocretacea (Bertels)						—
W. vicinalis (Musacchio)				---	---	
Cyamocytheridea? sp.						—
Neocyprideis? zampalensis Musacchio				---	---	
Ovocytheridea? rionegrensis Musacchio						—
Perissocytheridea informalis Musacchio						—
P. sp.						—
Allenocytheridea lobulata Ballent						—
Cophinia? alleniensis Ballent						—
Cytheropteron aff. *carinoa latum* Bate						—
C. inops Bertels						—
C. sp.						—
Cytherura argentinensis Bertels						—
C. ? *jaguelensis* Bertels						—
C. ? sp.						—
Hemicytherura? ranquelensis Bertels						—
H. rionegrensis Bertels						—
Mehesella? apolloi Bertels						—
Orthonotacythere? intensosulcata Bertels						—
Metacytheropteron? sp.					---	
Paracytheropteron? pellegrinensis Ballent						—
Semicytherura? similis Bertels						—
Pacaytheridea rionegrina Bertels						—
?Allaruella racedense García & Proserpio						—
"Gomphocythere" payunensis Musacchio						—
Cytheromorpha? flexuosa Bertels						—
Munseyella minima Bertels						—
Sphaeroleberis? abnormis Bertels						—
Tumidoleberis australis Bertels						—
Mosaeleberis? argentinensis Bertels						—
Bertelsiana escobari García & Proserpio						—
Majungaella australis García & Proserpio						—
Amphicytherura? sp.						—
Wichmannella araucana Bertels					+	—
W. cretacea Bertels						—
W. magna Bertels						—
Trachyleberis noviprinceps Bertels						—
T. princeps Bertels						—
Repandocosta? sp.					+	—
Actinocythereis tuberculata Bertels						—
Anebocythereis chubutiana García & Proserpio						—
A. cretacica García & Proserpio						—
A. spinosa García & Proserpio						—
Cythereis? excellens Bertels						—
C. ? *incerta* Bertels						—
C. ? *rionegrensis* Bertels						—
?Cythereis sp.						—
Henryhowella splendida Bertels						—
Acanthocythereis abundans (Bertels)						—
Protocosta spinosa Bertels						—
Bradleya? argentinensis Bertels						—
B. attilai Bertels						—
B. ? *patagonica* Bertels						—
Anticythereis arcana Bertels						—
A. ? *attenuata* Bertels						—
A. venusta Bertels						—
Platycythereis? velata Bertels						—
Occultocythereis? sp.						—
Oertliella sp. A						—
? *O.* sp. B						—
Togoina cretacea Bertels						—
Veenia (Nigeria) argentinensis Bertels						—
V. (N.) inornata Bertels						—
V. (N.) jaguelensis Bertels						—
V. (N.) punctata Bertels						—
V. (N.) tumida Bertels						—
Cytherella araucana Bertels						—
C. terminopunctata Holden						—
C. utilis Bertels						—
CALCAREOUS NANNOPLANKTON						
Abmuellerella octoradiata (Gorka)						—
Arkhangelskiella cymbiformis Vekshina						—
Actinozygus regularis (Gorka)						—
Braarudosphaera bigelowi (Grand & Braarud)						—
Cribrosphaera ehrenbergi Arkhangelsky						—
Chiastozygus litterarius (Gorka)						—
Cretarhabdus conicus Bramlette & Martini						—
Eiffellithus turriseiffeli (Deflandre & Fert)						—
Glaukolithus diplogrammus (Deflandre & Fert)						—
Lithraphidites quadratus Bramlette & Martini						—
Markalius inversus (Deflandre & Fert)						—
Micula staurophora (Gardet)						—
Staurolithes crux (Deflandre & Fert)						—
Watznaueria barnesae (Black)						—
Zigodiscus spiralis Bramlette & Martini						—

(After: Bertels, 1968, 1969, 1970 a, b, c, 1972 a, b, 1974, 1975, 1980; Musacchio, 1973; Uliana and Musacchio, 1979; Angelozzi, 1980, Ballent, 1980; Rossi de García and Proserpio 1980; Malumian et al. 1984).

TABLE 11. Cretaceous Charophytes and Invertebrates Described from Aimara and Parana Basins

TAXA	AIMARA BASIN (MAASTR.)		PARANA BASIN CRET.	
	ARG.	BOLIV.	LOWER	UPPER
CHARACEA				
Amblyochara sp.	●	●		
Chara elliptica Fritzsche	●	●		
Platychara compressa Knowltown	●			
"P." cruciana (Horn af Rantzien)	●			
Porochara gildemeisteri Koch & Blissenbach	●			
P. ovalis (Fritzsche)	●	●		
Praechara barbosai (Petri)				●
Rhoadsia sp.		●		
Tectochara sp. A	●			
T. sp. B.	●			
Triportheus sp.		●		
FORAMINIFERIDA				
Miliolinella sp.	●			
BIVALVIA				
?Anodontites freitasi Mezzalira				●
?A. paulistanensis Mezzalira				●
A. pricci Mezzalira				●
Corbicula dormitator Pilsbry		●		
C. aff. dorsata (Dunker)		●		
C. cf. exarata (Dunker)		●		
C. aff. nuculaeformis (Roem.)		●		
C. aff. venulina (Dunker)		●		
C. aff. zimmermanni (Dunker)	●	●		
?Diplodon arrudai Mezzalira				●
Florenceia peiropolensis Mezzalira				●
Itaimbeia priscus (Ihering)				●
Kidodia picardi Frenguelli	●			
Lima cf. galloprovincialis Math.		●		
Monocondylaea cominatoi Mezzalira				●
Pisidium sp, indet		●		
GASTROPODA				
Anoptychia sp. indet.	●			
Asthenotoma comonensis Fritzsche		●		
Asthenotoma globosa Fritzsche		●		
Brachycerithium dupliciornatum Bonarelli	●			
B. evolutum Bonarelli	●			
B. intermedium Bonarelli	●			
B. lacrymigerum Bonarelli	●			
B. majus Bonarelli	●			
B. microstoma Bonarelli	●			
B. minus Bonarelli	●			
B. ornatissimum Bonarelli	●			
B. pauciornatum Bonarelli	●			
B. recticosta Bonarelli	●			
B. seminudum Bonarelli	●			
Carbajalia binotata (Bonarelli)	●			
C. juvenis (Fritzsche)	●			
"Cerithium" miraflorense Fritzsche		●		
"C." pucacnse Fritzsche		●		
Doryssa(?) andicola Pilsbry		●		
Eocerithium holmbergi Bonarelli	●			
E. potosense (d'Orb.)	●	●		
Gerthia brackebuschi (Bonarelli)	●			
Gonioconcha nodosa Bonarelli	●			
G. n. evolutior Bonarelli	●			
G. spinata Bonarelli	●			
G. s. brachyspira Bonarelli	●			
G. striata Bonarelli	●			
G. striatospinata Bonarelli	●			
Goniospira ameghinorum Bonarelli	●			
Hadraxon bolivianum Fritzsche		●		
?Hydrobia prudentinensis Mezzalira				●
Katosira brachyspira Bonarelli	●			
K. carbajalensis Bonarelli	●			
K. c. alta Bonarelli	●			
K. c. angulicosta Bon.	●			
K. c. baculoidea Bon.	●			
K. c. ornatissima Bon.	●			
K. clathrata Bonarelli	●			
K. communis Bonarelli	●			
K. conoidea Bonarelli	●			
K. groeberi Bonarelli	●			
K. longobardii Bonarelli	●			
K. nagerai Bonarelli	●			
K. obliquicosta Bonarelli	●			
K. percrassa Bonarelli	●			
K. pseudovetusta Bonarelli	●			
K. rapidecrescens Bonarelli	●			
K. recticosta Bonarelli	●			
K. zygopleuroides Bonarelli	●			
K. z. elatior Bonarelli	●			
K. z. intermedia Bonarelli	●			
Kittliconcha doelloi Bonarelli	●			

TAXA	AIMARA BASIN (MAASTR.)		PARANA BASIN CRET.	
	ARG.	BOLIV.	LOWER	UPPER
Melanoides bicarinata Fritzsche	●			
M. b. grandis Fritzsche	●			
Natica sp.	●	●		
Naticopsis andina Bonarelli	●	●		
Nerinea undulatocostata Fritzsche		●		●
Physa aridi Mezzalira				●
"Planorbis" boliviensis Fritzsche		●		
"P." molino Pilsbry		●		
Protofusus andinus Bonarelli	●			
P. carbajalensis Bonarelli	●			
P. convexigyrus Bonarelli	●			
P. saltensis Bonarelli	●			
Pseudomelania gr. seignettei Dumort. & Font.	●			
P. subandina Bonarelli	●			
Spirostylus elegans Bonarelli	●			
S. (Heligmostylus) bonarellii (Cossman)	●			
S. (Doellocosmia) doelloi (Bon.)	●			
Tilcaria ferughoi (Bonarelli)	●			
Tyrsoecus andinus Bonarelli	●			
T. disputabilis (Bonarelli)	●			
T. frequens (Bonarelli)	●			
T. micracantha (Bonarelli)	●			
T. moniliformis Bonarelli	●			
T. perarmatus Bonarelli	●			
T. tyrsoecina (Bonarelli)	●			
Valvata humilis Fritzsche	●	●		
V. satira Fritzsche	●			
V. yaviana Fritzsche	●	●		
Viviparus fluviorum (Sow.)				●
V. souzai Mezzalira				●
Windhauscnia dubia Mezzalira	●			
Xystrella armata (Goldf.)	●			
X. grandis (Fritzsche)	●			
Zygopleura argentina Bonarelli	●			
Z. cossmanni Bonarelli	●			
Z. crassinodis Bonarelli	●			
Z. decarlesi Bonarelli	●			
Z. gallardoi Bonarelli	●			
Z. insignis Bonarelli	●			
Z. maimarensis Bonarelli	●			
Z. multicosta Bonarelli	●			
Z. paucinodosa Bonarelli	●			
Z. polygyrica (Fritzsche)	●			
Z. pustulosa Bonarelli	●			
Z. scopesi Bonarelli	●			
Z. spinicosta Bonarelli	●			
Z. turgida Bonarelli	●			
Z. t. elatior Bonarelli	●			
Z. t. proversicosta Bon.	●			
CONCHOSTRACA				
Cyzicus (Euestheria) barbosai (Almeida)			●	
C. (E) mendesi (Almeida)			●	
C. (E.) ribeiropretensis Souza			●	
C. (E.) triangularis Souza			●	
C. (Lioestheria) elliptica (Jones)			●	
Palaeolimnadiopsis suarezi Mezzalira			●	
"Pseudestheria sp."				●
OSTRACODA				
Candonopsis pyriformis Almeida			●	
Cypridea sp.	●		●	
C. oblonga (?)			●	
Cypris boliviana Pilsbry		●	●	
Pachecoia acuminata Almeida			●	
P. rodriguesi Almeida			●	
Eucandona? cf. huantraiconensis Bertels	●			
Cypridopsis sp.	●			
Ilyocypris cf. wichmanni Musacchio	●			
I. sp.	●			
INSECTA				
Prosceliphron auroranormai (Roselli)				●
DECAPODA				
Dynomenopsis branisai Secretan		●		
ECHINOIDS				
Echinocorys (Ananychites) poznanskii (Berry)		●		
Holectypus sp.		●		
Phymosoma peruana (Brueggen)		●		
"Pseudodiadema" rotulare pucaense Fritz.		●		

(AIMARA BASIN, after: d'Orbigny, 1842; Bonarelli, 1921, 1927; Berry 1922, 1932; Fritzsche, 1924; Cossmann, 1925; Frenguelli 1937, 1945; Pilsbry, 1939; Ahlfeld and Branisa, 1960; Lohmann and Branisa, 1962; Branisa et al., 1966, 1969; Musacchio 1972 a; Secretan 1972; Mendez and Viviers, 1973; Moroni, 1982; Stach and Angelozzi, 1984; PARANA BASIN (Fossils listed under Lower Cretaceous include Jurassic material), after: Ihering 1913; Roselli, 1939; Frenguelli, 1946; Almeida, 1950; Petri, 1955; Mezzalira, 1966, 1974; Salamuni and Bigarella, 1967; Souza et. al., 1971; Francis, 1975; Langguth, 1978).

TABLE 12. Cretaceous Vertebrates described from Southern South America

TAXON	AUSTRAL	CHUBUT	ARG. (ANDEAN)	CHILE (ANDEAN)	ARAUCO	AIMARA	SAN LUIS	PARANA
CONDRICHTHYES								
Synechodus nitidus Smith Woodward					•			
Hexanchus dentatus (Smith Woodward)				•				
Scapanorhynchus subulatus Agassiz					•			
Isurus angustidens (Agassiz)					•			
I. triangularis (Egerton)					•			
Lamna arcuata Agassiz					•			
L. appendiculata Agassiz					•			
Squatirhina sp.						•		
Scymnus sp.					•			
Pristidae sp.					•			
"*Ischyrhiza chilensis*" Philippi					•			
I. hartenbergeri Cappetta						•		
Pucapristis branisai Schäffer						•		
Dasyatis branisai Cappetta						•		
D. molinoensis Cappetta						•		
D. shaefferi Cappetta						•		
Pucabatis hoffstetteri Cappetta						•		
"*Rhombodus* cf. *binkhorsti*" Dames						•		
Callorhynchus ? sp.					•			
Ptychodus sp.					•			
OSTEICHTHYES								
Actinopterygii indet.							●	
Semionotidae								●
Lepidotes mawsoni Wood.					•			
L. maximus Wagner				●				
L. patagonicus Ameghino	•							
Austrolepidotes cuyanus Bocchino							●	
Lepidotus pusillus Bocchino							●	
Neosemionotus puntanus Bocchino							●	
N. sp. indet.							●	
Coelodus toncoensis Benedetto & Sanchez						•		
Macromesodon agrioensis Bocchino				●				
Microdon chilensis Biese				●	●			
Paramicrodon volcanensis Schultze					•			
Protosphyraena ferox Leidy					•			
Teleostei s. l.					•			
Haplospondylus clupeoides Cabrera	●							
Tharrias shamani Dolgopol de Saez				•				
Gasteroclupea branisai Signeux						•		
"? *Portheus* sp."					•			
"*Apateodus* sp."					•			
"*Enchodus* sp."					•			
Molinichthys inopinatus Gayet						•		
Siluriformes indet.						•		
"*Hemilampronites hespericus*" Cockerell					•			
"*Beryx* sp."					•			
Ceratodus sp.	•		•					
AMPHIBIA								
Saltenia ibanezi						•		
REPTILIA								
Archisauripus australis Lull							●	
Parabatrachopus argentinus Lull							●	
CHELONIA								
Niolania argentina (Ameghino)	?							
Platemys schuenensis Ameghino	?							
Trionyx argentina Ameghino								•
Podocnemis brasiliensis Staesche								•
P. elegans Suarez								•
P. harrisi Pacheco								•
Roxochelys vilavilensis Broin						•		
R. wanderleyi Price								•
SQUAMATA								
Dicarlesia (*Carlesia*) *incognita* (Huene)						•		
Dinilysia patagonica Smith Woodward			•					
Pristiguana brasiliensis Estes & Price								•
Archosauria or Cynodontia								
A. ?							●	
CROCODILIA								
Crocodilia indet.						•		
cf. *Goniopholis* sp.								•
Meridiosaurus vallisparadisi Mones								●
"*Goniopholis paulistanus*" Roxo								•
Uruguaysuchus aznarezi Rusconi								•

TAXON	AUSTRAL	CHUBUT	ARG. (ANDEAN)	CHILE (ANDEAN)	ARAUCO	AIMARA	SAN LUIS	PARANA
U. terrai Rusconi								•
"*Brasileosaurus pachecoi*" Huene								•
Itasuchus jesuinoi Price								•
"cf. *Machimosaurus* sp."								•
"*Symptosuchus contortidens*"	•							
Notosuchus terrestris Smith Woodward			●					
Sphagesaurus huenei Price								•
Baurusuchus pachecoi Price								•
Cynodontosuchus rothi Smith Woodward			•					
C. sp.						•		
Peirosaurus torminni Price	?		•					•
PTEROSAURIA								
Pterodaustro guinazui Bonaparte			●				●	
Puntanipterus globosus Bonaparte & Sanchez							●	
Ornithocheiridae ?	•							
SAURISCHIA								
teeth							?	
Coelurosauria indet.								•
Dolichochampsa minima Gasparini & Buffetaut								•
Ornithomimidae indet.							•	
Noasaurus leali Bonaparte & Powell							•	
Carnosauria indet.							•	
Genyodectes serus Smith Woodward		•						
Loncosaurus argentinus Ameghino	•							
Abelisaurus comahuensis Bonaparte & Novas				•				
Chubutisaurus insignis del Corro		●						
Unquillosaurus ceibalii Powell						•		
Antarctosaurus sp.			•					
A. brasiliensis Arid & Vizzotto								•
A. giganteus Huene		?						
A. wichmannianus Huene			•					•
Argyrosaurus superbus Lydekker		•	•					
"*Campylodon ameghinoi*" Huene		?						
Laplatasaurus araukanicus Huene			•					•
L. sp.		?						
Titanosaurus australis Lydekker			•					•
T. robustus Huene			•					
cf. *Titanosaurus* sp.						•		
Saltasaurus loricatus Bonaparte & Powell						•	•	
Clasmodosaurus spatula Ameghino	•							
Sphaerovum erbeni Mones								•
ORNITHISCHIA								
Iguanodontidae indet.								•
Iguanodonnichnus frenkii Casamiquela				●				
Camptosaurischus fasolae Casamiquela				●				
Hadrosauridae indet.			•					
Hadrosaurichnus australis Alonso						•		
Secernosaurus koerneri Brett-Surman		•						
Loricosaurus scutatus Huene			•					
Notoceratops bonarellii Tapia		•						
Tacuaremborum oblongum Mones								?
Kritosaurus sp.			•					
SAUROPTERYGIA								
Plesiosauria						•		
"*Pliosaurus chilensis*" Gay						•		
"*Cimoliasaurus andium*" Deecke						•		
C. sp.						•		
Aristonectes parvidens Cabrera						•		
A. sp.						•		
Stretosaurus sp.			●					
ICHTHYOSAURIA								
Myobradypterygius hauthali Huene	●							
"*Ichthyosaurus saladensis*" Rusconi	•							
AVES								
Aves indet.							•	
Neogaeornis wetzeli Lambrecht						•	•	
Enantiornis leali Walker						•		
MAMMALIA								
Brasilichnium elusivum Leonardi								●
Mesungulatum houssayi Bonaparte & Soria			•					

(AUSTRAL BASIN, after: Ameghino, F., 1899, 1900, 1903, 1906; Ameghino, C.,1916; Cabrera, 1927; Huene, 1927, 1929; d'Erasmo, 1935; Pascual and Bondesio, 1976; CHUBUT BASIN, after: Cabrera, 1941; del Corro, 1966, 1975; Bonaparte, 1978; Bonaparte and Gasparini, 1979; ANDEAN BASIN (including the Upper Cretaceous of Northern Patagonia), after: Lydekker, 1893; Dolgopol de Saez, 1957; Casamiquela et. al., 1969; Estes et al., 1970; Gasparini, 1973,1982; Gasparini and Dellape, 1976; Bocchino, 1977; Bonaparte, 1978, 1984; Cione and Laffite, 1980; Aramayo, 1981; Bonaparte and Novas, 1985; Bonaparte and Soria, 1985; ARAUCO BASIN, after Steinmann et al., 1895; Oliver Schneider, 1923; Lambrecht, 1929; Broili, 1930; Wetzel, 1930; Casamiquela, 1969 b; AIMARA BASIN, after: Reig, 1959; Ibañez, 1960; Parodi and Kraglievich, 1960; Parodi et. al., 1960; Parodi, 1962; Schaeffer, 1963; Branisa et al., 1964; Bonaparte and Bossi, 1967; Leanza, 1969; Wenz, 1969; Benedetto and Sanchez, 1971, 1972; Broin, 1973; Cappetta, 1975; Bonaparte et al., 1977; Powell 1979; Alonso, 1980; Bonaparte and Powell, 1980; Gasparini and Buffetaut, 1980; Walker, 1981; Baez, 1982; Gayet, 1982 a, b, c; Cione et al., 1985; SAN LUIS BASIN, after: Bonaparte, 1970, 1971, 1978, 1981; Bocchino, 1973, 1974; Sanchez, 1973; Bonaparte and Sanchez, 1975; PARANA BASIN, after: Ihering, R. 1909; Huene, 1929 a, 1931, 1934 a, b; Pacheco, 1931; Walther, 1932; Rusconi, 1933; Roxo, 1936; Staesche, 1937, 1944; Price, 1945, 1950 a, b, 1951, 1953, 1954, 1955, 1959; Arid and Vizzotto, 1966, 1971, 1975; Suarez, 1969; Estes and Price, 1973; Leonardi, 1980, 1981; Mones, 1980). ● Lower Cretaceous,   • Upper Cretaceous

TABLE 13. Tithonian Ammonite Zones/Assemblages from the Andean Basin

		EUROPEAN STANDARD ZONES	WEST CENTRAL ARGENTINA, CENTRAL CHILE
TITHONIAN	Upper	Paraulacosphinctes transitorius	Substeueroceras koeneni, *Aulacosphinctes azulensis, A. mangaensis, Pectinatites (?) striolatus, Berriasella fraudans inflata, B. inaequicostata, Parodontoceras calistoides, Aspidoceras longaevum, Substeueroceras exstans, Blanfordiceras vetustum, Himalayites andinus, Spiticeras acutum*
			Corongoceras alternans, *C. lotenoense, C. mendozanum, C. rigali, Aulacosphinctes mangaensis, Lytohoplites burckhardti Micracanthoceras tapiai, M. lamberti, B. pastorei, B. australis, B. krantzi, B. bardensis B. (?) delhaesi*
	Middle	Micracanthoceras ponti	Windhauseniceras internispinosum, *Wichmanniceras mirum, Pachysphinctes, americanensis, Hemispiticeras* aff. *steinmanni, Subdichotomoceras araucanense, S. windhauseni, Parapallasiceras* aff. *pseudocolubrinoides, P.* aff. *recticosta, Aulacosphinctoides* aff. *hundesianus, Aspidoceras euomphalum, Corongoceras lotenoense*
		Semiformiceras fallauxi	Aulacosphinctes proximus, *Subdichotomoceras* sp. indet. *Pseudhimalayites steinmanni, Aspidoceras andinum, A. neuquensis*
		Semiformiceras semiforme	Pseudolissoceras zitteli, *P. pseudoolithicum, Aspidoceras cieneguitense, Glochiceras steueri, Hildoglochiceras wiedmanni, Parastreblites comahuensis*
	Lower	Franconites vimineus	Virgatosphinctes mendozanus, *Pseudinvoluticeras douvillei, P. windhauseni, P. (?) wilfridi, Choicensisphinctes choicensis Ch. choicensis sutilis, Ch. erinoides, Virgatosphinctes andesensis, V. mexicanus, V. burckhardti, V. denseplicatus rotundus, V. evolutus, Subplanites malarguensis*
		Neochetoceras mucronatum	
		Hybonoticeras hybonotum	

TABLE 14. Lower Cretaceous Ammonite Zones/Assemblages from the Andean Basin

		EUROPEAN STANDARD ZONES	WEST CENTRAL ARGENTINA, CENTRAL CHILE
BARREMIAN	Upper	Colchidites sp. Heteroceras astieri Hemihoplites feraudi "Emericiceras" barremense	[*Silesites* aff. *vulpes, Spitidiscus seunesi*]
BARREMIAN	Lower	Moutoniceras sp. Pulchellia compressissima Spitidiscus hugii	Crioceratites andinus, *C. diamantensis, C. perditus, C. schlagintweiti, C. apricus,* "*Hoplitocrioceras*" *gentili*
HAUTERIVIAN	Upper	Pseudothurmannia angulicostata Plesiospitidiscus ligatus Subsaynella sayni	Holcoptychites neuquensis, *H. compressus, H. demissus, Weavericeras vacaensis*
HAUTERIVIAN	Lower	Lyticoceras nodosoplicatum Olcostephanus jeannoti Crioceratites loryi Acanthodiscus radiatus	Lyticoceras pseudoregale, *L. australe, Neocomites crassicostatus, Pseudofavrella angulatiformis, P. garatei, Acanthodiscus* spp., *Teschenites* sp.
VALANGINIAN	Upper	Teschenites callidiscus Himantoceras trinodosum Saynoceras verrucosum	Olcostephanus curacoensis, *O. atherstoni, O. (Lemurostephanus) araucanus, O. (L.) mingrammi, O. (L.) permolestus, Lissonia riveroi, Karakaschiceras attenuatus*
VALANGINIAN	Lower	Thurmanniceras campylotoxum Thurmanniceras pertransiens Thurmanniceras otopeta	Neocomites wichmanni, *Thurmannia pertrasiens, Acantholissonia gerthi, Lissonia riveroi, Sarasinella crassicostata*
BERRIASIAN	Upper	Fauriella boissieri Tirnovella occitanica	Spiticeras damesi, *S. andium, S. bodenbenderi, S. fraternum, S. mammatum, S. singulare, Cuyaniceras groeberi, C. inflatum, C. raripartitum, C. transgrediens, Thurmanniceras duraznense, T. keideli, T. neogaeus, Neocomites regularis, N.* aff. *occitanicus, Pseudoblanfordia australis*
BERRIASIAN	Lower	Pseudosubplanites euxinus	Argentiniceras noduliferum, *A. bituberculatum, Berriassella laxicosta, Frenguelliceras magister, F. simplex, Groebericeras bifrons, Hemispiticeras steinmanni, Substeueroceras disputabile, Thurmanniceras discoidale*

TABLE 15. Lower Cretaceous Ammonite Zones/Assemblages from Patagonia

		EUROPEAN STANDARD ZONES	PATAGONIAN ASSEMBLAGES AND ZONES
ALBIAN	U.	Stoliczkaia dispar	*Mariella patagonica*
		Mortoniceras inflatum	Puzosia vegaensis, *Labeceras singulare.,Myloceras andinum, Anades-moceras constrictum, "Parasilesites" desmoceratoides, Mortoniceras tarense*
	M.	Euhoplites lautus / Euhoplites loricatus / Hoplites dentatus	Sanmartinoceras patagonicum, *Feruglioceras piatnizkyi / Beudanticeras rollerii, Cleoniceras santacrucense / Aioloceras argentinum, Pseudosaynella bonarellii, Rossalites imlayi, Anapuzosia sp., Sinzovia leanzai, Dimitobelus aff. stimulus*
	L.	Douvilleiceras mammillatum / Leymeriella tardefurcata	
APTIAN	U.	Diadochoceras nodosocostatum	Pelt ocrioceras deeckei, *Helicancylus patagonicus, Toxoceratoides spp., Ptychoceras sp., "Parasilesites" turici, "P." russoi, "Parasilesites" cf. desmoceratoides, Feruglioceras piatnizkyi, Tetragonites? heterosulcatus, Sanmartinoceras cardielense, Sanmartinoceras sp.*
	M.	Epicheloniceras subnodosocostatum	
		Aconeceras nisus	
	L.	Deshayesites deshayesi	*Australiceras sp., Tropaeum sp., Toxoceratoides nagerai, Helicancylus bonarellii, Sanmartinoceras africanum, Sanmartinoceras walshense, Sinzovia piatnizkyi*
BARREMIAN	U.	Colchidites sp. / Heteroceras astieri / Hemihoplites feraudi / "Emericiceras" barremense	*Colchidites vulanensis, Heteroceras elegans, Sanmartinoceras africanum insignicostatum*
	L.	Moutoniceras sp. / Pulchellia compressissima / Spitidiscus hugii	Hatchericeras patagonense, *H. santacrucense, Criptocrioceras yrigoyeni, Pseudohatchericeras argentinense, Hatchericeras semilaeve*
HAUTERIVIAN	U.	Pseudothurmannia angulicostata / Plesiospitidiscus ligatus / Subsaynella sayni	"Favrella" wilckensi, *Protaconeceras patagoniense*
	L.	Lyticoceras nodosoplicatum / Olcostephanus jeannoti / Crioceratites loryi / Acanthodiscus radiatus	Favrella americana, *Belemnopsis patagoniensis, B. madagascariensis, Aegocrioceras sp.*
VALANGINIAN	U.	Teschenites callidiscus / Himantoceras trinodosum / Saynoceras verrucosum	*Olcostephanus sp.*
	L.	Thurmanniceras campylotoxum / Thurmanniceras pertransiens / Thurmanniceras otopeta	
BERRIASIAN	U.	Fauriella boissieri / Tirnovella occitanica	*Jabronella aff. michaelis, Neocosmoceras sp., Neocosmoceras ornatum, Delphinella sp., Thurmanniceras sp., Berriasella inaequicostata, Phyllopachyceras aureliae, Bochianites? sp.*
	L.	Pseudosubplanites euxinus	
JURASSIC		TITHONIAN (KIMMERIDGIAN)	*Berriasella cf. behrendseni, Protacanthodiscus feruglioi, Himalayites peregrinus, Corongoceras mendozanum, Lytohoplites burckhardti, Aspidoceras cf. haupti, Virgatosphinctes aff. andesensis, Torquatisphinctes (?) sp., Pseudosimoceras patagonicum, Aulacosphinctoides (?) sp.*

TABLE 16. Upper Cretaceous Zones/Assemblages from Patagonia

		EUROPEAN STANDARD ZONES	PATAGONIAN ASSEMBLAGES AND ZONES
MAASTRICHTIAN	U.	Pachydiscus llarenai	
	L.	Pachydiscus neubergicus	[*Eubaculites lyelli, E. ootacodensis, Pachydiscus quiriquinae, Hoploscaphites quiriquinensis, Kitchinites darwini, Naefia neogaeia*] Gunnarites kalika, *G. antarcticus, Maorites* spp., *Diplomoceras australe, Grossouvrites gemmatus*
CAMPANIAN	U.	Bostrychoceras polyplocum	Hoplitoplacenticeras plasticum, *Saghalinites kingianus involutior, Gaudryceras varagurense patagonicum, Hypophylloceras nera, Pseudophyllites peregrinus, Baculites duharti*
		Trachyscaphites spiniger	
	L.	Delawarella delawarensis	Karapadites centinelaensis; *Natalites hauthali, Neograhamites morenoi*
		Scaphites hippocrepis	
SANTONIAN	U.	Placenticeras syrtale Eupachydiscus isculensis	
	L.	Texanites texanus	
CONIACIAN	U.	Parabevahites emscheri Protexanites, Paratexanites Texanites pseudotexanus	*Baculites* sp., *Polyptychoceras* sp., *Anagaudryceras* sp. *Gauthiericeras santacrucense*
	L.	Barroisiceras haberfellneri	
TURONIAN	U.	Subprionocyclus neptuni Romaniceras deverianum	
	M.	Romaniceras ornatissimum Romaniceras kallesi Kamerunoceras turoniense	"Anapachydiscus" steinmanni, "A". patagonicus, "A." hauthali, "Patagiosites" (?) amarus, Parabinneyites paynensis, Parapuzosia magellanica, Placenticeras viedmaense, P. santacrucense, P. washbournei, P. patagonicum, "Canadoceras" megasiphon, Argentoscaphites mutantibus, Sciponoceras santacrucense
	L.	Mammites nodosoides Watinoceras coloradoense	
CENOMANIAN	U.	Calycoceras naviculare	*Calycoceras* sp.
	M.	Acanthoceras rhotomagense	*Desmoceras floresi*
	L.	Mantelliceras mantelli	*Hypoturrilites* cf. *gravesianus, Sciponoceras* sp.

species seem to be pandemic, whereas the ostracods are usually endemic, although with south African affinities (Musacchio, 1978, 1979b). Throughout northern Patagonia, from Neuquén to the Colorado Basin, Upper Cretaceous foraminifers and ostracods (Table 10) are known only from the Maastrichtian, as most marine sediments are restricted to that age (Figs. 19, 20). The foraminifers have affinities with faunas from eastern North America, whereas they differ from those known from southern Patagonia (Bertels, 1979, 1987).

VERTEBRATES

Most Cretaceous vertebrates described and figured from southern South America are listed in Table 12. The Lower Cretaceous record is relatively less important than the Upper Cretaceous, and is similar to that of the Jurassic (Chong and Gasparini, 1976; Bonaparte, 1978; Pascual et al., 1978; Gasparini, 1979). The fauna consists of pycnodontiforms, crocodilia, pterosaurs, ichthyosaurs, and a few ornithischians. In the Aimará Basin of northern Argentina there are anurans with African affinities. Mammalian remains have been found (Bonaparte, 1986b) in Barremian-Albian strata of northwestern Patagonia (see under Neuquén). These fossils come from marine and continental sediments and represent relatively isolated findings within the most

important and large Cretaceous basins of southern South America.

Vertebrates are better known from Upper Cretaceous strata (Huene, 1929b; Chong and Gasparini, 1976; Bonaparte, 1978; Bonaparte and Gasparini, 1979), a fact related to the major importance of continental over marine sedimentation at that time. Nevertheless, in a restricted area such as that of the Arauco Basin (Fig. 20), a large variety of marine taxa has been found (e.g., ellasmobranchs, osteichthyes, plesiosaurs). Continental facies are characterized by turtles, crocodilia, and titanosaurs, which seem to be fairly abundant in central and northern Patagonia, as well as in the Paraná and Aimará Basins. Mammalian remains, indicating the presence of eotherians, allotherians, and therians, have been described from northern Patagonia (Bonaparte and Soria, 1985; Bonaparte, 1986a,b; Goin et al., 1986; Bonaparte and Pascual, 1987; Bonaparte and Rougier, 1987). According to Bonaparte (1979), most Cretaceous crocodiles and chelonids occur north of 35° south latitude, whereas most dinosaurs are found south of that latitude. On that basis, Bonaparte (1979) suggested the existence of two biogeographic provinces during Late Cretaceous time. For that time, ornithischian evidence also suggests faunal connection with North America.

The biogeography and the relationships of the Cretaceous continental vertebrates of South America have been analyzed by Bonaparte (1986c).

MAGMATISM

Cretaceous igneous activity in southern South America has been primarily studied in central and northern Chile by Ruiz et al. (1961), Oyarzún and Villalobos (1969), Vistelius et al. (1970), Farrar et al. (1970), Clark et al. (1970), Levi (1973), Aguirre et al. (1974), Clark et al. (1976), Montecinos and Oyarzún (1976), Tilling (1976), Huete et al. (1977), Dávila et al. (1979), Montecinos (1979), Vergara and Drake (1979a,b), Damm and Pichowiak (1981), Coira et al., (1982), Munizaga and Vicente (1982), Berg and Breitkreuz (1983), Berg and Baumann (1985), Montecinos (1985), Montecinos and Helle (1985), Rivano et al. (1985), and Damm et al. (1986).

The magmatic evolution of the southern Andes has been summarized by Saunders and Tarney (1982) and Ramos et al. (1982). General information can be found in Ramos and Ramos (1979), Zeil (1979, 1980), Thorpe et al. (1980), Zeil et al. (1980), Malumián and Ramos (1984), and Avila-Salinas (1986).

A maximum of Cretaceous igneous activity seems to have occurred at 98 Ma, between two minima at 140 to 150 Ma and 65 to 70 Ma (Fig. 21; Ramos and Ramos, 1979). Between the two minima, three different suites of rocks are represented. An older part, 115 to 120 Ma, consists mainly of andesites, dacites, and rhyolites in the Andes, and of olivine basalts in the Paraná

Basin (Figs. 17, 21). A second group, about 90 to 110 Ma, represents the climax of Andean plutonism (Figs. 18, 21). Finally, the third and youngest Cretaceous magmatism includes the volcanic, mostly basaltic, rocks of northwest Argentina, basic stocks in the Paraná Basin, and post-tectonic granitic intrusions in the Andean Range of Patagonia and central and northern Chile (Figs. 19, 21).

In dealing with this subject it must be noticed that some K-Ar data of Mesozoic volcanic rocks in central and northern Chile are in error, owing to the effects of burial and contact metamorphism (Levi, 1969, 1970; Palmer et al., 1980).

VOLCANISM

As during Jurassic time (Riccardi, 1983a), in central and northern Chile, Cretaceous volcanic activity was mainly developed along the western margin of the Andean Basin (see Andean Basin). During each structural stage, the volcanism of calc-alkaline type was characterized by an initial rhyolitic and principally ignimbritic phase, followed by andesitic volcanism. Andesitic volcanism is well developed in the Lower Cretaceous, whereas acid volcanism began in the mid-Cretaceous with hypabyssal rocks, and developed fully during Cenozoic time (Huete et al., 1977).

From the Jurassic to the lower Tertiary, volcanic rocks exhibit a distinct chemical and temporal zonation with respect to the Perú-Chile trench, due to a gradual eastward shift of the volcanic belt (Farrar et al., 1970; Dostal et al., 1977).

Thousands of meters of rhyolitic volcanics intercalated with calcareous marine strata of the Lo Prado Formation, were succeeded by continental porphyric andesitic flows of the Veta Negra Formation (Fig. 11; Table 17). After the "Middle" Cretaceous diastrophic phase (see Tectonism), a continental volcanic and volcaniclastic event took place, and a thick succession of rhyolitic ignimbrites, rhyolitic breccias, and andesites—the Abanico, Viñita (and equivalent) formations (Fig. 11; Table 17)—was developed eastward of the older formations. Petrographic and chemical studies of these volcanics have been carried out, between 30° and 34° south latitude by Vergara (1969), and between 26° and 28° south latitude by Dostal et al. (1977).

Similarly, in southern Patagonia the southwestern margins of the Austral Basin were characterized by the Lower Cretaceous volcaniclastic rocks of the Hardy Formation (Table 17) that interfinger with marine rocks of the Yahgan Formation. Continental rhyolites, andesites, andesitic breccias, and pyroclastics of Early and Late Cretaceous age are widely distributed throughout the northern part of the Austral Basin (Fig. 17). Southward, they cover marine strata of progressively younger ages. The oldest and northernmost representatives of these rocks, with absolute ages ranging between 111 ± 5 and 115 ± 7 Ma, are probably related to a zone of sialic weakness coincident with the southern limit of the Nazca Plate (Ramos, 1979).

In the Paraná Basin, volcanic activity (Bossi and Fernández, 1963; Leinz et al., 1968; Bossi and Umpierre, 1975) extended northward along important northeastern striking fractures that were formed during the Late Jurassic. Basic flows at about 142 ± 10 Ma in age, present in northwestern Uruguay, are followed in northeastern Uruguay by andesitic, dacitic, and rhyolitic flows and syenitic intrusives that are dated at 120 to 131 Ma. At about the same time, between 125 and 145 Ma, a large outflow of basaltic lavas (the Serra Geral Formation) began, throughout the Paraná Basin. A final explosive alkaline magmatism from 51 to 82 Ma, was developed along the eastern margin of the Paraná Basin. Twelve Upper Jurassic to Early Cretaceous alkaline provinces are recognized in southern Brazil, central Uruguay, eastern Paraguay, and southeastern Bolivia (Beurlen, 1970; Herz, 1977; Asmus, 1982; Almeida, 1983; Avila-Salinas, 1986).

Lower Cretaceous basalts are also present in the Santos (see Fodor and Vetter, 1984), Santa Lucía, and Salado basins and in central Argentina (Fig. 17). In Sierra de Los Cóndores, six flows of picritic basalts, calc-alkaline trachytes, and trachybasalts have been documented, with ages ranging between 112 ± 6 and 129 ± 3 Ma. The paleomagnetism of these rocks was studied by Valencio (1972), and all information about the geology, chemistry, and radiometric ages has been summarized by Llambías and Brogioni (1981).

In the Aimará Basin, basaltic flows are present within the Pirgua Subgroup (Table 18), the Aroifilla Formation (Betanzos basalt), between the La Puerta and Tarapaya formations (the Maragua basalt), and within the Toro Toro Formation (the Tupiza basalt; Avila-Salinas, 1986). In the Pirgua Subgroup, two effusive episodes have been recognized. The first, 98 to $128 \pm$ Ma, is related to the El Cadillal and La Yesera formations; the second episode, 76.4 ± 3.5 to 78 ± 5 Ma, is related to the Las Curtiembres Formation.

PLUTONISM

The most important Cretaceous plutonism of southern South America is represented in the Andean Belt, along the international boundary of Argentina and Chile. In central and northern Chile (Figs. 17, 18), the plutons form an almost continuous belt to the east of the Jurassic intrusives of the Coastal Cordillera, and to the west of the Tertiary intrusives of the main Andean Range. This belt appears to represent an eastward migration of intrusive foci (Munizaga and Vicente, 1982), coupled, after the "Middle" Cretaceous, with a continuous west–east decrease in initial $^{87}Sr/^{86}Sr$ isotope ratio (McNutt et al., 1975; Berg and Baumann, 1985).

The best exposures of Cretaceous intrusives are present between 26° and 36° south latitude (Figs. 6, 10), where about 33 percent of the area is formed by plutonic rocks. At 31°-32° south latitude, Cretaceous intrusives have been included in the Illapel "Superunit" (divided in the Chalinga and Limahuida "Units") and the San Lorenzo "Unit" (Rivano et al., 1985). K-Ar isotopic analysis indicates ages of 133 ± 4 to 85.9 ± 2.2 Ma for the Chalinga "Unit", and 65.3 ± 3.1 Ma for the San Lorenzo "Unit".

Between Chañaral and Taltal, a large batholith with a granitic to dioritic composition, i.e., the Cerro del Pingo Plutonic Group (Naranjo and Puig, 1984), has K-Ar and $^{39}Ar/^{40}Ar$ isotopic ages ranging between 124 ± 4 and 109 ± 2 Ma. North of Taltal (Fig. 10), several batholiths occur in the Coastal Cordillera, e.g., the Sierra Minillas, Pastenes, and Tocopilla batholiths (Naranjo, 1978, 1981; Palacios and Espinoza, 1982), and the Esmeralda "Superunit" (Damm and Pichowiak, 1981; Pichowiak et al., 1982).

In the Patagonian Cordillera (Fig. 2), Cretaceous plutonism has been studied less (Nordenskjold, 1905; Quensel, 1913; Kranck, 1932), although intrusives of different ages are quite well represented as far south as Cape Horn (55° south latitude; Fig. 1). The most common petrographic types are granodiorites, diorites, granites, and tonalites, with subordinate quartz-monzonites, adamellites and gabbros. The modal composition of these rocks can be found in Vistelius et al. (1970), Ramos and Palma (1981), Palacios and Espinoza (1982), Stern and Stroup (1982), and Haller (1985). A northward decrease in age of these plutons has been suggested by Ramos and Palma (1981). Evidence for an eastward decrease in age seems, however, to be conflicting (Dalziel and Elliot, 1973; Halpern, 1973; Halpern et al., 1975; Skarmeta and Charrier, 1976; González Díaz, 1982; Munizaga et al.,

1984). In central and northern Chile, Pbα and K-Ar ages range between 66 and 136 Ma (Levi et al., 1963; Thomas, H., 1967; Segerstrom, 1967; Farrar et al., 1970; Corvalán and Munizaga, 1972; Aguirre et al., 1974; Munizaga and Vicente, 1982; Montecinos, 1985; Hervé et al., 1985a,b). Most values are clustered around 90 and 110 Ma. In the Patagonian Cordillera, radiometric ages, mostly determined by Rb-Sr, indicate at least three intrusive phases. They appear to have occurred between 10 and 50, 75 and 100, and 120 and 155 Ma (Halpern, 1962, 1972, 1973; Halpern and Carlin, 1971; Halpern and Rex, 1972; Toubes and Spikerman, 1974; Hervé et al., 1974; Skarmeta and Charrier, 1976; Skarmeta, 1978; Ramos, 1979; Pesce, 1979a; Thiele et al., 1979; González Díaz and Valvano, 1979; Sepúlveda and Viera, 1980; Ramos and Palma, 1981; Munizaga et al., 1984; Haller, 1985; Suárez et al., 1985b, 1986). The climax of the intrusive cycle also occurred close to the Late Albian, i.e., about 98 ± 4 Ma (Ramos and Ramos, 1979; Ramos, 1979). Plutonic intrusives along the eastern margin of the Andean Range, south of 36°25′ south latitude, seem to have been intruded at about the same time (Sillitoe, 1977; Turner, 1979; Zanettini, 1979; González Díaz and Nullo, 1980; González Díaz, 1982).

The maxima of plutonic activity along the Andean Cordillera are in agreement with the time interval ascribed to the "Middle" Cretaceous diastrophic phase (see Tectonism; Fig. 21). The Late Cretaceous cluster of isotopic ages, which in northern Chile is between 62 and 70 Ma, seems to agree with the timing of the Cretaceous-Tertiary diastrophic phase. It has also been suggested that the maxima in plutonic activity were coincident with regressions, whereas the minima were related to transgressions (Piraces, 1979). This conclusion is in agreement with the available data (Fig. 21).

Cretaceous intrusives of granitic composition have also been recorded in northwestern Argentina—the Aguilar, Abra Laite, Rangel, Tusaquillas, and Papachacra formations, and in southern Bolivia, the Mecoya-Rejará Formation (Fig. 13). Isotopic analysis indicates ages ranging from 96 ± 5 and 155 ± 10 Ma (Spencer, 1950; Halpern and Latorre, 1973; Linares and Latorre, 1975; Méndez, 1975; Turner and Salfity, 1977; Turner et al., 1979; Navarro García, 1984).

In central-eastern Bolivia, central Argentina, and the Paraná Basin are mostly basic intrusives that seem to be related to basaltic volcanism represented in the Paraná Basin. In eastern Bolivia, seven subcircular syenitic intrusions arranged with a southwest–northeast–trending strike constitute the Velasco alkaline province (Darbyshire and Fletcher, 1979; Almeida, 1983; Avila Salinas, 1986). Rb-Sr and K-Ar isotopic ages gave values between 134 and 143 ± Ma. Similar and probably related basic intrusives are present in central Bolivia near Cochabamba (Fig. 13): the Ayopaya alkaline province (Ahlfeld, 1966; Fletcher and Litherland, 1981; Avila-Salinas, 1986).

Basic intrusives with K-Ar absolute ages ranging between 116 ± 17 Ma and 76 Ma—the Candelaria Alkaline Complex—are also represented in eastern Bolivia, south of the Velasco Alkaline Province (Avila-Salinas, 1986).

In central Argentina, trachybasaltic dikes have absolute ages ranging between 63 ± 5 and 68 ± 5 Ma, and between 129 ± 10 and 150 ± 10 Ma (Gordillo and Lencinas, 1967; Linares and Valencio, 1974).

In Uruguay, northeast of Montevideo, syenite and trachyte intrusives are about 120 ± 5 m.y. old. Cretaceous alkaline intrusives are present in several areas of southern Brazil, along the margin of the Paraná Basin and in the Santos Basin (Figs. 12, 17; Umpierre and Halpern, 1971; Bossi and Umpierre, 1975; Asmus, 1982; Almeida, 1983).

TECTONISM

The existence of different Cretaceous diastrophic phases has been partially or totally dealt with by Groeber (1953), Stipanicic and Rodrigo (1970), Charrier and Vicente (1972), Vicente et al. (1973), Charrier (1973), Rutland (1974), Chotin (1977), Jensen (1979), Frutos (1980), Zambrano (1981), Moya and Salfity (1982), Salfity (1982), Malumián and Ramos (1984), and Salfity et al. (1984).

According to Stipanicic and Rodrigo (1970), five diastrophic phases occurred between the Late Jurassic and the Late Cretaceous. Most of them have been named differently for central-west Argentina, central and northern Chile, and Patagonia. They are, in ascending order: (1) the Araucana (Chiza, Santa Cruz) Phase, (2) the Catan Lil (Santiago, Chenque) Phase, (3) the Avilé Phase, (4) the Mirano (Antofagasta, Caleta Cordova) Phase, and (5) the "Laramic" or Early Magallanian Phase. The clearest evidence for these proposed diastrophic phases exists for the first, fourth, and fifth.

The late Oxfordian Araucana, or Araucanian, Phase produced a general uplift in central-west Argentina and central and northern Chile. This uplift resulted in an extensive marine regression and in the deposition of thick sequences of conglomerates during succeeding Kimmeridgian time. In Patagonia the Araucana Phase is related to the initial formation of the Austral or Magallanes Basin and is expressed in the disconformable relationship between Jurassic volcanics and the Upper Jurassic–Cretaceous marine sediments (Riccardi and Rolleri, 1980). The Araucanian diastrophism is probably related to the first rifting stages leading to the opening of the South Atlantic, and the development of a series of basins along the eastern margin of southern South America. It was also associated with a magmatic episode (see Magmatism).

The Early Valanginian Catan Lil (or Infra-Neocomian; Jensen and Vicente, 1977) Phase has been proposed on the basis of lithologic changes existing between the Vaca Muerta–Quintuco and Mulichinco formations (Fig. 9) of west-central Argentina. These changes indicate an unconformity and could be related to a eustatic fall of sea level (Gulisano et al., 1984; Legarreta and Kozlowski, 1984; Mitchum and Uliana, 1985).

The Late Hauterivian Avilé Phase of Stipanicic and Rodrigo (1970) has been postulated on the basis of a conspicuous sandstone level (Avilé Sandstone) within the Agrio Formation of the Neuquén Embayment (see Neuquén). It could represent the initial subphase of the "Middle" Cretaceous Mirano Phase.

The Mirano Phase (approximately Albian-Turonian in age), is also known as the Subhercynian, Intersenonian (Groeber, 1953), or Peruvian (Steinmann, 1929), Caleta Cordova, or Patagonian (Stipanicic and Rodrigo, 1970) Phase; it represents a major event in the history of the Andes. It has been developed as at least three subphases, i.e., the Initial, Intermediate, and Principal. Although the main subphase of this tectonic episode is now thought to have occurred in Cenomanian-Turonian time, all evidence of diastrophism recorded between Albian and Campanian time has been usually ascribed to this phase. Precise dating is made difficult by the continental character of most Upper Cretaceous sequences. Nevertheless, some authors (see Zambrano, 1986) differentiate Aptian (Initial Mirano or Patagonian), Cenomanian (Main Mirano or Patagonian), and Campanian (Peruvian or Subhercynian) phases.

The successive movements of the Mirano Phase determined a change in the evolution of the central Argentine-Chilean Andes. At first it produced an uplift of the Andean domain which ended Lower Cretaceous marine sedimentation, and it initiated "Middle" Cretaceous continental sedimentation, which is represented by the Colimapu, Huitrín (and equivalent) formations (Table 17). Overthrusting and increased faulting also affected the Lower Cretaceous and older units. Subsequently it was responsible for the unconformable relationship of the Rayoso and Neuquén groups in the Neuquén Embayment (Table 17), as well as for the unconformity existing between the volcanic and volcanoclastic formations, i.e., Cerrillos, Viñita, Augusta Victoria, and all oldest units, in central and northern Chile (Table 17). The Mirano Phase is associated with burial metamorphism and the main Andean plutonism.

In the Austral Basin, the Mirano Phase is primarily recorded by folding in the southwest (Cecioni, 1960), and in the more widespread regressive facies found in the northern part of the Austral Basin. A plutonic episode has also been dated as being between 75 and 100 Ma, and the "early Andean" orogeny is thought to have occurred between 84 and 100 Ma (Hervé et al., 1981). Nevertheless, an important continuous marine succession was being deposited at that time in the central regions of the basin.

In central Patagonia the Mirano Phase could be represented by the unconformity found at the top of the Las Heras Group (Table 17).

In the Aimará Basin, the Mirano or Peruvian Phase is evident in the unconformity existing at the top of La Puerta Formation, where the Maragua Basalt is present (Table 18). The same phase is represented by the Betanzos and Las Conchas basalts, the unconformity between the lower and upper parts of the Aroifilla "formation," and the unconformable relationship of the Tacurú and equivalents with all older units. Similarly, in the Paraná Basin

the Mirano Phase is expressed in the unconformity found at the base of the Bauru Formation and equivalents (Table 18).

The "Laramide" Phase—also known as Early Magallanian Phase (Katz, 1973), First Phase of the First (Tertiary) Movement (Groeber, 1951), or Ranquel Phase (Salfity et al., 1984)—is represented in central and northern Chile by an angular unconformity between Upper Cretaceous and lower Tertiary volcanic sedimentary units. Thus, in central Chile the Farellones Formation rests unconformably on the Abanico and Coya Machalí formations, whereas farther north the same relationship is found between the Los Elquinos and Viñita formations (Table 17).

In the Aimará Basin, the Ranquel Phase is evident in major changes in facies related to the deposition of the Balbuena Subgroup and El Molino Formation (Table 18; Salfity et al., 1984; Sempere, 1986). In southern Patagonia the Early Magallanian Phase is documented by the unconformity found at the base of the Río Bueno Formation, as well as by other local unconformities and strong facies changes in marginal areas. In other parts of the basin, the Cretaceous-Tertiary boundary could be transitional despite sharp microfaunal changes (Macellari, 1985b). A similar transition seems to be present in northern Patagonia, according to the continuity of the sedimentary sequence, although a Late Maastrichtian faunal hiatus has been recorded (Bertels, 1969; Camacho, 1972).

GEOLOGIC EVOLUTION AND PALEOGEOGRAPHIC SYNTHESIS

INTRODUCTION

General aspects of the paleogeography of the Cretaceous of southern South America, mainly from Argentina and Chile, have been presented; authors include Gerth (1932, 1955, 1960), Weaver (1942), Weeks (1947), Groeber (1953, 1959, 1963), Harrington (1962), Herrero Ducloux (1963), Radelli (1964), Ruiz et al. (1965), Borrello (1969), Cecioni (1970), Aubouin et al. (1973), Chotin (1976), Wiedmann (1980), Urien et al. (1981), Zambrano (1980b, 1981, 1986), Thiele and Nasi (1982), Malumián et al. (1983), Digregorio et al. (1984), Chong (1986), and Riccardi (1987), among others.

The present exposition is divided chronologically into four parts, following approximately the main stages of geologic evolution as expressed by major changes in paleogeography and sedimentary patterns: (1) Tithonian-Hauterivian, (2) Barremian-Cenomanian, (3) ?Turonian-Campanian, and (4) Maastrichtian.

The initial history of all Cretaceous sedimentary basins of southern South America is related to the initial break-up of Gondwanaland. Triassic to Late Jurassic intracontinental rifting, spreading, and subsidence culminated in the earliest Cretaceous with the opening of the South Atlantic. Crustal extension resulted in a series of northwest- to northeast-trending fault-bounded grabens and troughs.

Along the subduction zone of the Pacific coast of South America, Late Jurassic–Early Cretaceous marine basins were initiated with the development of volcanic arcs and ensialic troughs, in central Chile and west-central Argentina, the Andean Basin, and in southern Patagonia, the Austral Basin. At about the same time, several isolated rift basins, formed on the continent and along its Atlantic margin, were the site of continental volcanism and sedimentation.

BERRIASIAN-HAUTERIVIAN

By Kimmeridgian time, central-west Argentina and central and northern Chile had gone through a regional uplift that established continental conditions (Riccardi, 1983a). During Tithonian time, the area was again downwarped, and a volcanic arc and an ensialic trough were formed within the continental margin on eroded Jurassic rocks and granitic and metamorphic basement. A marine back-arc basin extended, in a north-northwest direction, from about 40° south latitude to the latitude of Copiapó and perhaps to Antofagasta (Fig. 17). In central-west Argentina, the marine basin developed a large eastward extension, the Neuquén Embayment.

During Berriasian-Hauterivian time, sedimentary patterns within the Andean Basin were related to episodic development of two distinctive volcanic arcs with intra-arc subbasins (Charrier, 1984; Ramos, 1986). Transgressive-regressive sequences (Fig. 21) were deposited during an Early Tithonian–Late Berriasian period of rapidly rising sea level (Mitchum and Uliana, 1985). In the southeastern margin of the Neuquén Embayment, shelf facies of the Quintuco, Loma Montosa, Pichi Picún Leufú, and Ortiz formations (Figs. 8, 9; Table 17) consist of sandstones, conglomerates, and shale with some carbonate. In the deep, west-central axial portion of the basin, basinal facies of the Vaca Muerta, Baños del Flaco, and Lo Valdés formations (Fig. 11, Table 17) consist of dark organic shales and calcareous rocks about 2,000 m thick. On the western side of the basin, marine turbidites and limestones are intercalated with volcanic and volcanoclastic sequences 2,000 to 4,000 m thick: the Arqueros, Lo Prado, and Bandurrias formations (Fig. 11, Table 17), derived from the western volcanic arc.

During Late Berriasian–Early Valanginian time, a short lived fall in sea level (Fig. 21) took place within a complex paleogeographic setting, in which shelf to fluviatile sandstones and conglomerates, the Mulichinco Formation, and carbonate platforms with reef-like structures, the Chachao Formation, coexisted within the Neuquén Embayment (Mombrú et al., 1979) (Fig. 7; Table 17). In Late Valanginian–Hauterivian time, a progressive sea-level rise represented by a transgressive sequence (Fig. 21) culminated with the deposition of the pelites and limestones of the Agrio Formation. In the southeastern margin of the Neuquén Embayment, inner shelf to continental facies are represented by the Centenario, La Amarga, and Bajada Colorada formations (Figs. 8, 9; Table 17). On the western belt of the

Andean Basin (Fig. 17), a volcanic pile 5,000 to 8,000 m thick, was formed, consisting of basaltic flows in the lower and middle parts, and andesitic flow breccias in the upper part.

South and north of Antofagasta (Fig. 17), and also between Iquique and Arica (Figs. 1, 10), Berriasian-Hauterivian time is mostly represented by continental beds, although marine fore-arc strata, such as the Blanco and El Way formations (Table 17), are also present. In the eastern back-arc areas of northern Chile, continental beds with evaporites were also deposited during this time, and a shallow sea coming from the south reached 25° south latitude (Maksaev, 1984).

South of 42° south latitude, a northwest-southeast ensialic basin—the Austral or Magallanes Basin—began to develop during the Late Jurassic. This back-arc basin was formed as part of a large volcano-tectonic rift zone subsided, forming a submarine trough along the Pacific margin of the continent (Bruhn, 1979). South of 50° south latitude, basaltic magmas intruded the continental crust and led to the development of an oceanic spreading center (De Wit and Stern, 1981) and an intra-arc or back-arc marginal basin (Dalziel, 1974, 1984; Dalziel et al., 1974; Suárez and Pettigrew, 1976; Bruhn, 1979; Suárez, 1979a,b; Nelson et al., 1980; Farquharson, 1982; Aberg et al., 1984). The marginal basin crust consisted of ophiolitic rocks and blocks of rifted continental basement and Jurassic silicic volcanics. North of 50° south latitude, no oceanic crust was generated, and the ensialic trough of this part of the Austral Basin has been interpreted as an aborted marginal basin (Aberg et al., 1984).

The Austral Basin was separated from the northern Andean Basin by the Chubut or Concepción "Dorsal" or High (Fig. 17; Aubouin and Borrello, 1966; Aubouin et al., 1973; Aguirre et al., 1974). However, a possible marine connection between the Andean Basin of central-west Argentina and Chile and the Austral Basin cannot be ruled out (Thiele and Hein, 1979).

Late Jurassic subsidence in the Austral Basin was followed by marine transgression, which covered the southwestern margins of the Deseado Massif (Fig. 17) and resulted in continued sedimentation throughout the Late Jurassic and Cretaceous (Fig. 21, Table 17). At first, fluviatile to marine sandstones intercalated with shales—the Springhill Formation—were deposited in the northeastern platform area in topographic depressions of the Jurassic volcanic basement (El Quemado Complex) that floored most of the Austral Basin. Marine sedimentation, in pelagic and euxinic environments, characterized by low faunal diversity, continued during the Early Cretaceous, and the maximum advance of the sea was attained during late Early Hauterivian time (Fig. 21; Riccardi and Rolleri, 1980). In the cratonward bathyal submarine slope of the marginal basin (Fig. 3), marine pelites, siltstones, and turbidites—the Zapata Formation—grade laterally and southward into andesitic detritus-rich graywacke sandstones and shales—the Yahgan Formation—which were deposited closer to the rear flank of the volcanic arc. At about the same time, volcanic activity developed in the volcanic arc, and rhyolitic flows, basalts and pyroclastic rocks—the Hardy Formation—accumulated along the rear flank of the volcanic arc and in the

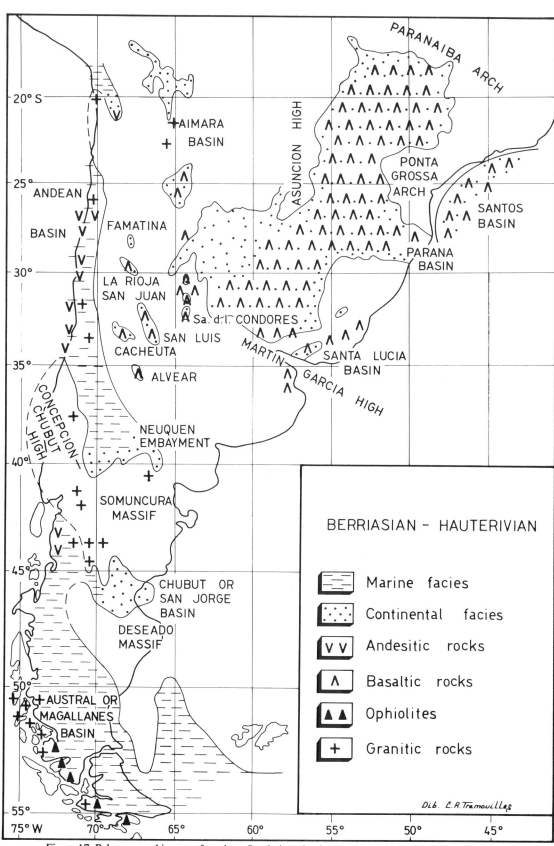

Figure 17. Paleogeographic map of southern South America for Berriasian-Hauterivian time (modified from Riccardi, 1987).

Pacific side of the back-arc basin. Marine rocks deposited on the platform area consist of dark laminated calcareous shales—the Río Mayer Formation.

Between 49° and 47° south latitude there is an Upper Jurassic–Lower Cretaceous volcanic gap. But the volcanic arc was active north of 47° south latitude, with marine sedimentation restricted to a narrow trough. North of 45° south latitude, an intra-arc basin was developed, and marine beds are interbedded with calc-alkaline volcanics (Ramos et al., 1982).

The Chubut or San Jorge Basin is an extensional basin that began to develop in central Patagonia (Fig. 17) during the Early Cretaceous. It was mostly filled with continental sediments, although the presence of marine microplankton in the western areas of the basin suggests a possible connection with the Austral Basin.

During the initial rift stage, the Chubut Basin was a lacustrine basin with slow deposition of suspended clay and organic matter, the Aguada Bandera-1 Formation (Table 17). Lacustrine sedimentation was followed by deposition of coarse clastics by braided fluvial systems, the Cerro Guadal Formation.

During the Late Jurassic, along eastern South America from the Colorado Basin in Argentina to northern Brazil, Precambrian lines of crustal weakness were reactivated by tensional forces, i.e., the Wealdean Reactivation (Almeida, 1972), resulting in a normal-faulted structural pattern. Within that framework a series of basins formed in the Precambrian-Paleozoic basement, such as the Paraná, Santa Lucía, Salado, Colorado Basins (Urien et al., 1975, 1981) (Figs. 17, 18).

The Paraná Basin was uplifted after the deposition of upper Paleozoic marine, and Triassic fresh-water sediments. This was followed by a new cycle of sedimentation in Jurassic–Early Cretaceous time, with deposition of the subaerial sediments of the Botucatu, San Cristóbal, and Tacuarembó formations (Table 18). Sedimentation patterns indicate a high degree of tectonic stability with a slow rate of uplift of source areas, culminating in an episode of intense volcanic activity (Soares et al., 1978). The Paraná Basin was subjected to intense fracturing. Large amounts of basic flows erupted through a system of deep fractures, and basaltic sills and dikes intruded Jurassic–Lower Cretaceous strata. Similar volcanic or volcanic-sedimentary sequences were formed in the Santa Lucía, Pelotas, and Santos Basins (Figs. 17, 18).

The volcanic rocks occurring in these basins are synchronous with those found along the southwest coast of Africa. All of them seem to have been related to the initial rifting of the South Atlantic about 120 to 130 Ma (Cordani, 1970; Larson and Ladd, 1973; Barker et al., 1976; Darbyshire and Fletcher, 1979; Fodor and Vetter, 1984), a hypothesis supported by paleomagnetic studies (Creer, 1962, 1964; Pacca and Hiodo, 1976; Ernesto et al., 1979).

The Lower Cretaceous successions represent the initial stages in the evolution of the taphrogenic basins developed along the South American continental margin (Campos et al., 1974; Ponte and Asmus, 1978). The earlier rifting stage characterized

by continental sedimentation was followed by a stage of rifting with basaltic flows.

Coarse clastic sediments intercalated with basaltic flows were also developed during the Early Cretaceous in a series of restricted tectonic basins—Intra-Pampean Basin—of central Argentina (Fig. 17).

During the Late Jurassic and Early Cretaceous, northern Argentina and Bolivia were also affected by tectonism, which gave rise to a block structure with northwest-southeast and northeast-southwest fractures. Thus, a major basin, the Aimará Basin, with a southeast-northwest trend and extending into southern Perú, began to develop between the Precambrian Brazilian Shield to the east and a positive high named the Calchaquí "Dorsal" (Aubouin et al., 1973) to the west (Figs. 14, 17).

Throughout Early Cretaceous time, the irregular relief of the Aimará Basin was filled by continental deposits; from north to south they include the La Puerta Formation, and the Pirgua Subgroup (Table 18).

During Late Jurassic–Early Cretaceous time, the continental region north of 35° south latitude was characterized by a warm and extremely arid climate. This part of the southern hemisphere arid zone has been identified by a paleodesert of regional dimensions (Botucatu, San Cristóbal, and Tacuarembó formations) in the Paraná Basin (Fig. 17).

South of 43° south latitude was an area characterized by a warm and humid climate with a cold season. This is suggested by sandstones composed of quartz and kaolinite—the Springhill Formation—which originated from Upper Jurassic volcanics, and by coal seams intercalated with clastic rocks.

Sea waters on the Pacific coast were also warm. Coral patch reefs and oolitic limestones with a rich and diverse invertebrate fauna, including crinoids, scleractinians, and large and thick-shelled bivalves, are present along the coast as far south as 45° south latitude. $\delta^{18}O$ - $\delta^{16}O$ isotope analyses on belemnites from the Austral Basin, at about 49° south latitude, indicate temperatures of 23.7° to 25.7°C, with seasonal variations of 3.9° to 9.6°C, for Berriasian-Hauterivian time (Bowen, 1961).

BARREMIAN-CENOMANIAN

In Barremian time, the magmatic arc of the Andean Basin (Fig. 18) had an important eastward migration, probably related to an increase in plate motion rates (Ramos, 1986). This migration was coincident with a low stand of sea level in the back-arc and intra-arc basins (Fig. 21) and the development of a single and expanded central arc. This resulted in the initial uplift of the Cordillera Principal of Argentina and Chile, which was to attain its present altitude by Tertiary time. In west-central Argentina and central Chile (Rivano, 1984), this uplift caused a complete reversal of the regional slope, and continental basins were developed on the site of the back-arc basin. Thus, at 35° to 40° south latitude, east of 70° west longitude, Barremian-Albian continental and lacustrine sediments accumulated under arid conditions, i.e., the Colimapu Formation (Table 17). Coarseness of sediments

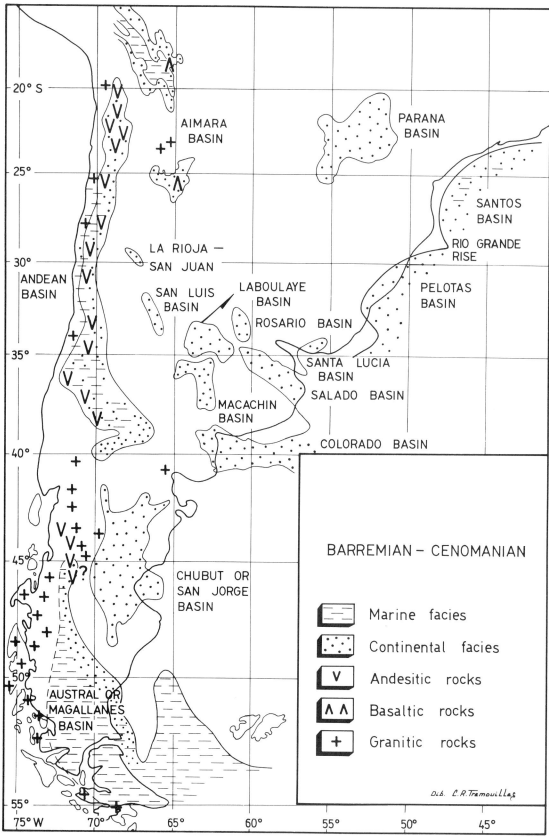

Figure 18. Paleogeographic map of southern South America for Barremian-Cenomanian time (modified from Riccardi, 1987).

decreased eastward; along the axis of the basin, evaporites of the Rayoso Group were deposited, marking the end of the regression (Table 17; Fig. 11). Meanwhile, andesitic volcanic activity continued in the west, and during Aptian-Albian time a volcanic cordillera began to develop, consisting of volcanic rocks and continental volcanoclastics, i.e., the Quebrada Marquesa and Veta Negra Formations (Table 17; Fig. 11).

During Barremian time a regressive sedimentary regime began (Fig. 21, Table 17) south of 45° south latitude (Fig. 18), with deposition of near-shore to continental sediments to the north of the Austral Basin—the Apeleg, Río Belgrano, Río Tarde, Kachaike, and Piedra Clavada formations—and marine sediments to the south—the Río Mayer Formation. Strong magmatism with acid volcanics and plutonic intrusives followed in the north of the basin, producing the Divisadero Formation (see Riccardi and Rolleri, 1980). South of 50° south latitude, closure of the marginal basin began in the Late Albian (Bruhn and Dalziel, 1977) and was accompanied by some uplift, which resulted in a northeast migration of the basinal axis. Flysch-like deposits, the Punta Barrosa Formation, were formed as a byproduct of the uplift of the "Paleo-Andes" to the west. These deposits have been considered as the first evidence of the foreland basin stage of development of the Austral Basin (Wilson and Dalziel, 1983).

By the beginning of Barremian time, the areas under sedimentation in the Chubut Basin (Fig. 18) had expanded to cover many of the regional highs located between the early Mesozoic graben troughs (Uliana and Biddle, 1987). The fan-delta system, of the D-129 Formation, subsided and was inundated by a deepening lacustrine environment (the Mina El Carmen Formation). Deposition persisted in the deeper and more restricted areas of the basin, while westward proximal fluvial and fan-delta systems were developed (the Castillo Formation) (Brown et al., 1982).

In the Parana Basin, the Lower Cretaceous volcanics were covered on the northwestern area by eolian sediments to form the Caiuá Formation (Table 18). Similar continental sequences were formed in all Atlantic basins, i.e., the Santa Lucía, Pelotas, Santos, Salado and Colorado Basins. The proto-oceanic stage is represented in the Santos Basin by Aptian evaporites. Continental beds were also deposited in a series of basins in central Argentina (Fig. 18).

"Middle" Cretaceous folding and uplift in northern Chile provided the source areas of continental sandstones in the Aimará Basin of northern Argentina and Bolivia (Fig. 18). The basalts found within these rocks are related to the same diastrophic phase (Avila-Salinas, 1986). During Cenomanian time, probable erosional leveling and tectonic lowering of the whole area was followed by a marine transgression coming from northwest Perú, characterized by shallow marine facies of the Ayavacas and Miraflores formations. The marginal areas were characterized by continental deposition, i.e., the Sucre Formation (Table 18).

Warm and arid climatic conditions persisted on the continental area north of 43° south latitude. These conditions are indicated by evaporites and red beds (Rayoso Group and the Huitrín, Colimapu, and Quebrada Marquesa formations) in the Andean Basin (Fig. 18; Table 17), from west-central Argentina to northern Chile, by sedimentological and palynological data in southern Brazil (Lima, 1983; Petri, 1983), and by saurischian remains in the northern part of the Chubut Basin (Fig. 18). In the latter area, conchostracans and fresh-water sponges suggest the seasonal presence of ponds and pools, probably in the "warmer" season of the year, which was also characterized by a strongly dry season.

Sea waters on the Pacific coast were increasingly warmer, as $\delta^{18}O$ - $\delta^{16}O$ analyses on belemnites from the Austral Basin at about 49° south latitude indicate 30.7° to 32.7°C for Albian time (Bowen, 1963). These results, however, seem anomalously high (Bowen, 1966; Waterhouse and Riccardi, 1971), especially since analyses of benthic and pelagic foraminifers suggest temperate to cold waters for the entire Austral Basin (Malumián, 1979).

?TURONIAN-CAMPANIAN

The "Middle" Cretaceous Mirano diastrophic phase caused the uplift of the western areas of central and northern Chile from about 37° to 18° south latitude (Fig. 19). Thick volcanic sequences were deposited to the west and east of the Abanico Paleovolcanic Range (Aubouin et al., 1973), along the western margin of the present Cordillera Principal. Volcaniclastic sediments accumulated in restricted back-arc continental basins, to form the Viñita, Augusta Victoria, Purilactis, and Tolar formations (Table 16).

In the eastern areas, south of 33° south latitude (Fig. 19), relatively thin, red continental sediments—the Neuquén Group—were deposited in the now-isolated basin of west-central Argentina.

Deepening of the Austral Basin was related to deformation and uplift within the Andean Cordillera, to the west. As a result, a regressive pattern continued in the north (Figs. 3, 4, 19), with the deposition of near-shore to continental sediments and pyroclastics, i.e., the Anita, Cardiel, and Pari Aike formations (Table 17). In the deep axial portion of the basin (Figs. 3, 4, 19), submarine fans were developed eastward and southeastward to form thick tongues of conglomerates in a sequence of turbidites, the Cerro Toro Formation (Table 17). Marine sedimentation was continuous in the southern part of the basin (Table 17). The main regressive pattern (Fig. 21) was punctuated by at least two short-lived transgressions associated with sea-level rises of probably Cenomanian-Turonian and Coniacian-Santonian ages (Arbe and Hechem, 1984).

In central and northern extra-Andean Patagonia (Fig. 19), continental beds, containing abundant pyroclastics and bearing dinosaur remains, were deposited, the Chubut and Neuquén Groups. In the Chubut Basin, renewed uplift—principally in the north and northwest—caused extensive southeastward progradation and aggradation of fluvial and fan-delta systems into lacustrine environments—the Comodoro Rivadavia and Cañadón Seco formations (Table 17) (Brown et al., 1982; Barcat et al., 1984a). Meanwhile, the central part of the basin was the site of

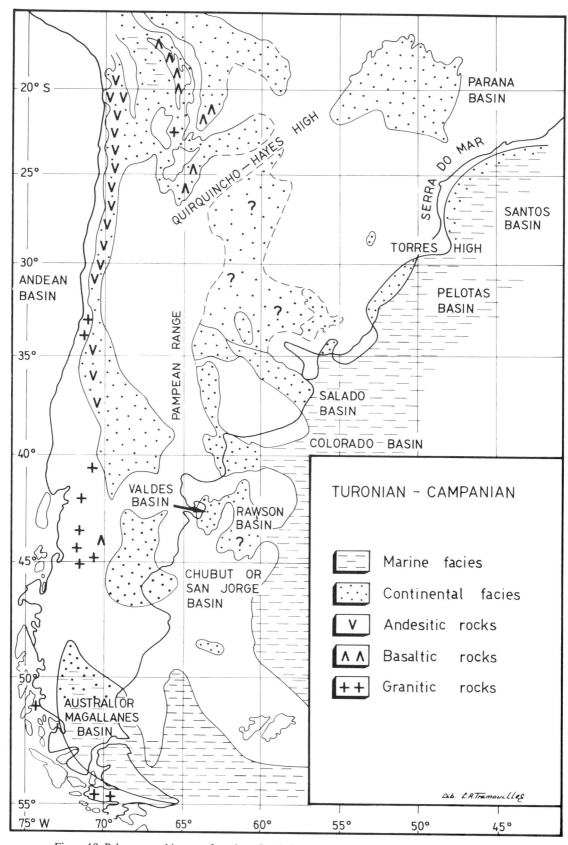

Figure 19. Paleogeographic map of southern South America for Turonian-Campanian time (modified from Riccardi, 1987).

extensive alluvial plains—the Bajo Barreal Formation (Brown et al., 1982).

Most of the Paraná Basin was emergent during early Late Cretaceous time, although vertical movements along the Atlantic continental margin resulted in deposition of conglomerates—the Bauru Group—in the southern and northern areas of the Paraná Basin (Fig. 19). The Pelotas and Santos Basins of southern Brazil were characterized by marine sedimentation. All other Atlantic basins, i.e., the Salado and Colorado basins, had mostly continental deposition. In northern Patagonia, however, from the Andes to the Colorado Basin, fresh- and brackish water sediments, represented by rocks of the Jagüel Formation, began to accumulate during latest Campanian time, to be followed in Maastrichtian time by an Atlantic transgression.

In the Aimará Basin of northern Argentina and Bolivia (Fig. 19), deposition of continental sediments with interbedded basalts continued, i.e., the Cotacucho, Moho and Toro Toro formations, Pirgua Subgroup, Betanzos, and Las Conchas basalts (Table 18), with some minor deposition of shallow-marine strata—the Chaunaca Formation. Continental rocks also accumulated farther east, in the Subandean Range and Chiquitanas Hills of Bolivia (Figs. 14, 19), producing the Tacurú, Beu, and El Portón formations (Table 18).

The widespread presence of red beds during Late Cretaceous time, from central Patagonia to Bolivia and southern Brazil (Fig. 19), suggest warm, dry, and arid conditions. In that area, vertebrates were relatively common, including chelonids, crocodiles, saurischia, and ornithischia (Bauru, Neuquén, Chubut groups and Pirgua Subgroup; Tables 12, 17, 18). Most chelonids and crocodiles are known north of 35° south latitude, while most dinosaurs are found south of 35° south latitude. This suggests (Bonaparte, 1979) a northern region with higher temperatures, more water, and plants, and a southern region with lower temperatures, less water, and open vegetation. However, most known plants have been described from central Chubut. This evidence suggests a northern belt, north of 43° south latitude, with a more continental climate and characterized by a strongly dry season, and a southern belt with temperate and humid climate. Progressive endemism of the invertebrate fauna of the Austral Basin suggest, in addition to paleogeographic factors, cool-water conditions (Thomson, 1981; Macellari, 1987).

MAASTRICHTIAN

As a result of the different subphases of "Middle" Cretaceous diastrophism (see under Tectonism), the continental basins of central and northern Chile disappeared due to folding and uplift, although volcanism was still present in most areas (Figs. 11, 20). At the same time, a few small fore-arc marine basins were developed along the western margin of the Coastal Cordillera, Navidad, and Arauco Basins (Fig. 20). They had a northwest orientation and were filled by detrital marine sediments to form the Quiriquina and Punta Topocalma formations (Table 18).

In the Austral Basin (Fig. 20), marine sedimentation, largely represented by retrogradational sedimentary units, became restricted to the southern areas. Uppermost Cretaceous and Tertiary units were derived from the south, west, and northwest, and show a progressive onlap geometry from west to east (Biddle et al., 1986). A short-lived(?) transgression, related to a sea-level rise, occurred during Late(?) Maastrichtian time.

In the Chubut Basin (Fig. 20), a final Late Cretaceous deltaic progradational episode, the Yacimiento Trébol Formation, closed the rift basin history (Brown et al., 1982).

During Early Maastrichtian time, subsidence of the Atlantic margin, coupled with sea-level rise, led to widespread marine sedimentation in the Colorado, Salado, Pelotas, and Santos basins (Fig. 20). Middle Maastrichtian maximum flooding in northern Patagonia (Fig. 21) occurred in a negative area located between two massifs, the Pampean Massif to the north and the Somuncurá Massif to the south. The transgression extended through the Colorado Basin, where it is represented by the Pedro Luro Formation, to the foothills of the Andes, i.e., the Jagüel, Loncoche, Malargüe, and Lefipán formations (Table 17). Regression proceeded from west to east during Late Maastrichtian–Early Paleocene time (Fig. 21). Existence of neritic marine sediments in the High Cordillera of west-central Argentina (Yrigoyen, 1979) suggests a possible Maastrichtian Atlantic-Pacific connection through the Abanico Paleovolcanic Chain, although facies changes within the Malargüe Group indicate an eastward paleoslope. The Salado Basin (Fig. 20) was probably connected to the north with the Aimará Basin, and to the south with the Macachín-Colorado basins.

In the Aimará Basin the maximum advance of the Maastrichtian sea is represented by the El Molino and Yacoraite formations. Alternative marine connections have been postulated with the Pacific through Chile or Perú, and with the Atlantic through the Chaco-Pampean Plain (Salfity, 1980; Salfity et al., 1985; Zambrano, 1986). At the end of Maastrichtian times the sea withdrew, and a continental sequence was deposited during early Tertiary times—the Santa Bárbara Subgroup.

In the Paraná Basin, cratonic subsidence produced continental sedimentation on a broad, stable platform. Sediments containing dinosaur remains—the Bauru Group—were deposited in a northern depression extending from 48° to 56° west longitude. Similar sediments were probably deposited in western Uruguay. At the end of Cretaceous time, a general uplift resulted in a new erosive cycle.

The ?Turonian-Campanian climatic zonation persisted during Maastrichtian time, although a cooling trend, within a tropical to subtropical climate was probably present in southern Brazil (Lima, 1983; Petri, 1983). Presence of limestones and stromatolites suggests that the Maastrichtian interior seas of the Aimará Basin (Fig. 20) had shallow and warm waters. A similar sea was present in northern Patagonia (Fig. 20), where planktonic foraminifers indicate temperate to warm waters (Bertels, 1979). The microfauna of the Austral Basin has different affinities, is less diverse than that of northern Patagonia, and indicates temperate

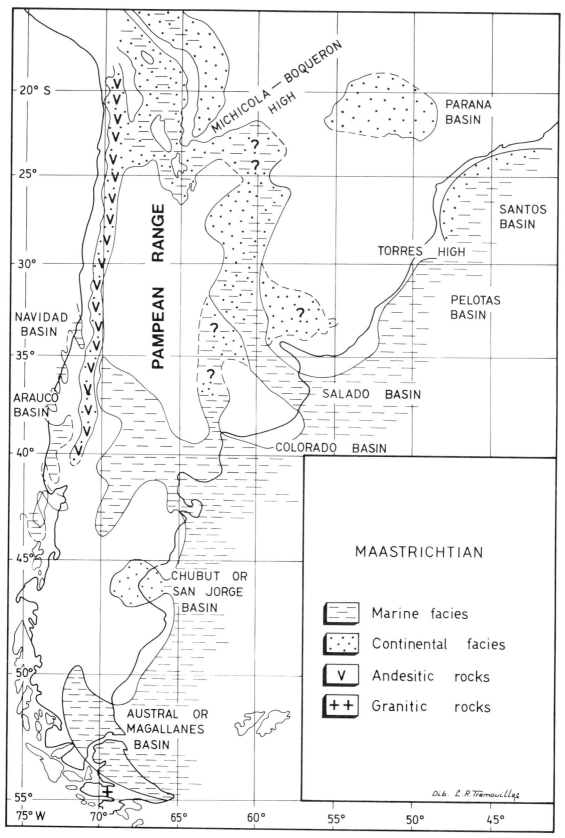

Figure 20. Paleogeographic map of southern South America for Maastrichtian time (modified from Riccardi, 1987).

to cold waters (Malumián, 1979). This conclusion is supported by macrofauna endemism (Macellari, 1985b).

CRETACEOUS TRANSGRESSIONS AND REGRESSIONS

In the last decade it has become common to correlate transgressive-regressive depositional sequences of local marine successions with tectono-eustatically controlled global changes of sea level (but see Jeletzky, 1977, 1978). The results are usually expressed by transgressive-regressive graphs or sea-level curves, and compared with one or several of the proposed curves of global sea-level changes.

For the Cretaceous of southern South America, this effort at correlation has been discussed or attempted only for restricted areas and stages (Reyment and Tait, 1972; Reyment, 1972, 1977, 1979, 1980; Charrier and Malumián, 1975; Matsumoto, 1977, 1980; Reyment and Mörner, 1977; Wiedmann, 1980; Malumián and Ramos, 1984; Hallam et al., 1986; Petri, 1987; Riccardi, 1987). The stratigraphic synthesis of the Cretaceous of the entire area, presented above, now provides a good basis on which to attempt a more comprehensive analysis of this subject. The objective is to compare the marine sequences of different basins and to examine the relative roles played by eustatic and tectonic events in the development of transgressive-regressive sedimentary patterns.

The analysis (Fig. 21) is restricted to those basins including Cretaceous marine strata. All of these basins are located on the Pacific (the Austral, Andean, Arauco, Navidad, and Aimará basins) or Atlantic (the Colorado, Salado, Pelotas, and Santos basins) margins of the South American subcontinent (Figs. 17, 20).

Differences in detail of published data for the sedimentary sequences of these basins are obvious from the stratigraphic synthesis presented above. Nevertheless, an attempt has been made (Fig. 21) to document, as precisely as possible, the transgressive and regressive Cretaceous histories of the different areas, and to compare them with the curve of global sea-level changes of Vail et al. (1977), and against the time scale proposed by Odin and Kennedy (1982).

Pre-Aptian marine Cretaceous beds in southern South America are restricted to the Austral and Andean basins (Fig. 21). In the Andean Basin, a late–Early Tithonian transgressive sequence was succeeded by a Middle Tithonian–Early Valanginian regressive sequence. The early Tithonian transgression corresponds to the widespread development of the Vaca Muerta Formation, whereas the Middle Tithonian–Early Valanginian regression is mostly represented by the prograding pattern of the Quintuco and Mulichinco formations. A condensed interval or marine hiatus at the Tithonian-Berriasian boundary has been interpreted as evidence of a rapid and pronounced sea-level rise (Mitchum and Uliana, 1985). A Late Valanginian–Hauterivian transgression interrupted by two short-lived regressions (Mendiberri, 1985) culminated with deposition of the lower part of the Agrio Formation. According to Mitchum and Uliana (1985), this transgression corresponds to a general sea-level rise punctuated by a Late Valanginian sea-level fall. Finally, the Andean Basin experienced a Hauterivian-Aptian regression, which resulted in a complete withdrawal of the Pacific sea, although two marine levels of the Huitrín Formation attest to an Early Aptian high stand of sea level (Legarreta, 1986). Part of this area, in west-central Argentina, was covered again by the sea in the course of a widespread Maastrichtian transgression coming from the Atlantic (see below).

In the Austral Basin a Tithonian–late Early Hauterivian transgression (Fig. 21) is represented by the basal sandstones of the Springhill Formation and the following widespread development of a pelitic interval characterized by the ammonite *Favrella americana* (Favre) (Table 15; Plate 8, Figs. 3, 4). On a finer scale, a Valanginian fall of sea level could be represented by a faunal break(?) and stratigraphic hiatus (Biddle et al., 1986). Installation of a Barremian–Early Aptian regressive pattern is indicated by prograding shallow marine facies represented in the basin's northern margin. A pause in the main regressive pattern, coincident with deepening of the basin, suggests a Late Aptian–Middle Albian sea-level rise. This rise of sea level ended the previously restricted euxinic marine regime of the Austral Basin. Widespread flooding and open marine circulation fostered migration of a more diversified extra-basinal ammonite fauna (Table 15). Regressive sequences characterize most of the Late Cretaceous history of the Austral Basin, although at least two short-lived rises of sea level probably occurred between the Cenomanian and Early Campanian (Arbe and Hechem, 1985). A Maastrichtian transgressive sequence is represented by the Calafate and Man Aike formations (Fig. 21).

Comparison of the transgressive-regressive patterns described for the Andean and Austral basins with the curve of global sea-level changes of Vail et al. (1977) indicates several discrepancies. This situation is clearly evident for the Andean Basin where Tithonian–Early Valanginian and Hauterivian-Aptian major regressive sequences were formed at times of relatively high global sea levels. Regressions resulted from an excess of clastic supply in relation to subsidence and sea-level rise. Thus, the transgressive pattern of the Andean Basin appears to have been controlled by regional tectonics in an area in which local vertical movements were greater than global sea-level changes. Regional tectonics and magmatism at plate edge positions seem to have also controlled the access of oceanic waters to the back-arc depocenters. Local low and high stands of sea level were probably related to changes in plate motion rates and compressive regimes (Ramos, 1986). Thus, maxima in plutonic activity were coincident with regressions, whereas minima were related to transgressions (Piraces, 1979).

The late Early Cretaceous regression is a consequence of high plate convergence rates associated with a maximum in igneous activity, major uplift, and eastward migration of the magmatic arc (Ramos, 1986) (Fig. 21).

As in the Andean Basin, the Pacific margin of the Austral

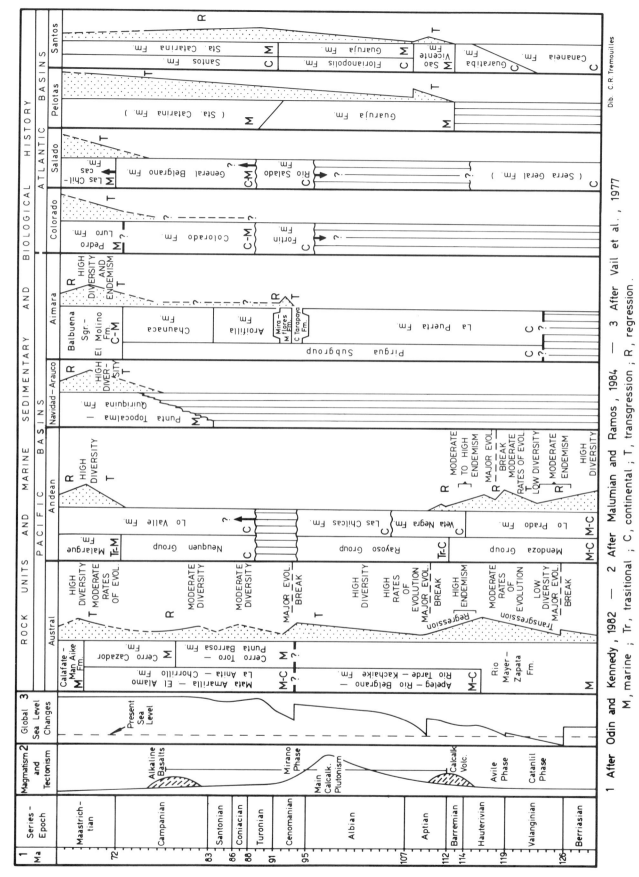

Figure 21. Summary of geological and biological events, marine Cretaceous rocks, southern South America.

1 After Odin and Kennedy, 1982 — 2 After Malumian and Ramos, 1984 — 3 After Vail et al., 1977

M, marine ; Tr, trasitional ; C, continental ; T, transgression ; R, regression.

Dib C.R. Tremouilles

Basin was tectonically active during Early Cretaceous time. However, the tectonic and paleogeographic settings differed as southern Patagonia was characterized by the opening of a marginal basin, and sedimentation was mostly controlled by a tectonically more stable area, the Deseado Massif, located on the northeastern margin of the Austral Basin. Therefore, a different transgressive-regressive pattern emerged, as local tectonic movements were not large enough to completely overprint the record of global changes of sea level. Thus, the major Tithonian-Hauterivian transgressive pattern of the Austral Basin is in agreement with a global rise of sea level (Fig. 21). Similarly, closure of the marginal basin at the end of Early Cretaceous time, coupled with the initial uplift of the Andean Cordillera and the deepening of the Austral Basin, could not completely overprint the Albian global rise in sea level (Fig. 21).

Only for the Early Cenomanian is there coincidence between the Late Cretaceous regressive pattern of sedimentation of the Austral Basin and the curve of global changes of sea level. Late Cenomanian–Early Campanian global sea-level changes are not clearly evident in southern Patagonia, although for Late Campanian–Early Maastrichtian time there is again a good match between sedimentary patterns and the widespread transgression recorded in many parts of the earth. Most of the Late Cretaceous transgressive-regressive marine history of the Austral Basin was probably the result of both local uplift of the Andes and global changes of sea level.

Late Campanian–Early Maastrichtian strata are also present in the Arauco, Navidad, and Aimará basins (Fig. 20) on the Pacific margin of the South American subcontinent. They correspond to transgressive sequences that match a worldwide transgressive peak.

Evidence for another important Cretaceous global sea-level rise is found in the Aimará Basin. This transgression corresponds to the Ayavacas and Miraflores formations of Perú and Bolivia and has been dated, on the occurrence of the ammonite genus *Neolobites* Fischer (Plate 15, Figs. 1-3), as Late Cenomanian, although generic affinities could indicate Middle Albian time.

Marine sedimentation of the Atlantic basins of southern South America was restricted in time to the late Early and Late Cretaceous. No evidence is known to support a Late Valanginian marine transgression across northern Patagonia, as was postulated by Reyment and Tait (1972, Fig. 2). Faunal similarities between southern Africa and west-central Argentina can be accounted for by interconnection through the Austral Basin, where a similar fauna has also been recorded (Riccardi, 1976, 1977; Nullo et al., 1981a).

The first evidence of marine sedimentation in the Atlantic basins of southern South America analyzed herein is found in the Pelotas and Santos basins (Fig. 18); it consists of Aptian limestones and salt beds. That Lower Cretaceous marine strata are absent in the epicratonic basins of the Argentinian continental shelf is probably due to restricted water circulation in the opening South Atlantic Ocean (Sclater et al., 1977; Thiede and van Andel, 1977; Reyment, 1980; Urien et al., 1981). The existence of Lower Cretaceous marine sequences in the sedimentary wedges buried under the continental slope off Argentina is very likely considering the accepted sequence of opening of the South Atlantic and the marine sediments recorded in the Malvinas (Falkland) Plateau. The presence of neritic and bathyal facies in the Pelotas Basin and of salt beds north of the Walvis–Rio Grande Ridge in the Santos Basin could be related to a global rise of sea level with intermittent northward flooding over the ridge barrier. A rise in sea level continued during Albian–Early Cenomanian time, when shelf and/or bathyal facies were deposited in the Pelotas and Santos Basins (Asmus, 1981, 1984). Throughout Late Cretaceous time, marine sedimentation was continuous in both the Pelotas and Santos Basins. However, in the Santos Basin a post-Middle Turonian regressive pattern emerged due to the large volume of sediments shed from the west by the rising Serra do Mar hinge line (Williams and Hubbard, 1984). This pattern differs from a picture of continued marine transgressions, as represented in the Pelotas Basin and in other northern Brazilian basins. In spite of these differences, a Late Campanian–Maastrichtian transgression appears to be present in both the Pelotas and Santos basins.

During Late Cretaceous time, subsidence of the northern part of the Argentinian Atlantic margin, coupled with a sea-level rise, led to widespread shallow-marine sedimentation in the Colorado and Salado basins (Fig. 20). Lack of accurate biostratigraphic dating precludes a detailed assessment of the initial stages of these successions. Nevertheless, the earliest marine encroachment (?Turonian) is recorded in the Colorado Basin, probably as a result of subsidence of a basement ridge that fringes the Patagonian segment of South America (Uliana and Biddle, 1987). Subsequently, a Late Campanian–Middle Maastrichtian transgression, coming from the Atlantic through the Colorado Basin, is well documented in northern Patagonia, the Salado Basin, and the Chaco-Pampean plain (Fig. 20).

A. C. Riccardi

Plate 1

Figures 1–2. *Pseudinvoluticeras douvillei* Spath, phragmocone, Lower Tithonian, Picún Leufú, Neuquén Province, Argentina (MLP 8210); 1, lateral view; 2, ventral view; × 1.

Figures 3–4. *Choicensispinctes choicensis* (Burckhardt), phragmocone and body chamber, Lower Tithonian, Cerro Lotena, Neuquén Province, Argentina (MLP 6342); 1, ventral view; 2, lateral view; × 1.

Plate 2

Figures 1–2. *Hemispiticeras* aff. *steinmanni* (Steuer), phragmocone, Middle Tithonian, Cañada de Leiva, Mendoza Province, Argentina (MLP 14941); 3, lateral view; 4, ventral view; × 1.

Figures 3–5. *Aspidoceras andinum* Steuer, phragmocone, Middle Tithonian, Río Diamante, Mendoza Province, Argentina (MLP 3568); 3, ventral view; 4, lateral view; 5, apertural view; × 1.

Figures 6–7. *Aulacosphinctes proximus* (Steuer), phragmocone and body chamber, Middle Tithonian, Mina La Eloisa, Mendoza Province, Argentina (MLP 14629); 3, apertural view; 4, lateral view; × 1.

Plate 3

Figures 1–2. *Substeueroceras koeneni* (Steuer), phragmocone, Upper Tithonian, Mina de Rafaelita, Mendoza Province, Argentina (MLP 3362); 1, lateral view; 2, ventral view; × 1.

Figures 3–4. *Parodontoceras calistoides* (Behrendsen), complete phragmocone, Upper Tithonian, Tril west, Neuquén Province, Argentina (MLP 9924); 4, ventral view; 5, lateral view; × 1.

Figures 5–6. *Spiticeras damesi* (Steuer), incomplete phragmocone and body chamber, Upper Berriasian, Arroyo del Yeso, Mendoza Province, Argentina (MLP 19373); 5, lateral view; 6, ventral view.

Figures 7–8. *Cuyaniceras transgrediens* (Steuer), phragmocone, Upper Berriasian, Mendoza Province, Argentina (MLP 4511); 7, ventral view; 8, lateral view; × 1.

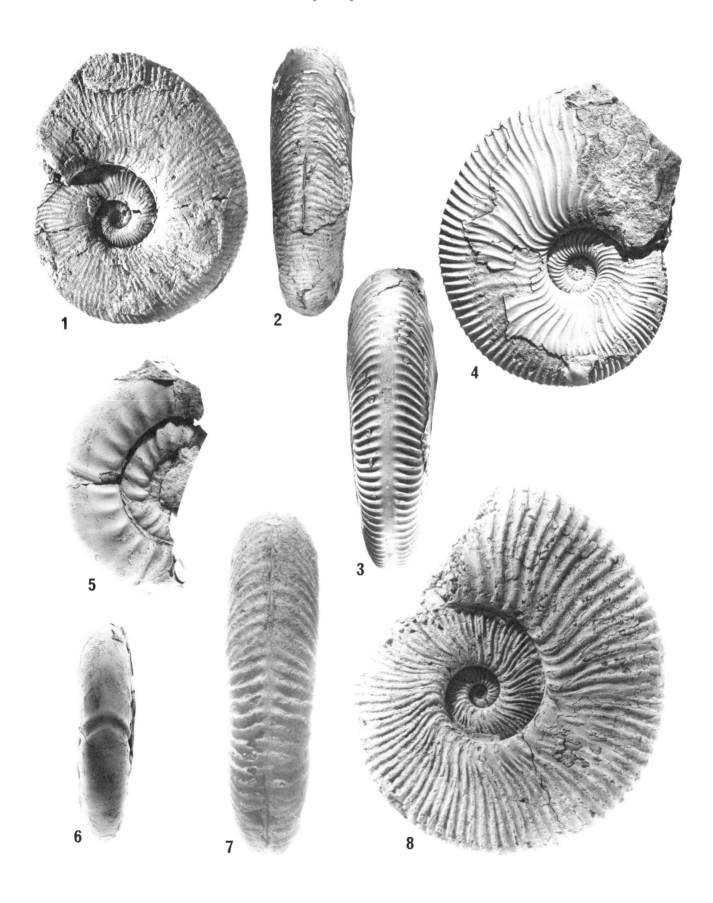

A. C. Riccardi

Plate 4

Figures 1–2. *Thurmanniceras discoidale* (Gerth), phragmocone, Lower Berriasian, Loncoche, Mendoza Province, Argentina (MLP 14940); 1, lateral view; 2, ventral view; × 1.

Figures 3–4. *Frenguelliceras* sp., phragmocone, Lower Berriasian, Cañada de Leiva, Mendoza Province, Argentina (MLP 14944); 3, lateral view; 4, ventral view; × 1.

Figures 5–6. *Karakaschiceras attenuatus* (Behrendsen), phragmocone, Upper Valanginian–Lower Hauterivian, Cerro Pitren, Neuquén Province, Argentina (MLP 11071); 5, lateral view; 6, ventral view; × 1.

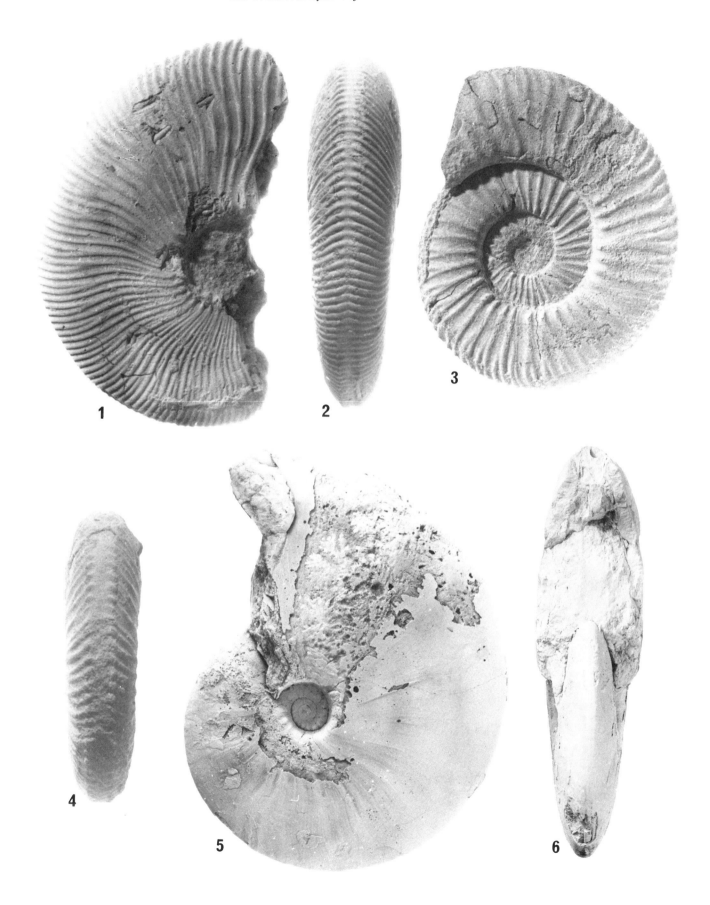

Plate 5

Figures 1–2. *Olcostephanus atherstoni* (Sharpe), phragmocone, Upper Valanginian–Lower Hauterivian, Cerro Pitren, Neuquén Province, Argentina (MLP 10956); 1, lateral view; 2, ventral view; × 1.

Figures 3–4. *Lissonia riveroi* (Lisson), phragmocone with body chamber, Valanginian, Cañada de Leiva, Mendoza Province, Argentina (MLP 14945); 3, ventral view; 4, lateral view; × 1.

Figures 5–6. *Pseudofavrella angulatiformis* (Behrendsen), lectotype, body chamber, Lower Hauterivian, Cerro Pitren, Neuquén Province, Argentina (GPIG 498-24); 5, lateral view; 6, ventral view; × 1.

Plate 6

Figure 1. *Pseudofavrella garatei* Leanza and Leanza, phragmocone and body chamber, Lower Hauterivian, Bajada del Agrio, Neuquén Province, Argentina (MLP 17367); lateral view; × 1.

Figures 2–3. *Lyticoceras pseudoregale* (Burckhardt), phragmocone, Lower Hauterivian, Arroyo de la Yesera, Mendoza Province, Argentina (MLP 3590); 2, apertural view; 3, lateral view; × 1.

Figures 4–5. *Acanthodiscus wichmanni* Gerth, phragmocone, Lower Hauterivian, Cerro Guanacos, Mendoza Province, Argentina (MLP 3657); 4, lateral view; 5, ventral view; × 1.

Figures 6–7. *Holcoptychites neuquensis* (Douvillé), phragmocone and body chamber, Upper Hauterivian, Loma del Rayoso, Neuquén Province, Argentina (MLP 3618); 6, lateral view; 7, ventral view; × 1.

Plate 7

Figures 1–2. *Crioceratites diamantensis* (Gerth), phragmocone and body chamber, Upper Hauterivian–Lower Barremian, Planicie Negra, Neuquén Province, Argentina (FCEN 5150); 1, lateral view; 2, ventral view; × 1.

Figures 3–5. *Crioceratites apricus* (Giovine), holotype, phragmocone, Upper Hauterivian–Lower Barremian, cerro Curaco, Neuquén Province, Argentina (FCEN 5320); 3, lateral view; 4, ventral view; 5, apertural view; × 1.

A. C. Riccardi

Plate 8

Figures 1–2. *Crioceratites schlagintweiti* (Giovine), holotype, phragmocone, Upper Hauterivian–Lower Barremian, Estancia Gallardo, Neuquén Province, Argentina (FCEN 5147); 1, lateral view; 2, apertural view; × 1.

Figures 3–4. *Favrella americana* (Favre), plaster cast of paralectotype, phragmocone, Lower Hauterivian, Lago Belgrano, Santa Cruz Province, Argentina (GPIF 17600); 3, ventral view; 4, lateral view; × 1.

Figure 5. *Belemnopsis madagascariensis* (Besairie), Hauterivian, Lago San Martín, Santa Cruz Province, Argentina (MLP 21302); ventral view; × 1.

Plate 9

Figures 1–2. *Protaconeceras patagoniense* (Favre), lectotype, phragmocone and body chamber, Upper Hauterivian, lago Belgrano, Santa Cruz Province, Argentina (GPIF 1820); 1, lateral view; 2 apertural view; × 1.

Figures 3–4. *Protaconeceras patagoniense* (Favre), allotype, phragmocone and body chamber, Upper Hauterivian, lago Belgrano, Santa Cruz Province, Argentina (MLP 17100); 3, lateral view; 4, ventral view; × 1.

Figures 5–6. *Belemnopsis patagoniensis* (Favre), holotype, Hauterivian, lago Belgrano, Santa Cruz Province, Argentina (GPIF 40); 5, cross section; 6, ventral view; × 1.

Figures 7–8. *Hatchericeras patagonense* Stanton, holotype, phragmocone, Lower Barremian, lago Pueyrredón, Santa Cruz Province, Argentina (PU 66); 1, lateral view; 2 apertural view; × 0.50.

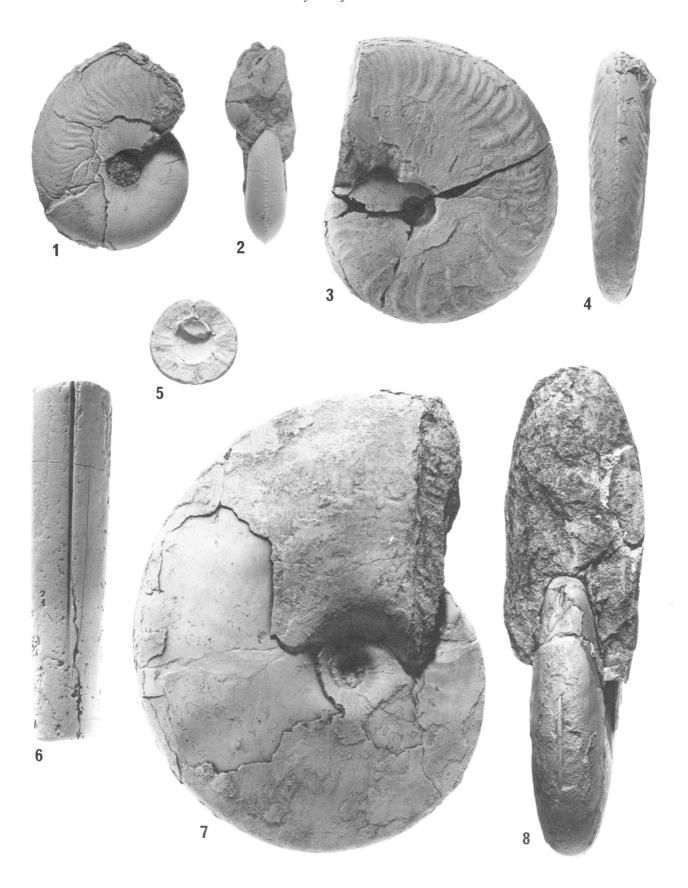

Plate 10

Figures 1–2. *Pseudohatchericeras argentinense* (Stanton), lectotype, phragmocone, Lower Barremian, lago Pueyrredón, Santa Cruz Province, Argentina (PU 69); 1, lateral view; 2, ventral view; × 1.

Figures 3–4. *Hatchericeras pueyrrydonensis* Stanton, holotype, phragmocone, Lower Barremian, lago Pueyrredón, Santa Cruz Province, Argentina (PU 74); lateral view; 4, apertural view, × 1.

Figure 5. *Ptychoceras* sp., Upper Aptian, lago San Martín, Santa Cruz Province, Argentina (MLP 17490); lateral view; × 1.

Figures 6–7. *Hatchericeras tardense* Stanton, holotype, phragmocone, Lower Barremian, lago Pueyrredón, Santa Cruz Province, Argentina (PU 73); 3, lateral view; 4, apertural view; × 1.

Figures 8–9. *Colchidites vulanensis australis* Klinger, Kakabadze and Kennedy, phragmocone and body chamber, Upper Barremian, Tucu Tucu, Santa Cruz Province, Argentina (MLP 20571); 8, lateral view; 9, ventral view; × 1.

Figure 10. *Helicancylus bonarellii* (Leanza), Upper Aptian, lago San Martín, Santa Cruz Province, Argentina (MLP 17094); lateral view; × 1.

Plate 11

Figure 1. *Peltocrioceras deeckei* (Favre), phragmocone, Upper Aptian, lago San Martín, Santa Cruz Province, Argentina (MLP 15209); lateral view; × 1.

Plate 12

Figures 1–2. *Sinzovia leanzai* Riccardi et al., holotype, phragmocone and body chamber, Lower Albian, lago San Martín, Santa Cruz Province, Argentina (UNC 4340); 1, lateral view; 2, ventral view; × 1.

Figures 3–4. *Sinzovia piatnitzkyi* Riccardi et al., holotype, phragmocone and body chamber, Upper Aptian, lago San Martín, Santa Cruz Province, Argentina (FCEN 11111); 3, ventral view; 4, lateral view; × 1.

Figures 5–6. *Dimitobelus* aff. *stimulus* Whitehouse, Lower Albian, Lago San Martín, Santa Cruz Province, Argentina (MLP 21303); 5, ventral outline; 6, left profile; × 1.

Figures 7–9. *Rossalites imlayi* (Leanza), body chamber, Lower Albian, lago San Martín, Santa Cruz Province, Argentina (MLP 17369); 7, lateral view; 8, ventral view of shaft; 9, ventral view of hook; × 1.

Plate 13

Figure 1. *Sanmartinoceras patagonicum* Bonarelli, plaster cast of the holotype, phragmocone and body chamber, Albian, lago San Martín, Santa Cruz Province, Argentina (SGN 9302); lateral view; × 1.

Figures 2–3. *Aioloceras argentinum* (Bonarelli), phragmocone and body chamber, Lower Albian, lago San Martín, Santa Cruz Province, Argentina (MLP 17368); 1, apertural view; 2, lateral view; × 1.

Figures 4–9. *Dimitobelus* aff. *stimulus* Whitehouse, Lower Albian, lago San Martín, Santa Cruz Province, Argentina; 4-5, ventral outline and right profile of specimen MLP 21304; 6-7, dorsal outline and left profile of specimen MLP 21305; 8-9, ventral outline and right profile of specimen MLP 21306; × 1.

Figure 10. *Labeceras singulare* (Leanza), Upper Albian, Estancia La Vega, Santa Cruz Province, Argentina (MLP 20601). Lateral view; × 1.

Figures 11–12. *Labeceras crassetuberculatum* Klinger, Upper Albian, Estancia La Vega, Santa Cruz Province, Argentina (MLP 16961); 11, lateral view; 12, ventral view; × 1.

Figures 13–14. *Myloceras andinum* (Leanza), holotype, pragmocone, Upper Albian, Estancia La Vega, Santa Cruz Province, Argentina (UNC 4349); 13, lateral view; 14, apertural view; × 1.

Figures 15–16. *Myloceras andinum* (Leanza), phragmocone with part of body chamber, Upper Albian, Estancia La Vega, Santa Cruz Province, Argentina (MLP 20638); 15, lateral view; 16, apertural view; × 1.

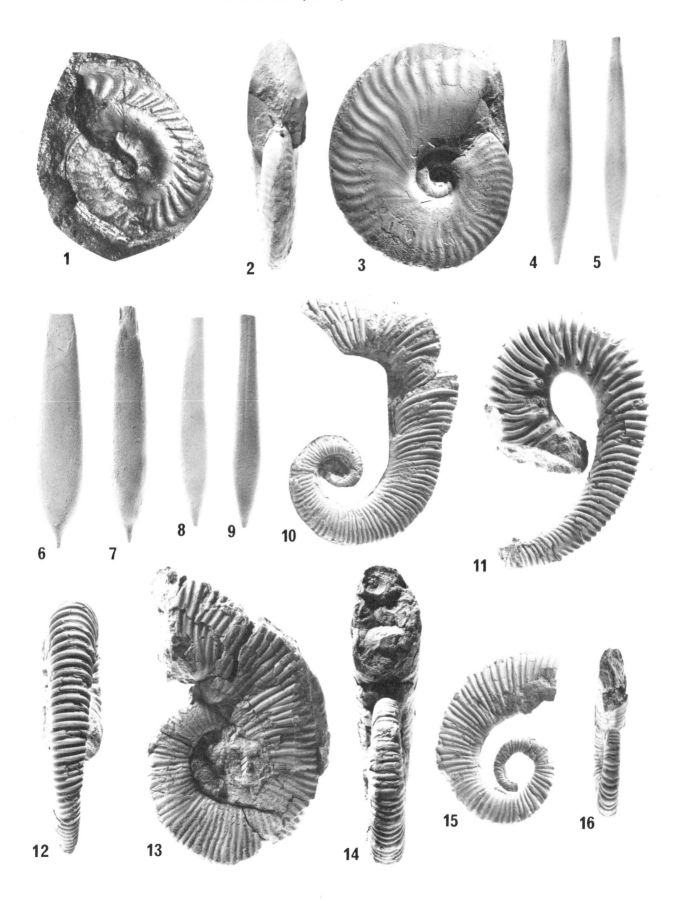

A. C. Riccardi

Plate 14

Figures 1–2. *Puzosia vegaensis* Leanza, holotype, phragmocone and body chamber, Upper Albian, Estancia La Vega, Santa Cruz Province, Argentina (SGN 12476); 1, lateral view; 2, ventral view; × 1.

Figures 3–4. *Placenticeras patagonicum* Leanza, paratype, phragmocone, Upper Cenomanian–Turonian, Lago Viedma, Santa Cruz Province, Argentina (USNM 132587, courtesy Dr. R. W. Imlay); 3, ventral view; 4, lateral view; × 1.

Figure 5. *Calycoceras* sp., phragmocone and body chamber, Cenomanian, lago Argentino, Santa Cruz Province, Argentina (SGN s/n); lateral view; × 0.25.

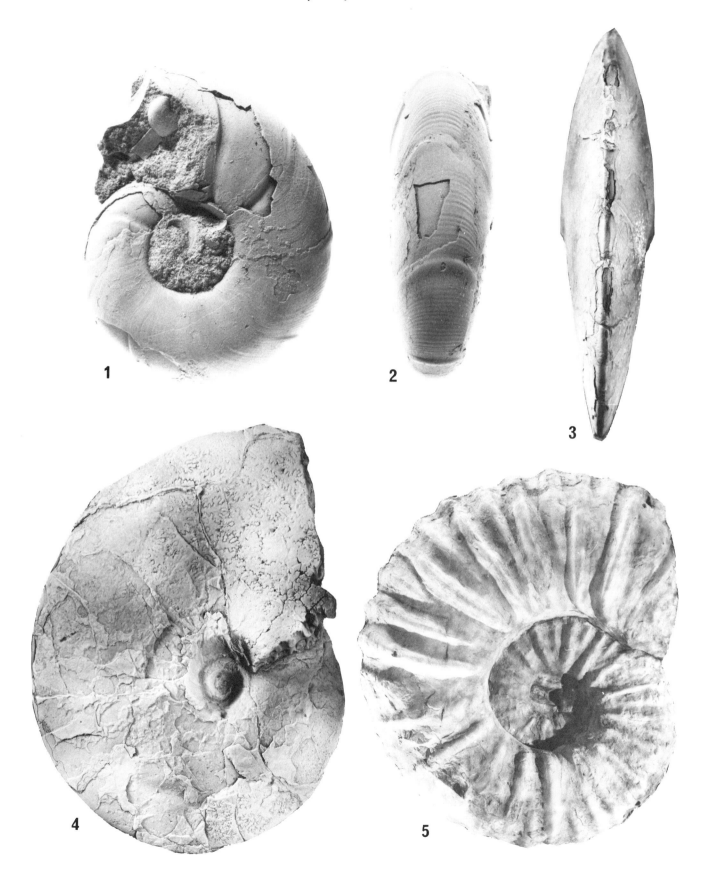

Plate 15

Figures 1–3. *Neolobites* sp., Cenomanian, Miraflores, Bolivia (Muséum National d'Histoire Naturelle, Paris, France, courtesy Prof. J. Sornay); 1, 2, 3. lateral views; × 1.

Figures 4–5. *Gauthiericeras santacrucense* (Leanza), holotype, phragmocone, Coniacian, cerro Indice, Santa Cruz Province, Argentina (UNC 4390, courtesy Prof. M. Hünicken); 4, lateral view; 5, ventral view, × 1.

A. C. Riccardi

Plate 16

Figures 1–3. *Karapadites centinelaensis* (Blasco et al.), holotype, phragmocone and body chamber, Lower Campanian, lago Argentino, Santa Cruz Province, Argentina (SGN 15498); 1, lateral view; 2, ventral view; 3, apertural view; × 1.

Figures 4–6. *Hoplitoplacenticeras plasticum* Paulcke, complete specimen, Campanian, cerro Cazador, Santa Cruz province, Argentina (author's collection); 4, lateral view; 5, apertural view; 6, ventral view; × 1.

Plate 17

Figures 1–2. *Natalites* cf. *hauthali* (Paulcke), phragmocone and body chamber, Lower Campanian, lago Argentino, Santa Cruz Province, Argentina (SGN 15499); 1, lateral view; 2, ventral view; × 1.

Figures 3–4. *Maorites seymourianus* (Kilian and Reboul), phragmocone and body chamber, Maastrichtian, Seymour Island, Antarctica (MLP 13579); 3, lateral view; 4, ventral view; × 1.

Figures 5–6. *Grossouvrites gemmatus* (Huppe), phragmocone, Maastrichtian, Seymour Island, Antarctica (MLP 13551); 5, ventral view; 6, lateral view; × 1.

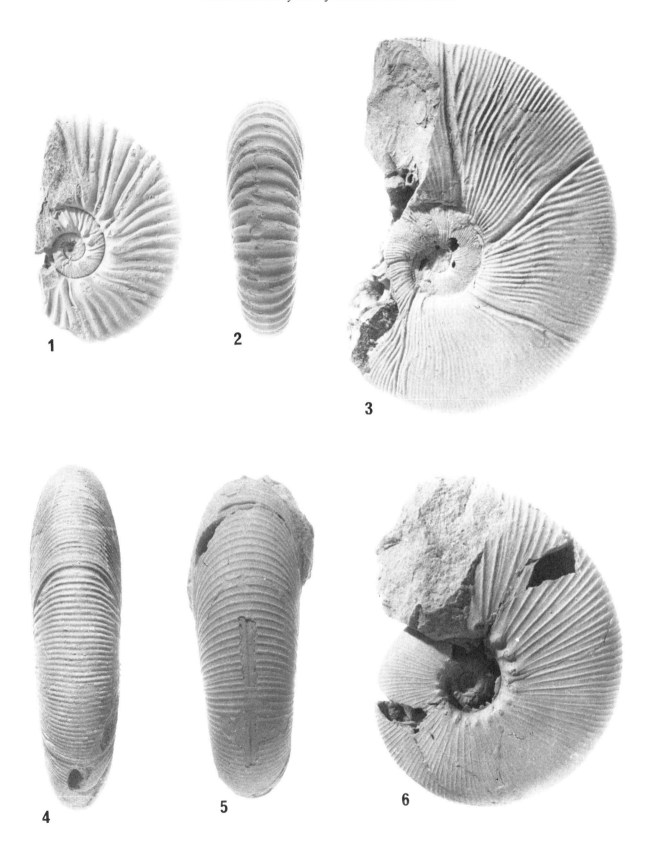

1

2

3

4

5

6

A. C. Riccardi

Plate 18

Figures 1–3. *Eubaculites ootacodensis* (Stoliczka), body chamber, Maastrichtian, El Cain, Río Negro Province, Argentina (MLP 12093); 1, lateral view; 2, ventral view; 3, cross section; × 1.

Plate 19

Figure 1. Upper Jurassic volcanics, in the foreground, are overlain by marine Lower Cretaceous pelites of the Zapata Formation, on the upper right. Austral Basin. Cerro Hobler, northwestern margin of Lago Argentino, southern Argentina. (Photograph: A. C. Riccardi).

Figure 2. Marine Upper Cretaceous turbidites of the Cerro Toro Formation. Austral Basin. Cerro Horqueta, northern margin of Lago Argentino, southern Argentina. (Photograph: A. C. Riccardi).

A. C. Riccardi

Plate 20

Figure 1. To the right, Tithonian–Valanginian marine strata of the Vaca Muerta and Muli-chinco formations, capped to the left by Hauterivian–Barremian marine beds of the Agrio Formation. In the background Aptian–Albian beds of the Huitrín Formation. Andean Basin, Looking south in Pampa de Tril, south of Buta Ranquil, centralwest Argentina (Photograph: A. C. Riccardi).

Figure 2. Continental red beds of the Aptian-Albian Rayoso Group. Andean Basin. North of Río Agrio, centralwest Argentina (Photograph: A. C. Riccardi).

A. C. Riccardi

Plate 21

Figure 1. Basaltic flows of the Serra Geral Formation, Paraná Basin, east of São Joaquim, Santa Catarina State, Brazil (Photograph: courtesy of R. Andreis).

Figure 2. To the right Maastrichtian limestones of the Yacoraite Formation, capped to the left by Lower Tertiary marls of the Mealla Formation. Aimará Basin. Tonco brachysyncline, Salta Province, northern Argentina (Photograph: courtesy of R. Palma).

REFERENCES CITED

Abad, E., 1976, Las Formaciones Cerrillos y Hornitos al Norte de Vallenar, Provincia de Atacama, Chile. Primer Congreso Geológico Chileno, Actas 1, p. A97–A114.

———, 1980, Cuadrángulos Estación Algarrobal, Yerbas Buenas, Cerro Blanco, Merceditas y Tres Morros, Región de Atacama. Instituto Investigaciones Geológicas Chile, Carta Geológica 38, p. 1–48.

Aberg, G., Aguirre, L., Levi, B., and Nyström, J. O., 1984, Spreading-subsidence and generation of ensialic marginal basins: An example from the early Cretaceous of Central Chile. In: Kokelaar, B. P., and Howells, M. F., eds., Marginal Basin Geology. Geological Society of London Special Publication 16, p. 185–193.

Aceñolaza, F. G., 1968, Geología estratigráfica de la región de la Sierra de Cajas, Dpto. Humahuaca, Pcia. de Jujuy. Asociación Geológica Argentina, Revista 23(3), p. 207–222.

Aceñolaza, F. G., and Toselli, A. J., 1981, Geología del Noroeste Argentino. Facultad Ciencias Naturales, Universidad Nacional Tucumán, Publicación Especial 1287, p. 1–212.

Aguirre, L., 1960, Geología de los Andes de Chile Central, Provincia de Aconcagua. Instituto Investigaciones Geológicas Chile, Boletín 9, p. 5–70.

Aguirre, L., and Egert, L., 1965, Cuadrángulo Quebrada Marquesa, Provincia de Coquimbo. Instituto Investigaciones Geológicas Chile, Carta Geológica 15, p. 1–92.

———, 1970, Cuadrángulo Lambert (La Serena), Provincia de Coquimbo. Instituto Investigaciones Geológicas Chile, Carta Geológica 23, p. 1–14.

Aguirre, L., and Levi, B., 1964, Geología de la Cordillera de los Andes de las provincias de Cautín, Valdivia, Osorno y Llanquihue. Instituto Investigaciones Geológicas Chile, Boletín 17, p. 1–37.

Aguirre, L., Charrier, R., Davidson, J., Mpodozis, A., Rivano, S., Thiele, R., Tidy, E., Vergara, M., and Vicente, J. C., 1974, Andean magmatism: Its paleogeographic and structural setting in the central part (30°–35°S) of the southern Andes. Pacific Geology, v. 8, p. 1–38.

Aguirre Urreta, M. B., 1983, Crustáceos Decápodos Barremianos de la Región del Tucu-Tucu, Provincia de Santa Cruz. Ameghiniana, v. 19(3-4), p. 303–317.

———, 1985a, Ancylocerátidos (Ammonoidea) Aptianos de la Cordillera Patagónica Austral, Provincia de Santa Cruz, Argentina. Academia Nacional Ciencias, Córdoba, Boletín 56(3-4), p. 135–257.

———, 1985b, Nuevos Glyfeoideos de Argentina y una reseña de otros Crustáceos Decápodos Cretácicos. Noveno Congreso Brasileiro Paleontología (Fortaleza, 1985) (in press).

———, 1986, Aptian Ammonites from the Argentinian Austral Basin, The Subfamily Helicancylinae Hyatt, 1894. Annals South African Museum, v. 96(7), p. 271–314.

Aguirre Urreta, M. B., and Klinger, H. C., 1986, Upper Barremian Heteroceratinae (Cephalopoda, Ammonoidea) from Patagonia and Zululand, with comments on the Systematics of the Subfamily. Annals South African Museum, v. 96(8), p. 315–358.

Aguirre Urreta, M. B., and Ramos, V. A., 1981a, Estratigrafía y Paleontología de la Alta cuenca del Río Roble, Cordillera Patagónica, Provincia de Santa Cruz. Octavo Congreso Geológico Argentino, Actas 3, p. 101–138.

———, 1981b, Crustáceos Decápodos del Cretácico inferior de la Cuenca Austral, Provincia de Santa Cruz, Argentina. Cuencas Sedimentarias Jurásico y Cretácico América del Sur, v. 2, p. 599–623.

Aguirre Urreta, M. B., and Suárez, M., 1985, Belemnites de una secuencia turbidítica volcanoclástica de la Formación Yahgan. Titoniano-Cretácico inferior del extremo sur de Chile. IV Congreso Geológico Chileno, Actas 1(1), p. 1–16.

Ahlfeld, F., 1946, Geología de Bolivia. Museo La Plata, Geología, Revista (n.s.), v. 3, p. 5–370.

———, 1956, Bolivia. In: Jenks, W. F., ed., Handbook of South American Geology. Geological Society America Memoir, v. 65, p. 169–186.

———, 1959, Correlación del Horizonte Calcáreo de Miraflores con el de Ayavacas. Yacimientos Petrolíferos Fiscales Bolivianos (YPFB), Boletín Técnico 1(3), p. 7–12.

———, 1966, Geologische Untersuchungen in der Provinz Ayopaya (Bolivien). Neues Jahrbuch Mineralogie, Abhandlungen, B. Bd. 69:388–417.

———, 1970, Zur Tektonik des Andinen Bolivien: Geologische Rundschau, v. 59(3), p. 1124–1140.

Ahlfeld, F., and Branisa, L., 1960, Geología de Bolivia. Instituto Boliviano del Petróleo, La Paz, 245 p., 90 figs., 12 lám.

Alarcón, F. B., and Vergara, M. M., 1964, Nuevos antecedentes sobre la Geología de la Quebrada El Way. Instituto de Geología Universidad de Chile, Publicación 26, p. 1–128.

Albino, A. M., 1986, Nuevos Boidae Madtsoiinae en el Cretácico tardío de Patagonia (Formación Los Alamitos, Rio Negro, Argentina). IV Congreso Argentino de Paleontología y Bioestratigrafía, Actas 2, p. 15–21.

Aliste, N., Pérez, E., and Carter, W., 1960, Definición y Edad de la Formación Patagua, Provincia de Aconcagua, Chile. Revista Minerales, v. 15, p. 40–50.

Almeida, F.F.M., 1950, Uma fanula de crustáceos bivalvos do arenito de Botucatu no Est. de São Paulo. Ministerio da Agricultura, Departamento Nacional de Produção Mineral, Divisão Geologia e Mineralogia, Boletim 134, p. 1–24.

———, 1953, Botucatu, a Triassic desert of South America. XIX Congres Géologique International, Comptes Rendus, p. 9–24.

———, 1964, Grupo São Bento. In: Geologia do Estado de São Paulo. Instituto Geografico e Geologico, Boletim 41, p. 85–101.

———, 1972, Tectono-Magmatic Activation of the South American Platform and Associated Mineralization. XXIV International Geological Congress, 3, p. 339–346.

———, 1976, The system of continental rifts bordering the Santos Basin, Brazil. In: International Symposium on Continental Margins of Atlantic Type. Academia Brasileira de Ciências, Anais 48 (Suppl.), p. 15–26.

———, 1983, Relações Tectônicas das Rochas Alcalinas Mesozoicas da Região Meridional de Plataforma Sul-American. Revista Brasileira de Geociências, v. 13(3), p. 139–158.

Almeida, F.F.M., and Barbosa, O., 1953, Geologia das quadriculas de Piracicaba e Rio Claro, Estado de São Paulo. Ministerio de Agricultura, Departamento Nacional Produção Mineral, Divisão Geologia e Mineralogia, Boletim 143, p. 1–96.

Alonso, R. N., 1980, Icnitas de Dinosaurios (Ornithopoda, Hadrosauridae) en el Cretácico superior del Norte de Argentina. Acta Geológica Lilloana, v. 15(2), p. 55–63.

Alonso, R. N., and Marquillas, R. A., 1986, Nueva localidad con huellas de Dinosaurios y primer hallazgo de huellas de aves en la Formación Yacoraite (Maastrichtiano) del Norte Argentino. IV Congreso Argentino de Paleontología y Bioestratigrafía, Actas 2, p. 33–41.

Alonso, R., Viramonte, J., and Gutiérrez, R., 1984, Puna Austral-Bases para el subprovincialismo geológico de la Puna Argentina. Noveno Congreso Geológico Argentino, Actas 1, p. 43–63.

Amaral, G., Cordani, U. G., Kawashita, K., and Reynolds, J. H., 1966, Potassium-argon dates of basaltic rocks from southern Brazil. Geochimica et Cosmochimica Acta, v. 30, p. 155–189.

Amaral, G., Busher, J., Cordani, U. G., Kawashita, K., and Reynolds, J. H., 1967, Potassium-argon ages of alkaline rocks from southern Brazil. Geochimica et Cosmochimica Acta, v. 31(2), p. 117–142.

Ameghino, C., 1890, Exploraciones geológicas en la Patagonia. Instituto Geográfico Argentino Boletín 11(43), p. 1–46.

———, 1916, Sobre "Ceratodus Iheringi" de la Formación guaranítica de la Patagonia (comun.). Physis 2(10), p. 169.

Ameghino, F., 1897, Notes on the geology and paleontology of Argentina. Geological Magazine, v. 4(391), p. 4–118.

———, 1899, Nota preliminar sobre el "Loncosaurus argentinus", un representante de la familia de los Megalosauridae en la República Argentina. Sociedad Científica Argentina, Anales 47, p. 61–62.

———, 1900, Mamíferos del Cretácico inferior de la Patagonia (Formación de las areniscas abigarradas). Museo Nacional de Buenos Aires, Comunicaciones 6, p. 197–206.

——, 1903–1904, Nuevas especies de mamíferos cretáceos y terciarios de la República Argentina. Sociedad Científica Argentina, Anales 56, p. 193–208; 57, p. 162–175, 327–341; 58, p. 35–41, 56–71, 182–192, 225–291.

——, 1906, Les Formations sédimentaires du Crétacé supérieur et du Tertiaire de Patagonie. Museo Nacional de Buenos Aires, Anales (3) 15 (8), p. 1–568.

Amos, A. J., and Rocha Campos, A. C., 1970, A Review of South American Gondwana Geology 1967–1969. Second Gondwana Symposium, Proceedings and Papers, p. 1–13.

Andreis, R. R., Iñíguez, A. M., Lluch, J. J., and Sabio, D. A., 1974, Estudio Sedimentológico de las Formaciones del Cretácico superior del área del lago Pellegrini (Provincia de Río Negro, República Argentina). Asociación Geológica Argentina, Revista 29(1), p. 85–104.

Angelozzi, G. N., 1980, Dos nuevas especies de *Ilyocypris* (Ostracoda) de ambiente salobre del Cretácico superior en la Cuenca del Neuquén, República Argentina. Ameghiniana, v. 17(2), p. 163–167.

Aramayo, S., 1981, Hallazgo de *Lepidotes maximus* Wagner (Pisces) en el Titoniano de la Provincia de Neuquén, Argentina. II Congreso Latino-Americano de Paleontología, Anais I, p. 321–328.

Arbe, H. A., and Hechem, J. J., 1984, Estratigrafía y Facies de depósitos marinos profundos del Cretácico Superior, Lago Argentino, Provincia de Santa Cruz. Noveno Congreso Geológico Argentino, Actas 5, p. 7–41.

——, 1985, Estratigrafía y Facies de depósitos continentales, litorales y marinos del Cretácico superior, Lago Argentino, Provincia de Santa Cruz. Noveno Congreso Geológico Argentino, Actas 7, p. 124–158.

Archangelsky, S., 1963a, Notas sobre la flora fósil de la zona de Ticó, Provincia de Santa Cruz. Ameghiniana, v. 3(2), p. 57–63.

——, 1963b, Notas sobre la flora fósil de la zona de Ticó Provincia de Santa Cruz, 2. Tres nuevas especies de *Mesosingeria*. Ameghiniana, v. 3(4), p. 113–122.

——, 1963c, Notas sobre la flora fósil de la zona de Ticó, provincia de Santa Cruz, 3. *Ruflorinia pilifera* n. sp., 4. *Equisetites* sp. Ameghiniana, v. 3(8), p. 221–6.

——, 1963d, Notas sobre la flora fósil de la zona de Ticó, provincia de Santa Cruz, 5. *Sphenopteris* cf. *goepperti* Dunker, 6. *Cladophlebis* sp. Ameghiniana, v. 3(9), p. 280–284.

——, 1964, Dos nuevas especies de megasporas. Ameghiniana, v. 4(2), p. 52–55.

——, 1965, Fossil Ginkgoales from the Ticó Flora, Santa Cruz province, Argentina. British Museum (Natural History), Geology, Bulletin 10(5), p. 122–137.

——, 1967a, Estudio de la Formación Baqueró, Cretácico inferior de Santa Cruz, Argentina. Museo La Plata, Paleontología, Revista 5(32), p. 63–171.

——, 1967b, Notas sobre la flora fósil de la zona de Ticó, Provincia de Santa Cruz, 8. Seis especies del género *Sphenopteris*. Ameghiniana, v. 5(4), p. 149–157.

——, 1970, Fundamentos de Paleobotánica. Facultad de Ciencias Naturales y Museo, Universidad Nacional de La Plata, Serie Técnica y Didáctica 10, p. 1–347.

——, 1977a, Vegetales fósiles de la Formación Springhill, Cretácico en el Subsuelo de la Cuenca Magallánica, Chile. Ameghiniana, v. 13(2), p. 141–158.

——, 1977b, *Balmeiopsis,* Nuevo nombre genérico para el palinomorfo *Inaperturopollenites limbatus* Balme, 1957. Ameghiniana, v. 14(1-4), p. 122–126.

——, 1978, Megafloras Fósiles. In: Relatorio Geología y Recursos Naturales del Neuquén. Séptimo Congreso Geológico Argentino, p. 187–192. Buenos Aires.

Archangelsky, S., and Baldoni, A., 1972a, Revisión de las Bennettitales de la Formación Baqueró (Cretácico inferior). Museo La Plata, Paleontología, Revista 7(44), p. 195–265.

——, 1972b, Notas sobre la flora fósil de la zona de Ticó, provincia de Santa Cruz. X, Dos nuevas especies de *Pseudoctenis* (Cycadales). Ameghiniana,

v. 9(3), p. 241–264.

Archangelsky, S., and Gamerro, J. C., 1964, Estudio palinológico de la Formación Baqueró (Cretácico), Provincia de Santa Cruz, I. Ameghiniana, v. 4(5), p. 159–167.

——, 1966a, Estudio palinológico de la Formación Baqueró (Cretácico), Provincia de Santa Cruz, II. Ameghiniana, v. 4(6), p. 201–209.

——, 1966b, Estudio palinológico de la Formación Baqueró (Cretácico), Provincia de Santa Cruz, III. Ameghiniana, v. 4(7), p. 229–236.

——, 1966c, Estudio palinológico de la Formación Baqueró (Cretácico), Provincia de Santa Cruz, IV. Ameghiniana, v. 4(10), p. 363–369.

——, 1967, Spores and pollen types of the Lower Cretaceous in Patagonia (Argentina). Review of Palaeobotany and Palynology, v. 1, p. 211–217.

Archangelsky, S., and Romero, E. J., 1974, Polen de Gimnospermas (Coníferas) del Cretácico superior y Paleoceno de Patagonia. Ameghiniana, v. 11(3), p. 217–236.

Archangelsky, S., and Seiler, J., 1980, Algunos resultados de la Perforación UN-Oil OS 1, SO de la provincia de Chubut. Segundo Congreso Argentino de Paleontología y Bioestratigrafía, Actas 5, p. 215–225.

Archangelsky, S., Baldoni, A., Gamerro, J. C., Palamarczuk, S., and Seiler, J., 1981, Palinología Estratigráfica del Cretácico de Argentina Austral. Diagramas de Grupos Polínicos del Sudoeste de Chubut y Noroeste de Santa Cruz. Octavo Congreso Geológico Argentino, Actas 4, p. 719–742.

Archangelsky, S., Baldoni, A., Gamerro, J. C., and Seiler, J., 1983, Palinología Estratigráfica del Cretácico de Argentina Austral. II. Descripciones Sistemáticas. Ameghiniana, v. 20(3-4), p. 199–226.

——, 1984, Palinología Estratigráfica del Cretácico de Argentina Austral. III Distribución de las Especies y conclusiones. Ameghiniana, v. 21(1), p. 15–33.

Arid, F. M., 1977, Paleogeographical evidences from Bauru Basin (Upper Cretaceous) of southern Brazil. Naturalia, v. 3, p. 7–13.

Arid, F. M., and Vizotto, L. D., 1966, Un quelônio fossil de Sao José do Río Preto. Ciência e Cultura, v. 18(4), p. 422.

——, 1971, *Antarctosaurus brasiliensis* um novo saurópode do Cretáceo superior do sul do Brasil. XXV Congresso Brasileiro Geología, Anais, p. 298–305.

——, 1975, Crocodilideos fosseis nas proximidades de Santa Adélia (SP). Ciência e Cultura, v. 17(2), p. 138–139.

Asmus, H. E., 1981, Geologia das Bacias Marginais Atlânticas Mesozóicas-Cenozóicas do Brasil. In: Cuencas Sedimentarias del Jurásico y Cretácico de América del Sur, v. 1, p. 127–155.

——, 1982, Geotectonic Significance of Mesozoic-Cenozoic Magmatic rocks in the Brazilian Continental Margin and Adjoining Emerged Area. Quinto Congreso Latinoamericano de Geología (Argentina 1982), Actas 3, p. 761–779.

——, 1984, Geologia da Margem Continental Brasileira. In: Schobbenhaus, C., Almeida, D. de, Derze, C. R., and Asmus, H. E., eds., Geologia do Brasil, Ministerio das Minas e Energia, Brasilia, p. 443–472.

Asmus, H. E., and Ponte, F. C., 1973, The Brazilian marginal basins. In: Nairn, A.E.M., and Stehli, F. G., eds., The Ocean Basins and Margins, v. 1, The South Atlantic, p. 87–133. Plenum Publishing Corporation, New York.

Asmus, H. E., and Porto, R., 1972, Classificacão das Bacias Sedimentares Brasileiras segundo a Tectônica de Placas. XXVI Congresso Brasileiro Geologia, Anais 2, p. 67–90.

Aubouin, J., and Borrello, A. V., 1966, Chaînes Andines et Chaînes Alpines: Regard sur la Géologie de la Cordillère des Andes au Parallèle de l'Argentine Moyenne. Société géologique de France, Bull. (7) 8, p. 1050–1070.

Aubouin, J., Borrello, A. V., Cecioni, G., Charrier, R., Chotin, P., Frutos, J., Thiele, R., and Vicente, J. C., 1973, Esquisse Paléogéographique et structurale des Andes Méridionales. Revue de Géographie Physique et Géologie Dynamique (2) 15 (1-2), p. 11–72.

Audebaud, E., Capdevila, R., Dalmayrac, B., Debelmas, J., Laubacher, G., Lefevre, C., Marocco, R., Martínez, C., Mattauer, M., Megard, F., Paredes, J., and Tomasi, P., 1973, Les Traits géologiques essentiels des Andes Centrales (Pérou Bolivie). Revue Géographie Physique et Géologie

A. C. Riccardi

Dynamique (2) 15 (1-2), p. 73–114.

Avila-Salinas, W. A., 1986, El Magmatismo Cretácico en Bolivia. Cretácico de America Latina, IGCP 242, Primer Simposio (La Paz, 1986), p. 52–70.

Báez, A. M., 1982, Redescription and relationships of *Saltenia ibanezi*, a Late Cretaceous pipid frog from northwestern Argentina. Ameghiniana, v. 18(3-4), p. 127–154.

——, 1985, Anuro Leptodáctilido en el Cretácico superior (Grupo Baurú) de Brasil. Ameghiniana, v. 22(1-2), p. 75–79.

Baker, C. L., 1923, The lava field of the Parana Basin, South America. Journal of Geology, (3) 31, p. 66–79.

Baldoni, A. M., 1975, Revisión de las Bennettitales de la Formación Baqueró (Cretácico inferior), Pcia. de Santa Cruz. II, Bracteas. Ameghiniana, v. 11(4), p. 328–356.

——, 1977a, *Ptilophyllum ghiense* n. sp., una nueva Bennettital de Paso Roballos, Provincia de Santa Cruz. Ameghiniana, v. 14(1-4), p. 53–58.

——, 1977b, Nota sobre *Sueria rectinervis* Menéndez del Cretácico inferior de la Formación Baqueró, Provincia de Santa Cruz. Ameghiniana, v. 14(1-4), p. 301–304.

——, 1979, Nuevos elementos paleoflorísticos de la Tafoflora de la Formación Spring Hill, Límite Jurásico-Cretácico, Subsuelo de Argentina y Chile Austral. Ameghiniana, v. 16(1-2), p. 103–119.

——, 1980, Análisis de algunas Tafofloras Jurásicas y Eocretácicas de Argentina y Chile. Segundo Congreso Argentino de Paleontología y Bioestratigrafía, Actas 5, p. 41–65.

——, 1981, Tafofloras Jurásicas y Eocretácicas de América del Sur. In: Cuencas Sedimentarias Jurásicas y Cretácicas de América del Sur, v. 2, p. 359–391.

Baldoni, A. M., and Archangelsky, S., 1983, Palinología de la Formación Springhill (Cretácico inferior), Subsuelo de Argentina y Chile Austral. Revista Española de Micropaleontología, v. 14(1), p. 47–101.

Baldoni, A. M., and De Vera, H., 1981, Plantas fósiles de la Formación Apeleg (Cretácico) en la zona del lago Fontana, Provincia de Chubut. Ameghiniana, v. 17(4), p. 289–296.

Baldoni, A. M., and Ramos, V. A., 1981, Nuevas localidades con plantas fósiles cretácicas en la Cordillera Patagónica. Octavo Congreso Geológico Argentino, Actas 4, p. 743–759.

Baldoni, A. M., and Taylor, T. N., 1985, Megasporas Cretácicas de la Formación Springhill en el Subsuelo de Argentina y Chile Austral. Ameghiniana, v. 21(2-4), p. 151–167.

Ballent, S. C., 1980, Ostrácodos de ambiente salobre de la Formación Allen (Cretácico superior) en la Provincia de Río Negro (República Argentina), v. 17(1), p. 67–82.

Banks, L. M., and Díaz de Vivar, V., 1975, Exploration in Paraguay reactivated. Oil and Gas Journal, v. 73(40), p. 160, 164, 166–8.

Barbosa, A. F., Wohlers, A., Almeida, F.F.M., Guimarães, J.E.P., Paolielo, P. C., Freitas, R. O. de, Mezzalira, S., Petri, S., and Knecht, T., 1964, Geología do Estado de São Paulo. Instituto Geográfico e Geológico, Boletim 41, p. 5–263. São Paulo.

Barcat, C., Cortiñas, J. S., Nevistic, V. A., and Vásquez, J. R., 1984a, Turbiditas Lacustres en el Cretácico superior del Noreste de Santa Cruz. Noveno Congreso Geológico Argentino, Actas 5, p. 274–284.

Barcat, C., Cortiñas, J. S., Nevistic, V. A., Stach, N. H., and Zucchi, H. E., 1984b, Geología de la región comprendida entre los Lagos Musters, Colhue Huapi y la Sierra Cuadrada, Departamentos Sarmiento y Paso de Indios, Provincia del Chubut. Noveno Congreso Geológico Argentino, Actas 2, p. 263–282.

Barcha, S. F., and Arid, F. M., 1977, Parametros granulometricos de estruturas sedimentares e interpretacão do ambiente dinamico da Formacão Bauru. Naturalia, v. 3, p. 15–34.

Barker, P., et al., 1976, Evolution of the southwestern Atlantic Ocean Basin: Results of Leg 36, Deep Sea Drilling Project. Initial Report DSDP, v. 36, p. 993–1014.

Barth, W., 1973, Zur Stratigraphie des Altpaläozoikums und der Kreide bei Tarabuco und Icla im Depto. Chuquisaca, Bolivien. Muenstersche Forschungen zur Geologie und Palaeontologie, v. 31/32, p. 207–231.

Behrendsen, O., 1891–1892, Zur Geologie des Ostabhanges der argentinischen Cordillere. Deutsche Geologische Gesellschaft, 43 (1891), p. 369–420; 44 (1892), p. 1–42.

Bell, C. M., and Suarez, M., 1985, Formación Quebrada Monardes: Depositación fluvial en un ambiente árido, Jurásico-Cretácico, Atacama. IV Congreso Geológico Chileno, Actas 1, p. 29–37.

Bellieni, G., Brotzu, P., Comin-Chiaramonti, P., Ernesto, M., Melfi, A. J., Pacca, I. G., Piccirillo, E. M., and Stolfa, D., 1983, Petrological and Paleomagnetic Data on the Plateau Basalt to Rhyolite Sequences of the Southern Parana Basin (Brazil). Academia Brasileira de Ciências, Anais 55(4), p. 355–383.

Benavides Cáceres, V. E., 1956, Cretaceous System in Northern Peru. American Museum Natural History, Bulletin 108(4), p. 353–494.

Benedetto, J. L., and Sánchez, T. M., 1971, El hallazgo de peces Pycnodontiformes (Holostei) en la Formación Yacoraite (Cretácico superior) de la Provincia de Salta (Argentina) y su importancia paleoecológica. Acta Geológica Lilloana 11(8), p. 153–175.

——, 1972, *Coelodus toncoensis* nov. sp. (Pisces, Holostei, Pycnodontiformes) de la Formación Yacoraite (Cretácico superior) de la Provincia de Salta. Ameghiniana, v. 9(1), p. 59–71.

Berg, K., and Baumann, A., 1985, Plutonic and metasedimentary rocks from the Coastal Range of northern Chile: Rb-Sr and U-Pb isotopic systematics. Earth and Planetary Science Letters 75(2-3), p. 101–115.

Berg, K., and Breitkreuz, C., 1983, Mesozoische Plutone in der Küstenkordillere: Petrogenese, Geochronologie, Geochemie und Geodynamik mantelbetonter Magmatite. Geotektonische Forschungen, v. 66, p. 1–107.

Bernasconi, I., 1954, Nota sobre una nueva especie de Equinoideo fósil de Tierra del Fuego. Physis, v. 20(59), p. 397–400.

Berry, E. W., 1906, A note on Mid-Cretaceous geography. Science, v. 23, p. 509–510.

——, 1916, The Upper Cretaceous Floras of the World. Maryland Geological Survey, Special Publication, p. 181–313.

——, 1918, Age of certain plant-bearing beds and associated marine formations in South America. Geological Society of America Bulletin, v. 29, p. 637–648.

——, 1922, Marine Upper Cretaceous and a new Echinocorys from the Altaplanicie of Bolivia. Journal of Geology, v. 30(3), p. 227–231.

——, 1924, Mesozoic plants from Patagonia. American Journal of Science, (5)7, p. 473–482.

——, 1932, Sketch of the geology of Bolivia. Panamerican Geologist, v. 57(4), p. 241–262.

——, 1937, An Upper Cretaceous flora from Patagonia. Johns Hopkins University Studies in Geology, v. 12, p. 11–31.

——, 1939, The fossil flora of Potosí, Bolivia. Johns Hopkins University Studies in Geology, v. 13, p. 9–67.

Bertels, A., 1964, Micropaleontología del Paleoceno de General Roca (Prov. de Rio Negro). Museo La Plata, Paleontología, Revista (N.S.), v. 4(23), p. 125–184.

——, 1968, Micropaleontología y Estratigrafía del límite Cretácico-Terciario en Huantrai-co (Provincia de Neuquén). Ostracoda. Pt. I. Ameghiniana, v. 5(8), p. 279–298.

——, 1969, Estratigrafía del límite Cretácico-Terciario de la Patagonia Septentrional. Asociación Geológica Argentina, Revista 24(1), p. 41–54.

——, 1970a, Micropaleontología y Estratigrafía del límite Cretácico-Terciario en Huantrai-có (Provincia del Neuquén). Pt. II. Ameghiniana, v. 6(4), p. 253–290.

——, 1970b, Los Foraminíferos Planctónicos de la Cuenca Cretácico-Terciaria en Patagonia Septentrional (Argentina), con consideraciones sobre la Estratigrafía de Fortín General Roca (Provincia de Río Negro). Ameghiniana, v. 7(1), p. 1–56.

——, 1970c, *Hiltermannia* n. gen. (Foraminiferida) del Cretácico superior (Maastrichtiano) de Argentina. Ameghiniana, v. 7(2), p. 167–172.

——, 1972a, Buliminacea y Cassidulinacea (Foraminiferida) guias del Cretácico superior (Maastrichtiano Medio) y Terciario Inferior (Daniano Inferior) de la República Argentina. Revista Española de Micropaleontología, v. 4(3), p. 327–353.

——, 1972b, Ostrácodos de agua dulce del miembro inferior de la Formación Huantrai-có (Maastrichtiano inferior), Prov. del Neuquén, Rep. Argentina. Ameghiniana, v. 9(2), p. 173–182.

——, 1974, Upper Cretaceous (lower Maastrichtian?) ostracodes from Argentina. Micropaleontology, v. 20(4), p. 385–397.

——, 1975, Upper Cretaceous (middle Maastrichtian) ostracodes of Argentina. Micropaleontology, v. 21(1), p. 97–130.

——, 1979, Paleobiogeografía de los Foraminíferos del Cretácico superior y Cenozoico de América del Sur. Ameghiniana, v. 16(3-4), p. 273–356.

——, 1980, Estratigrafía y Foraminíferos (Protozoa) Bentónicos del límite Cretácico-Terciarico en el área tipo de la Formación Jagüel, provincia del Neuquén, República Argentina. Segundo Congreso Argentino de Paleontología y Bioestratigrafía, Actas 2, p. 47–80.

——, 1987, Los Foraminíferos del Cretácico de la República Argentina: sus tendencias paleobiogeográficas. Academia Nacional de Ciencias Exactas, Físicas y Naturales, Anales 37, p. 265–305.

Bettini, F. H., Pombo, R. A., Mombrú, C. A., and Uliana, M. A., 1979, Consideraciones sobre el Diastrofismo Andino en la Vertiente Oriental de la Cordillera Principal, entre los 34° 30′ y los 37° de Latitud Sur. Séptimo Congreso Geológico Argentino, Actas 1, p. 671–683.

Beurlen, K., 1956, Paläogeographie und Morphogenese des Parana Beckens. Deutsche Geologische Gesellschaft, Zeitschrift 106, p. 519–537.

——, 1970, Geologie von Brasilien. Gebrüder Borntraeger. Berlin, Stuttgart.

Beurlen, K., Sena Sobrinho, M., and Martins, E. A., 1955, Formacões gondwânicas do Rio Grande do Sul. Museum Nacional, Rio de Janeiro, Geologia, Boletim, N.S., v. 22, p. 1–55.

Bianchi, J. L., 1984, Interpretación Tectogenética y Paleoambiental de la Cuenca de Rawson, Plataforma Continental Argentina. Noveno Congreso Geológico Argentino, Actas 3, p. 47–60.

Bianucci, H. A., Acevedo, O. M., and Cerdan, J. J., 1981, Evolución Tectosedimentaria del Grupo Salta en la Subcuenca Lomas de Olmedo (Provincias de Salta y Formosa). Octavo Congreso Geológico Argentino, Actas 3, p. 159–172.

Biddle, K. T., Uliana, M. A., Mitchum, R. M., Fitzgerald, M. G., and Wright, R. C., 1986, The stratigraphic and structural evolution of the Magallanes Basin, southern South America. Special Publications International Association Sedimentology, v. 8, p. 41–61.

Biese, W., 1942, La distribución del cretáceo inferior al sur de Copiapó. Primer Congreso Panamericano de Minas y Geología, Anales 2, p. 429–466.

——, 1958, *Microdon* del Aptiano de Copiapó. XX Congreso Geológico Internacional 7, p. 235–238.

Bigarella, J. J., 1972, Continental drift and paleocurrent analysis (a comparison between Africa and South America). Boletim Paranaense de Geociências, v. 30, p. 73–97.

——, 1973a, Paleocorrentes e Deriva Continental (Comparação entre América do Sul e Africa). Boletim Paranaense de Geociências, v. 31, p. 140–224.

——, 1973b, Textural characteristics of the Botucatu sandstone. Boletin Paranaense de Geociências, v. 31, p. 85–94.

Bigarella, J. J., and Salamuni, R., 1961, Early Mesozoic wind patterns as suggested by dune bedding in the Botucatu sandstone of Brazil and Uruguay. Geological Society of America Bulletin, v. 72, p. 1089–1106.

——, 1964, Paleowind patterns in the Botucatú sandstone (Triassic-Jurassic) of Brazil and Uruguay. In: Nairn, A.E.M., ed., Problems in Paleoclimatology, J. Wiley & Sons Ltd., London.

——, 1967a, Some palaeogeographic and palaeotectonic features of the Paraná Basin. In: Bigarella, J. J., Becker, R. D., and Pinto, I. D., eds., Problems in Brazilian Gondwana Geology, p. 236–301. Curitiba, Paraná.

——, 1967b, A review of South American Gondwana geology. In: Reviews Prepared for the First Symposium on Gondwana Stratigraphy (Mar del Plata, Argentina, 1967), Committee for the Study of Geological Documentation. International Union of Geological Sciences, p. 7–137.

Biró, L., 1976, *Titanites chilensis* n. sp. en la Formación Lo Valdes, Titoniano-Neocomiano, Provincia de Santiago, Chile. Primer Congreso Geológico Chileno, Actas 3, p. L11–L19.

——, 1979, Geología de la Franja Costera entre Cocholgue y Coronel, Provincia de Concepción (36°35′–37° Lat. Sur), Chile. Seminario/Taller sobre Desarrollo e Investigación de los Recursos Marinos de la Octava Región, Chile (1978), Actas: 20–30.

——, 1980, Algunos ammonites nuevos en la Formación Lo Valdes, Titoniano-Neocomiano, Provincia de Santiago (33°50′ Lat. Sur), Chile. Segundo Congreso Argentino de Paleontología y Bioestratigrafía. Actas 1, p. 223–235.

——, 1982a, *Hoploscaphites constrictus* (J. Sowerby) en la Formación Quiriquina, Campaniano-Maastrichtiano, Región del Bio-Bio, Chile, Sudamerica (36°30′–36°45′ Lat. Sur). III Congreso Geológico Chileno, Actas 1, p. A1–A16.

——, 1982b, Contribución al conocimiento de *Naefia neogaeia* Wetzel, Coleoidea, en la Formación Quiriquina, Campaniano-Maastrichtiano, Región del Bio-Bio, Chile, Sudamérica (36°30′–36°45′ Lat. Sur). III Congreso Geológico Chileno, Actas 1, p. A17–A28.

——, 1982c, Revisión y Redefinición de los "Estratos de Quiriquina," Campaniano-Maastrichtiano, en su localidad tipo, en la Isla Quiriquina, 36°37′ Lat. Sur, Chile, Sudamerica, con un perfil complementario en Cocholgüe. III Congreso Geológico Chileno, Actas, 1, p. A29–A63.

Bischoff, G., 1957, Stratigraphie, Tektonik und Magmatismus des Permus und Mesozoikums im Gebiet von Jacarèzinho (Nordparaná). In: Beiträge zur Geologie von Brazil. Beihefte zum Geologischen Jahrbuch 25, p. 81–104.

Björnberg, A.J.S., and Landim, P.M.B., 1966, Sôbre os arenitos da serra de Mantiqueira e os arenitos da Formação Botucatu (Eocretáceo). Boletim Paranaense de Geografía, v. 18/20, p. 19–24.

Björnberg, A.J.S., Landim, P.M.B., and Bosio, N. J., 1970, Observações sôbre a deposição do sedimento Bauru na regiao centroocidental do Estado de São Paulo. Sociedade Brasileira Geologia Boletim 19(1), p. 79–90.

Blasco, G., Nullo, F., and Proserpio, C., 1979, *Aspidoceras* en Cuenca Austral, Lago Argentino, Prov. de Santa Cruz. Asociación Geológica Argentina. Revista 34(4), p. 282–293.

Blasco, G., Nullo, F. E., and Ploszkiewicz, J., 1980, El género *Colchidites* Djanelidze, 1926 y la posición estratigráfica del género *Hatchericeras* Stanton, 1901, en la Estancia TucuTuca, Provincia de Santa Cruz. Asociación Geológica Argentina, Revista 35(1), p. 41–58.

Bocchino, A., 1973, Semionotidae (Pisces, Holostei, Semionotiformes) de la Formación Lagarcito (Jurásico superior?), San Luis, Argentina. Ameghiniana, v. 10(3), p. 254–268.

——, 1974, *Austrolepidotes cuyanus* gen. et sp. nov. y otros restos de peces fósiles de la Formación Lagarcito (?Jurásico superior), San Luis, Argentina. Ameghiniana, v. 11(3), p. 237–248.

——, 1977, Un nuevo Gyrodontidae (Pisces, Holostei, Pycnodontiformis) de la Formación Agrio (Cretácico inferior) de la Provincia de Neuquén, Argentina. Ameghiniana, v. 14(1-4), p. 175–185.

——, 1979, Revisión de los Osteichthyes fósiles de la República Argentina. I, Identidad de *Tharrias feruglioi* Bordas 1943 y *Oligopleurus groeberi* Bordas 1943. Ameghiniana, v. 15(3-4), p. 301–320.

Bodenbender, G., 1892, Sobre el Terreno Jurásico y Cretácico en los Andes Argentinos entre el río Diamante y el río Limay. Academia Nacional de Ciencias. Córdoba, Boletin 13, p. 5–44.

——, 1922, El Nevado de Famatina. Dirección General de Minas Geología e Hidrología, Anales 16(1), p. 7–68.

——, 1924, El Calchaqueño y los Estratos de la Puna de Penck. Academia Nacional de Ciencias, Córdoba, Boletin 27, p. 405–468.

Boll, A., and Hernández, R., 1985, Area Tres Cruces, Provincia de Jujuy. Análisis Estratigráfico-Estructural. Evaluación como Objetivo Exploratorio. Informe Yacimientos Petrolíferos Fiscales, Buenos Aires (unpublished), p. 1–67.

——, 1986, Sedimentitas cretácicas de la Subcuenca de Tres Cruces, Provincia de Jujuy. Primera Reunión Argentina de Sedimentología, Resúmenes Expandidos, p. 130–133.

Bonaparte, J. F., 1970, *Pterodaustro guinazui,* pterosaurio de la Formación Lagarcito, Provincia de San Luis, Argentina y su significado en la geología regional. Acta Geológica Lilloana, v. 10(10), p. 209–225.

——, 1971, Descripción del Cráneo y mandíbulas de *Pterodaustro guinazui*

(Pterodactyloidea-Pterodaustriidae nov.) de la Formación Lagarcito, San Luis, Argentina. Museo Ciencias Naturales Mar del Plata, Publicación 1(9), p. 263–272.

———, 1978, El Mesozoico de América del Sur y sus Tetrápodos. Opera Lilloana, v. 26, p. 1–596.

———, 1979, Faunas y Paleobiogeografía de los Tetrápodos Mesozoicos de América del Sur. Ameghiniana, v. 16(3-4), p. 217–238.

———, 1981, Los Fósiles Mesozoicos, In: Geología y Recursos Minerales de la Provincia de San Luis. Relatorio Octavo Congreso Geológico Argentino, p. 97–99.

———, 1984, Nuevas pruebas de la conexión física entre Sudamerica y Norteamérica en el Cretácico Tardío (Campaniano). III Congreso Argentino de Paleontología y Bioestratigrafía (Corrientes, 1982), Actas, p. 141–149.

———, 1986a, A new and unusual Late Cretaceous mammal from Patagonia. Journal of Vertebrate Paleontology, v. 6(3), p. 264–270.

———, 1986b, Sobre *Mesungulatum houssayi* y nuevos mamíferos Cretácicos de Patagonia, Argentina. IV Congreso Argentino de Paleontología y Bioestratigrafía, Actas 2, p. 48–61.

———, 1986c, History of the terrestrial Cretaceous vertebrates of Gondwana. IV Congreso Argentino de Paleontología y Bioestratigrafía, Actas 2, p. 63–95.

Bonaparte, J. F., and Bossi, G. E., 1967, Sobre la presencia de dinosaurios en la Formación Pirgua del Grupo Salta, y su significado cronológico. Acta Geológica Lilloana, v. 9, p. 25–44.

Bonaparte, J. F., and Gasparini, Z. B. de, 1979, Los Saurópodos de los Grupos Neuquén y Chubut y sus relaciones cronológicas. Séptimo Congreso Geológico Argentino, Actas 2, p. 393–406.

Bonaparte, J. F., and Novas, F. E., 1985, *Abelisaurus comahuensis,* n. g., n. sp., Carnosauria del Cretácico Tardío de la Patagonia. Ameghiniana, v. 21 (2-4), p. 259–265.

Bonaparte, J. F., and Pascual, R., 1987, Los Mamíferos (Eotheria, Allotheria y Theria) de la Formación Los Alamitos, Campaniano de Patagonia, Argentina. IV Congreso Latinoamericano de Paleontología (Bolivia, 1987), Memorias, v. 1, p. 361–378.

Bonaparte, J. F., and Powell, J. E., 1980, A continental assemblage of tetrapods from the Upper Cretaceous beds of El Brete, northwestern Argentina (Sauropoda-Coelurosauria-Carnosauria-Aves). Société Géologique de France, Mémoire, N.S., v. 139, p. 19–28.

Bonaparte, J. F., and Rougier, G., 1987, Mamíferos del Cretácico inferior de Patagonia. IV Congreso Latinoamericano de Paleontología (Bolivia, 1987), Memorias, v. 1, p. 343–359.

Bonaparte, J. F., and Sánchez, T., 1975, Restos de un Pterosaurio, *Puntaniptero globosus,* de la Formación La Cruz, Prov. de San Luis, Argentina. Primer Congreso Argentino de Paleontología y Bioestratigrafía, Actas 2, p. 105–114.

Bonaparte, J. F., and Soria, M. F., 1985, Nota sobre el Primer Mamífero del Cretácico Argentino, Campaniano-Maastrichtiano (Condylarthra). Ameghiniana, v. 21 (2-4), p. 177–183.

Bonaparte, J. F., Salfity, J., Bossi, G., and Powell, J., 1977, Hallazgo de dinosaurios y aves cretácicas en la Formación Lecho de El Brete (Salta), próximo al límite con Tucumán. Acta Geológica Lilloana, v. 14, p. 5–17.

Bonaparte, J. F., Franchi, M. R., Powell, J. E., and Sepúlveda, E. G., 1985, La Formación Los Alamitos (Campaniano-Maastrichtiano) del sudeste del Río Negro, con descripción de *Kritosaurus australis* n. sp. (Hadrosauridae). Significado paleogeográfico de los vertebrados. Asociación Geológica Argentina, Revista 39 (3-4), p. 284–299.

Bonarelli, G., 1914, La estructura geológica y los yacimientos petrolíferos del distrito minero de Orán (provincia de Salta). Dirección General Minas, Geología e Hidrología, Boletin B9, p. 1–43.

———, 1921, Tercera contribución al conocimiento geológico de las regiones petrolíferas subandinas del norte (provincia de Salta y Jujuy). Ministerio Agricultura de la Nación, Sección Geología, Mineralogía y Minería, Anales 15(1), p. 1–97.

———, 1927, Fósiles de la Formación Petrolífera o Sistema de Salta. Academia Nacional de Ciencias, Córdoba, Boletín 30, p. 51–116.

Bonarelli, G., and Nágera, J. J., 1921, Observaciones geológicas en las inmediaciones del lago San Martín (Territorio de Santa Cruz). Dirección General de Minas, Boletín 27B, p. 1–39.

Bordas, A., 1943, Peces del Cretáceo del río Chubut (Patagonia), Physis, v. 19(53), p. 313–318.

Boric, R., 1981, Cuadrángulo Estación Colupito y Toco, Región de Antofagasta. Instituto Investigaciones Geológicas Chile, Carta Geológica 49-50, p. 1–52.

Borrello, A. V., 1969, Los Geosinclinales de la Argentina. Dirección Nacional de Geología y Minas, Anales 14, p. 1–188.

———, 1972, Cordillera Fueguina. In: Leanza, A. F., ed., Geología Regional Argentina, p. 741–753. Academia Nacional Ciencias, Córdoba.

Bosio, N. J., and Landim, P.M.B., 1971, Un Estudo sedimentológico sobre a Formação Caiuá. Boletim Paranaense de Geociências, v. 28/29, p. 145–157.

Boso, M. A., Brandán, E. M., and Salfity, J. A., 1984, Estratigrafía y Paleoambientes del Subgrupo Pirgua (Cretácico) en la Comarca de Brealito, Provincia de Salta. Noveno Congreso Geológico Argentino, Actas 1, p. 108–123.

Bossi, G. E., 1969, Geología y Estratigrafía del sector sur del valle de Choranoro. Acta Geológica Lilloana, v. 10(2), p. 17–64.

———, 1977, La Formación Cerro Rajado, Provincia de La Rioja. Acta Geológica Lilloana, v. 14, p. 19–40.

Bossi, G. E., Mendes Piccoli, A. E., Pilatti, F., Thomaz, S. L., Chaiben Jabur, I., Rodrigues, M. A., and Ramos Madeiros, E., 1978, Paleocorrentes da Formação Botucatu nas Folhas de Montenegro, Novo Hamburgo, Taquara, Gravataí e São Leopoldo, RS. Acta Geologica Leopoldensia, v. 2(3), p. 83–109.

Bossi, G. E., and Wampler, M., 1969, Edad del Complejo Alto de las Salinas y Formación El Cadillal según el método K-Ar. Acta Geológica Lilloana, v. 10(7), p. 141–160.

Bossi, J., 1966, Geología del Uruguay. Universidad de la República, Montevideo, Uruguay.

Bossi, J., and Fernández, A., 1963, Evidencias de diferenciación magmática hacia el final del Gondwana uruguayo. Universidade Federale do Paraná, Geología, Boletim 9, p. 1–20.

Bossi, J., and Umpierre, M., 1975, Magmatismo Mesozoico de Uruguay y Río Grande del Sur; Sus recursos minerales asociados y potenciales. II Congreso Iberó-Americano de Geología Económica, v. 2, p. 119–140.

Bossi, J., Chebataroff, N., and Lopardo, J., 1963, Contribución al estudio del horizonte de Guichón. Facultad de Agronomía de Montevideo, Boletín 69, p. 1–17.

Bowen, R., 1961, Paleotemperature analyses of Belemnoidea and Jurassic paleoclimatology. Journal of Geology, v. 69, p. 309–320.

———, 1963, O^{18}/O^{16} Paleotemperature measurements on mesozoic Belemnoidea from Neuquén and Santa Cruz Provinces, Argentina. Journal of Paleontology, v. 37(3), p. 714–718.

———, 1966, Paleotemperature Analyses. Elsevier Publishing Co., Amsterdam.

———, 1972, Mid-Paleozoic to Recent structural history of the eastern border of the Parana Basin, South America. In: Continental Drift Emphasizing the History of the South Atlantic area. EOS (American Geophysical Union Transactions), v. 53(2), p. 172–3.

Bracaccini, O. I., 1970, Cuenca del Salado. In: Leanza, A. F., ed., Geología Regional Argentina, p. 407–417. Academia Nacional Ciencias, Córdoba.

———, 1980, Cuenca del Salado. Segundo Simposio de Geología Regional Argentina, v. 2, p. 879–918. Academia Nacional Ciencias, Córdoba.

Brackebusch, L., 1883, Estudios sobre la Formación Petrolífera de Jujuy. Academia Nacional Ciencias, Córdoba, Boletín 5 (2), p. 137–252.

———, 1891, Mapa Geológico del Interior de la República Argentina, escala 1:1,000,000. Gotha.

Brandmayr, J., 1945, Contribución al conocimiento geológico del extremo sud-sudoeste del Territorio de Santa Cruz (Región Cerro Cazador-alto Río Turbio), Boletín Informaciones Petroleras 256, p. 415–437.

Braniša, L., 1968, Hallazgo del amonite *Neolobites* en la caliza Miraflores y de huellas de dinosaurios en la formación El Molino y su significado para determinación de la edad del "Grupo Puca." Instituto Boliviano del Petróleo,

Boletín, v. 8(1), p. 16–29.

Braniša, L., Hoffstetter, R., and Signeaux, J., 1964, Additions à la faune ich-thyologique du Crétacé supérieur de Bolivie. Museum National d'Histoire Naturelle, Paris, Bulletin, Sér. 2, 36, p. 279–297.

Braniša, L., Hoffstetter, R., Freneix, S., Roman, J., and Sornay, J., 1966, Nouvelle contribution a l'etude de la paléontologie et de l'âge du groupe Puca (Crétacé de Bolivie). Museum National d'Histoire Naturelle, Paris, Bulletin, Sér. 2, 38, p. 301–310.

Braniša, L., Grambast, L., and Hoffstetter, R., 1969, Quelques précisions nouvelles, d'après les charophytes, sur l'âge du groupe Puca (crétacé-paléogène, Bolivie). Société Géologique de France, Comptes Rendus 8, p. 321–2.

Broili, F., 1930, Plesiosaurierreste von der Insel Quiriquina. Neues Jahrbuch für Geologie und Paläontologie, Abhandlungen 63, p. 497–514.

Broin, F. de, 1973, Une espece nouvelle de Tortue pleurodire (?*Roxochelys vilavilensis* n. sp) dans le Cretacé supérieur de Bolivie. Société Géologique de France Bulletin 13(3-4), p. 445–452.

Brown, L. F., Barcat, C., Fisher, W. L., and Nevistic, A., 1982, Seismic Strati-graphic and Depositional systems Analyses: New Exploration Approaches applied to the Golfo San Jorge Basin, Argentina. Primer Congreso Nacional de Hidrocarburos, Petróleo y gas, Conferencias, p. 127–156. Instituto Argen-tino del Petróleo, Buenos Aires.

Brüggen, J., 1950, Fundamentos de la Geología de Chile. Instituto Geográfico Militar, Santiago de Chile.

Bruhn, R. L., 1979, Rock structures formed during back-arc basin deformation in the Andes of Tierra del Fuego. Geological Society of America Bulletin, v. 90, p. 998–1012.

Bruhn, R. L., and Dalziel, I.W.D., 1977, Destruction of the Early Cretaceous marginal basin in the Andes of Tierra del Fuego. In: Talwani, M., and Pitman, W. C., eds., Island Arcs, Deep Sea Trenches and Back-Arc Basins. Maurice Ewing Series, 1, Am. Geophys. Union, p. 395–405.

Burckhardt, C., 1900a, Profils géologiques transversaux de la Cordillère Argentino-Chilienne. Museo La Plata, Sección Geológica y Mineralógica, Anales II, p. 1–136.

——, 1900b, Coupés géologiques de la Cordillère entre las Lajas et Curacautin. Museo La Plata, Sección Geológica y Mineralógica, Anales III, p. 1–102.

——, 1903, Beiträge zur Kenntniss der Jura-und Kreideformation der Cordillère. Palaeontographica 50, p. 1–144.

Busteros, A. G., and Lapido, O. R., 1984, Rocas Básicas en la vertiente Norocci-dental de la Meseta del Lago Buenos Aires, Povincia de Santa Cruz. Asocia-ción Geológica Argentina, Revista 38 (3-4), p. 427–436.

Cabrera, A., 1927, Sobre un pez fósil ("*Haplospondylus clupeoides*" gen. et sp. nov. del lago San Martín (Territorio de Santa Cruz). Museo La Plata, Revista, v. 30, p. 317–319.

——, 1941, Un plesiosaurio nuevo del Cretáceo de Chubut. Museo La Plata, Paleontología, Revista, n.s., v. 2(8), p. 113–130.

Camacho, H. H., 1949, La Fáunula Cretácica del Hito XIX (Tierra del Fuego). Asociación Geológica Argentina, Revista 4(4), p. 249–254.

——, 1967a, Las transgresiones del Cretácico superior y Terciario de la Argen-tina. Asociación Geológica Argentina, Revista 22(4), 253–280.

——, 1967b, Consideraciones sobre una fauna del Cretácico superior (Maestrichtiano) del Paso del Sapo, Curso medio del Río Chubut. Ameghiniana, v. 5(4), p. 131–134.

——, 1968, Acerca de la Megafauna del Cretácico superior de Huantrai-có, Provincia del Neuquén (Argentina). Ameghiniana, v. 5(9), p. 321–329.

——, 1969, Nota sobre fósiles del Cretácico superior de Mechanquil, Provincia de Mendoza (Argentina). Ameghiniana, v. 6(3), p. 219–222.

——, 1971, Nuevos fósiles del Cretácico superior (Maestrichtiano) del De-partamento Malargüe, Provincia de Mendoza, Argentina, Ameghiniana, v. 7(4), p. 329–334.

——, 1972, The Cretaceous-Tertiary Boundary in Argentina. XXIV Interna-tional Geological Congress, Proceedings 7, p. 490–495.

Camacho, H. H., and Olivero, E. B., 1985, El género *Steinmanella* Crickmay, 1930 (Bivalvia, Trigoniidae) en el Cretácico inferior del sudoeste

gondwánico. Academia Nacional de Ciencias Exactas, Físicas y Naturales, Anales 37, p. 41–62.

Camacho, H. H., and Riccardi, A. C., 1978, Invertebrados, Megafauna. Relatorio Geología y Recursos Naturales del Neuquén, Séptimo Congreso Geológico Argentino, p. 137–144.

Caminos, R., 1980, Cordillera Fueguina. In: Segundo Simposio de Geología Regional Argentina, v. 2, p. 1463–1501. Academia Nacional Ciencias, Córdoba.

Campos, C.W.M., Ponte, F. C., and Miura, K., 1974, Geology of the Brazilian continental margin. In: Burk, C. A. and Drake, C. L., eds., The Geology of Continental Margins, p. 447–461. Springer-Verlag, New York.

Caorsi, J. H., and Goñi, J. C., 1958, Geología Uruguaya. Instituto Geológico Uruguay, Boletín 37, p. 9–73.

Cappetta, H., 1975, Sur quelques Sélaciens nouveaux du Crétacé supérieur de Bolivie (Amérique du Sud). Geobios, v. 8(1), p. 5–24.

Carbajal, E., Pascual, R., Pinedo, R., Salfity, J. A., and Vucetich, M. G., 1977, Un nuevo mamífero de la Formación Lumbrera (Grupo Salta) de la Comarca de Carahuasi (Salta, Argentina). Edad y Correlaciones. Museo Ciencias Naturales Mar del Plata, Publicaciones 2(7), p. 148–163.

Carozzi, A., Baliña, M., Blanco, S., Corbari, S., Constanzo, J., Fernández, S., Gustavino, L., Orchuela, I., Pedrazzini, M., and Valenzuela, M., 1982, Análi-sis sedimentológico y Diagenético de los carbonatos del miembro superior de la Formación Quintuco en el Yacimiento Loma La Lata, Provincia del Neuquén. Primer Congreso Nacional Hidrocarburos, Petróleo y gas, Traba-jos Técnicos, p. 65–84.

Carozzi, A. V., Bercowski, F., Rodríguez, M., Sánchez, M., and Vonesch, T., 1981, Estudio de Microfacies de la Formación Chachao (Valanginiano), Provincia de Mendoza. Octavo Congreso Geológico Argentino, Actas 2, p. 545–565.

Carraro, C. C., Gamermann, N., Eick, N. C., Bortoluzzi, C. A., Jost, H., and Pinto, J. F., 1974, Mapa Geologico do Estado do Rio Grande do Sul, Escala 1:1,000,000. Instituto de Geociências, Universidade Federal Rio Grande do Sul, Brazil.

Carter, W. D., 1963, Unconformity marking the Jurassic-Cretaceous boundary in the La Ligua area, Aconcagua province, Chile. United States Geological Survey Professional Paper 450E, p. 61–63.

Carter, W. D., Pérez, E., and Aliste, N., 1961, Definition and age of Patagua Formation, Province of Aconcagua, Chile. American Association of Petro-leum Geologists Bulletin 45(11), p. 1892–6.

Casamiquela, R. M., 1964, Sobre un nuevo dinosaurio hadrosáurido de la Argen-tina. Ameghiniana, v. 3(9), p. 285–308.

——, 1969a, Historia geológica del valle de Huahuel Niyeu, área extra-andina del suroeste de la provincia de río Negro, República Argentina (con énfasis en el Pleistoceno). Asociación Geológica Argentina, Revista 24(3), p. 287–329.

——, 1969b, La presencia en Chile de *Aristonectes* Cabrera (Plesiosauria) del Maastrichtense del Chubut, Argentina. Edad y carácter de la transgresión "Rocanense." Cuartas Jornadas Geológicas Argentinas, Actas 1, p. 199–213.

——, 1979, La zona litoral de la transgresión Maastrichtense en el Norte de la Patagonia. Aspectos Ecológicos. Ameghiniana, v. 15(1-2), p. 137–148.

——, 1980, Considérations écologiques et zoogéographiques sur les Vertébrés de la zone littorale de la mer du Maastrichtien dans le Nord de la Patagonie. Société Géologique de France Mémoire, n.s., v. 139, p. 53–55.

Casamiquela, R. M., and Chong, G., 1980, La presencia de *Pterodaustro* Bonaparte, (Pterodactyloidea) del Neojurásico (?) de la Argentina, en los Andes del Norte de Chile. Segundo Congreso Argentino de Paleontología y Bioestratigrafía, Actas 1, p. 201–209.

Casamiquela, R. M., and Fasola, A., 1968, Sobre pisadas de dinosaurios del Cretácico inferior de Colchagua (Chile). Departamento Geología, Univer-sidad Chile, Publicación 30, p. 1–24.

Casamiquela, R. M., Corvalán, J., and Franquesa, F., 1969, Hallazgo de Dino-saurios en el Cretácico superior de Chile. Instituto Investigaciones Geo-lógicas Chile, Boletín 25, p. 5–31.

Cazau, L. B., and Uliana, M. A., 1973, El Cretácico superior continental de la

Cuenca Neuquina. Quinto Congreso Geológico Argentino, Actas 3, p. 131–163.

Cazau, L. B., Cellini, N., and Oliver Gascón, J., 1976, El Subgrupo Santa Bárbara (Grupo Salta) en la porción oriental de la provincia de Salta y Jujuy. Sexto Congreso Geológico Argentino, Actas 1, p. 341–355.

Cecioni, G., 1955a, Edad y Facies del Grupo Springhill en Tierra del Fuego. Facultad de Ciencias Físicas y Naturales, Universidad de Chile, Anales 12, p. 243–256.

——, 1955b, Distribuzione verticale di alcune Kossmaticeratidae nella Patagonia Chilena. Societa Geologica Italiana, Bollettino 74, p. 141–149.

——, 1956a, *Leopoldia? paynensis* Favre: Sua posizione stratigrafica in Patagonia. Societa Italiana Scienza Naturali e Museo Civico di Storia Naturale di Milano, Atti 95(1), p. 135–145.

——, 1956b, Significato della ornamentazione in alcune Kossmaticeratidae della Patagonia. Rivista Italiana di Paleontologia e Stratigrafia, v. 62, p. 3–8.

——, 1957a, Cretaceous flysch and molasse in Departamento Ultima Esperanza, Magallanes Province, Chile. American Association of Petroleum Geologists Bulletin, v. 41(3), p. 538–564.

——, 1957b, Eta della flora dei Cerro Guido e Stratigrafia del Departamento Ultima Esperanza, Provincia di Magellano, Cile. Societa Geologica Italiana, Bollettino 76, p. 1–18.

——, 1960, Orogénesis Subhercínica en el Estrecho de Magallanes. Facultad de Ciencias Físicas y Naturales, Universidad de Chile, Anales 17, p. 280–289.

——, 1961, El Titónico inferior marino en la provincia de Tarapacá y consideraciones sobre el arqueamiento central de los Andes. Departamento de Geología, Universidad de Chile, Comunicaciones 1(3), p. 1–19.

——, 1970, Esquema de Paleogeografía Chilena. Editorial Universitaria de Chile, Santiago.

——, 1978, Petroleum possibilities of the Darwin's Navidad Formation near Santiago, Chile. Museo Nacional de Historia Natural Publicación Ocasional 25, p. 3–28.

——, 1980, Darwin's Navidad Embayment, Santiago region, Chile, as a model of the southeastern Pacific Shelf. Journal of Petroleum Geology, v. 2(3), p. 309–321.

Cecioni, G., and Charrier, R., 1974, Relaciones entre la Cuenca Patagónica, la Cuenca Andina y el Canal de Mozambique. Ameghiniana, v. 11(1), p. 1–38.

Cecioni, G., and García, F., 1960a, Observaciones geológicas en la Cordillera de la Costa de Tarapacá. Instituto de Investigaciones Geológicas Chile, Boletín 6, p. 5–28.

——, 1960b, Stratigraphy of Coastal Range in Tarapacá Province, Chile. American Association of Petroleum Geologists Bulletin, v. 44(10), p. 1609–1620.

Chang, K. H., 1975, Unconformity-bounded units. Geological Society of America, Bulletin, v. 86, p. 1544–1552.

Charrier, R., 1973, Interruptions of spreading and the compressive tectonic phases of the meridional Andes. Earth and Planetary Science Letters, v. 20, p. 242–249.

——, 1981, Mesozoic and Cenozoic stratigraphy of the central Argentinian-Chilean Andes (32°-35° S) and chronology of their tectonic evolution. Zentralblatt Geologie Paläontologie, v. 1(3/4), p. 344–355.

——, 1982, La Formación Leñas-Espinoza: Redefinición, Petrografía y Ambiente de sedimentación. Revista Geológica Chile, v. 17, p. 71–82.

——, 1984, Areas subsidentes en el Borde Occidental de la Cuenca Tras-Arco Jurásico–Cretácica, Cordillera Principal Chilena entre 34° y 34°30'S. Noveno Congreso Geológico Argentino Actas 2, p. 107–124.

Charrier, R., and Lahsen, A., 1968, Contribution a l'étude de la limite Crétacé-Tertiaire de la Province de Magellan, Extreme-sud du Chile. Revue de Micropaleontologie, v. 11(2), p. 111–120.

Charrier, R., and Malumián, N., 1975, Orogénesis y epeirogénesis en la región austral de América del Sur durante el Mesozoico y el Cenozoico. Asociación Geológica Argentina, Revista 30(2), p. 193–207.

Charrier, R., and Vicente, J. C., 1972, Liminary and geosyncline Andes: Major orogenic phases and synchronical evolution of the central and austral sectors of the southern Andes. In: Conference on Solid Earth Problems, II: Symposium on the results of Upper Mantle Investigations with Emphasis on Latin

America. International Upper Mantle Project, Scientific Report 37-II, p. 451–470.

Charrier, R., Linares, E., Niemeyer, H., and Skarmeta, J., 1979, Edades Potasio-Argón de Vulcanitas Mesozoicas y Cenozoicas del Sector Chileno de la Meseta Buenos Aires, Aysen, Chile, y su significado geológico, Séptimo Congreso Geológico Argentino, Actas 2, p. 23–43.

Chebli, G. A., Nakayama, C., Sciutto, J. C., and Serraiotto, A. A., 1976, Estratigrafía del Grupo Chubut en la región central de la provincia homónima. Sexto Congreso Geológico Argentino, Actas 1, p. 375–392.

Cherroni, C., 1977, El Sistema Cretácico en la parte Boliviana de la Cuenca Cretácica Andina. Yacimientos Petrolíferos Fiscales Bolivianos, Revista Técnica 6(1-2), p. 5–46.

Chong, G., 1976, Las relaciones de los Sistemas Jurásico y Cretácico en la zona preandina de Chile. Primer Congreso Geológico Chileno, Actas 1, p. A21–A42.

——, 1977, Contribution to the knowledge of the Domeyko Range in the Andes of Northern Chile. Geologische Rundschau, v. 66(2), p. 374–404.

——, 1985, Hallazgo de restos óseos de dinosaurios en la Formación Hornitos-Tercera región de Atacama, Chile. IV Congreso Geológico Chileno, Actas 1(1), p. 152–161.

——, 1986, El Sistema Cretácico de los Andes de Chile entre su limite norte y los 27°00' Lat. Sur. In: Cretácico de América Latina, IGCP 242, Primer Simposio (La Paz, 1986), p. 75–86.

Chong, G., and Gasparini, Z. B. de, 1976, Los Vertebrados Mesozoicos de Chile y su aporte Geo-Paleontológico. Sexto Congreso Geológico Argentino, Actas 1, p. 45–67.

Chotin, P., 1976, Essai d'interpretation du Bassin Andin chiléno argentin mésozoique en tant que bassin marginal. Société Géologique du Nord, Annales 96(3), p. 177–184.

——, 1977, Les Andes Méridionales a la Latitude de Concepcion (Chili, 38° S): Portion Intracratonique d'une Chaine développée en bordure de la Marge active Est-Pacifique. Revue de Géographie Physique et de Géologie Dynamique (2) 19 (4), p. 353–376.

Cingolani, C. A., Varela, R., and Leguizamón, M. A., 1981, Las vulcanitas alcalinas cretácicas del Cerro Morado, Sierra de Mogna, Provincia de San Juan y su implicancia estratigráfica. Asociación Argentina Mineralogía, Petrología y Sedimentología, Revista v. 12(3-4), p. 53–70.

Cione, A. L., 1986, "*Haplospondylus*" *clupeoides* Cabrera, 1927, un Clupeomorfo (Actinopterygii, Teleostei) del Cretácico inferior de Patagonia. Ameghiniana, v. 22(3-4), p. 296–299.

Cione, A. L., and Laffite, G., 1980, El primer Siluriforme (Osteichthyes, Ostariophysi) del Cretácico de Patagonia. Consideraciones sobre el área de diferenciación de los Siluriformes. Aspectos Biogeográficos. Segundo Congreso Argentino de Paleontología y Bioestratigrafía, Actas 2, p. 35–46.

Cione, A. L., and Pereira, S. M., 1985, Los peces de la Formación Yacoraite (Cretácico tardío-Terciario, Noroeste Argentino) como indicadora de salinidad. Asociación Geológica Argentina, Revista, 40(1-2), p. 83–88.

Cione, A., Pereira, S. M., Alonso, R., and Arias, J., 1985, Los Bagres (Osteichthyes, Siluriformes) de la Formación Yacoraite (Cretácico tardío) del Noroeste Argentino. Consideraciones biogeográficas y bioestratigráficas. Ameghiniana, v. 21(2-4), p. 294–304.

Cisternas, M. E., and Oviedo, L., 1979, Perfil Tectónico-Estratigráfico en la Precordillera de Atacama y Flanco Occidental de la Cordillera Claudio Gay, en la Lat. 26°40', III Región, Chile. Segundo Congreso Geológico Chileno, Actas 1, p. B79–B97.

Cisternas, M. E., Diaz, L., Fontbote, LL., Mayer, Ch., and Amstutz, G. C., 1985, Nuevos antecedentes sobre la evolución de la Cuenca Neocomiana en la Zona Copiapó-Vallenar. IV Congreso Geológico Chileno, Actas 4, p. 599–612.

Clark, A. H., Farrar, E., Haynes, S. J., Quirt, G. S., Conn, H., and Zentilli, M., 1970, K-Ar Chronology of granite emplacement and associated mineralization, Copiapó Mining District, Atacama, Chile. Economic Geology, v. 65(6), p. 736.

Clark, A. H., Caelles, J. C., Farrar, E., Haynes, S. J., Lortie, R. B., McBride, S. L.,

Quirt, G. S., Robertson, R.C.R., and Zentilli, M., 1976, Longitudinal variations in the metallogenic evolution of the central Andes. In: Strong, D. F., ed., Metallogeny and Plate Tectonics. Geological Association of Canada Special Paper 14, p. 23–58.

Cloos, D., 1961, La presencia de "Cornaptychus" y "Laevilamellaptychus" (Cephalopoda-Ammonoidea) en Argentina. Asociación Geológica Argentina, Revista 16(1-2), p. 5–13.

——, 1962, Los "Aptychi" (Cephalopoda-Ammonoidea) de Argentina. Asociación Geológica Argentina, Revista 16(3-4), p. 117–141.

Cockerell, T.D.A., 1936, The fauna of the Sunchal (or Margas Verdes) Formation, northern Argentina. American Museum Novitates 886, p. 1–9.

Codignotto, J., Nullo, F., Panza, J., and Proserpio, C., 1979, Estratigrafía del Grupo Chubut entre Paso de Indios y Las Plumas, Provincia del Chubut, Argentina. Séptimo Congreso Geológico Argentino, Actas 1, p. 471–480.

Coira, B., Davidson, J., Mpodozis, C., and Ramos, V., 1982, Tectonic and magmatic evolution of the Andes of northern Argentina and Chile. Earth Science Reviews, v. 18, p. 303–332.

Comte, D., and Hasui, Y., 1971, Geochronology of eastern Paraguay by the potassium-argon method. Revista Brasileira de Geociências, v. 1(1), p. 33–43.

Cordani, U. G., 1970, Idade do vulcanismo no Oceano Atlântico Sul. Instituto Geociências e Astronomia Universidade São Paulo, Boletim 1, p. 9–76.

Cordani, U. G., and Vandoros, P., 1967, Basaltic rocks of the Parana Basin. In: Bigarella, J. J., Becker, R. D., and Pinto, I. D., Problems in Brasilian Gondwana Geology, p. 207–231. Curitiba.

Cordani, U. G., Sartori, P.L.P., and Kawashita, K., 1980, Geoquimica dos Isótopos de Estrôncio e a Evolução da Atividade Vulcânica na Bacia do Paraná (Sul do Brasil) Durante o Cretáceo. Academia Brasileira de Ciências, Anais 52(4), p. 811–818.

Corro, G. del, 1966, Un nuevo Dinosaurio carnívoro del Chubut (Argentina). Museo Argentino de Ciencias Naturales, Comunicaciones Paleontología 1(1), p. 1–4.

——, 1975, Un nuevo Saurópodo del Cretácico superior. *Chubutisaurus insignis* gen. et sp. nov. (Saurischia-Chubutisauridae nov.) del Cretácico superior (Chubutiano), Chubut, Argentina. Primer Congreso Argentino de Paleontología y Bioestratigrafía, Actas 2, p. 229–240.

Cortelezzi, C. R., and Cazeneuve, H., 1967, Estudio geocronológico de los Basaltos de Nogoyá (Prov. Entre Ríos) y su relación con las Rocas Efusivas del Sur de Brasil y Uruguay. Museo La Plata, Revista, Geología, n.s., v. 6(39), p. 19–32.

Cortelezzi, C. R., Traversa, L., and Pavlicevic, R. R., 1981, Estudio petrológico y ensayos físicos de las rocas alcalinas del sur de las provincias de Córdoba y San Luis. Octavo Congreso Geológico Argentino, Actas 4, p. 885–901.

Cortés, J. M., 1980, Senoniano marino en el flanco oriental del Macizo nordpatagónico. Asociación Geológica Argentina, Revista 35(3), p. 438–9.

Cortiñas, J. S., and Arbe, H. A., 1981, El Cretácico continental de la región comprendida entre los cerros Guadal y Ferrarotti, Departamento Tehuelches, Provincia del Chubut. Octavo Congreso Geológico Argentino, Actas 3, p. 359–372.

Corvalán, J., 1956, Uber marine Sedimente des Tithon und Neokom der Gegend von Santiago. Geologische Rundschau, v. 45, p. 919–926.

——, 1959, El Titoniano de Río Leñas, Prov. de O'Higgins. Instituto Investigaciones Geológicas Chile, Boletin 3, p. 5–65.

——, 1974, Estratigrafía del Neocomiano marino de la región al sur de Copiapó, Provincia de Atacama. Revista Geológica de Chile, v. 1, p. 13–36.

Corvalán, J., and Munizaga, F., 1972, Edades radiométricas de rocas intrusivas y metamórficas de la Hoja Valparaiso–San Antonio, Chile. Instituto Investigaciones Geológicas Chile, Boletín, 28, p. 5–40.

Corvalán, J., and Pérez, E., 1958, Fósiles guías chilenos Tithoniano-Neocomiano. Instituto Investigaciones Geológicas Chile, Manual 1, p. 1–48.

Corvalán, J., and Vergara, M., 1980, Presencia de fósiles marinos en las Calizas de Polpaico. Implicaciones paleoecológicas y paleogeográficas. Revista Geológica de Chile, v. 10, p. 75–83.

Cossmann, M., 1925, Description des gastropodes mesozoiques du Nord-Ouest de l'Argentine. Museum National d'Histoire Naturelle, Paleontologie Invertebrates Com., 2 (10), p. 193–209.

Coulon, F. K., Gamermann, N., and Formoso, M.L.L., 1973, Considerações sobre à gênese da Formação Tupanciretâ. Pesquisas, v. 2, p. 79–89.

Coutinho, J.M.V., Coimbra, A. M., Neto, M. B., and Rocha, G. A., 1982, Lavas Alcalinas analcimiticas associadas ao Grupo Bauru (Kb) No Estado de São Paulo, Brasil. Quinto Congreso Latinoamericano de Geología (Argentina 1982), Actas 2, p. 185–195.

Covacevich, V., Varela, J., and Vergara, M., 1976, Estratigrafía y Sedimentación de la Formación Baños del Flaco al sur del Río Tinguiririca, Cordillera de los Andes, Provincia de Curicó, Chile. Primer Congreso Geológico Chileno, Actas 1, p. A191–A211.

Creer, K. M., 1962, Paleomagnetism of the Serra Geral Formation. Royal Astronomical Society, Geophysical Journal, v. 7, p. 1–22.

——, 1964, Palaeomagnetism and the results of its application to South American rocks. Boletim Paranaense de Geografía, v. 10–14, p. 93–138.

Creer, K. M., Miller, J. A., and Gilbert Smith, A., 1965, Radiometric age of the Serra Geral Formation. Nature, v. 207, p. 282–3.

Criado Roque, P., 1979, Subcuenca de Alvear (Provincia de Mendoza). Segundo Simposio Geología Regional Argentina. Academia Nacional Ciencias, Córdoba, v. 1, p. 811–836.

Criado Roque, P., Mombrú, C. A., and Moreno, J., 1981, Sedimentitas Mesozoicas. In: Geología y Recursos Minerales de la Provincia de San Luis, Relatorio Octavo Congreso Geológico Argentino, p. 79–96.

Dalziel, I.W.D., 1974, Evolution of the margins of the Scotia Sea. In: Burk, C. A., and Drake, C. L., eds., The Geology of Continental Margins. Springer-Verlag, New York, p. 567–579.

——, 1984, The Scotia Arc: An international geological laboratory. Episodes, v. 7(3), p. 8–13.

Dalziel, I.W.D., and Elliot, D. H., 1973, The Scotia Arc and Antarctic margin. In: Nairn, A.E.M., and Stehli, F. G., eds., The Ocean Basins and Margins, v. 1, The South Atlantic, p. 171–246. Plenum Publishing Corporation, New York.

Dalziel, I.W.D., De Wit, M. J., and Palmer, K. F., 1974, Fossil marginal basin in the southern Andes. Nature, v. 250, p. 291–294.

Damm, K. W., and Pichowiak, S., 1981, Geodynamik und Magmengenese in der Küstenkordillere Nordchiles zwischen Taltal und Chañaral. Geotektonische Forschungen, v. 61, p. 1–166.

Damm, K. W., Pichowiak, S., and Todt, W., 1986, Geochemie, Petrologie und Geochronologie der Plutonite und des Metamorphen Grundgebirges in Nordchile. Berliner Geowissenschaftliche Abhandlungen, v. 66(1), p. 73–146.

Danieli, C., and Porto, J. C., 1968, Sobre la extensión austral de las formaciones Mesozoico-Terciarias de la Provincia de Salta, limítrofe con Tucumán. Terceras Jornadas Geológicas Argentinas, Actas 1, p. 77–90.

Darbyshire, D.P.F., and Fletcher, C.J.N., 1979, A Mesozoic alkaline province in eastern Bolivia. Geology, v. 7(11), p. 545–8.

Darwin, C., 1846, Geological Observations on the Volcanic Islands and Parts of America Visited during the Voyage of H.M.S. Beagle. Smith, Elder and Co., London.

Davidson, J., and Vicente, J. C., 1973, Características Paleogeográficas y Estructurales del Area fronteriza de las Nacientes del Teno (Chile) y Santa Elena (Argentina) (Cordillera Principal, 35° a 35°15′ de Latitud Sur). Quinto Congreso Geológico Argentino, Actas 5, p. 11–55.

Davidson, J. M., and Godoy, E., 1976, Observaciones sobre un perfil Geológico de los Andes Chilenos en la Latitud 25°40′ S. Sexto Congreso Geológico Argentino, Actas 1, p. 69–87.

Davila, A., Hervé, F., and Munizaga, F., 1979, Edades K/Ar en granitoides de la Cordillera de la Costa de la Provincia de Colchagua, VI Región, Chile Central. Segundo Congreso Geológico Chileno, Actas 1, p. F107–F120.

De Alba, E., 1970, Sistema del Famatina. In: Leanza, A. F., ed., Geología Regional Argentina, p. 143–184. Academia Nacional Ciencias, Córdoba.

Dedios, P., 1967, Cuadrángulo Vicuña, Provincia de Coquimbo. Instituto Investigaciones Geológicas Chile, Carta Geológica 16, p. 5–65.

—— , 1978, Cuadrángulo Rivadavia, Región de Coquimbo. Instituto Investigaciones Geológicas Chile, Carta Geológica 28, p. 1–19.

De Ferrariis, C., 1968, El Cretácico del Norte de la Patagonia. Terceras Jornadas Geológicas Argentinas, Actas 1, p. 121–144.

De Giusto, J. M., Di Persia, C. A., and Pezzi, E. E., 1980, Nesocratón del Deseado. In: Segundo Simposio de Geología Regional Argentina, v. 2, p. 1389–1430. Academia Nacional Ciencias, Córdoba.

Delaney, P.J.V., and Goñi, J., 1963, Correlação preliminar entre as Formações Gondwanicas do Uruguay e Rio Grande do Sul, Brasil. Boletim Paranaense de Geografía, v. 8–9, p. 3–21.

Dellapé, D., Pando, G., and Volkheimer, W., 1979, Estratigrafía y Palinología de las Formaciones Mulichinco, Agrio y Grupo La Amarga, al sur de Zapala. Séptimo Congreso Geológico Argentino, Actas 1, p. 593–607.

De Witt, M. J., and Stern, C. R., 1981, Variation in the degree of crustal extension during formation of a back-arc basin. Tectonophysics, v. 72, p. 229–260.

Días, H. A., and Massabié, A. C., 1974, Estratigrafía y Tectónica de las sedimentitas triásicas, Potrerillos, provincia de Mendoza. Asociación Geológica Argentina, Revista 29(2), p. 185–204.

Digregorio, J. H., 1972, Neuquén. In: Leanza, A. F., ed., Geología Regional Argentina, p. 439–505. Academia Nacional de Ciencias, Córdoba.

—— , 1978, Estratigrafía de las Acumulaciones Mesozoicas. Relatorio Geología y Recursos Naturales del Neuquén, Séptimo Congreso Geológico Argentino, p. 37–65.

Digregorio, J. H., and Uliana, M. A., 1975, Plano Geológico de la Provincia de Neuquén, Escala 1:500,000. II Congreso Ibero-Americano de Geología Económica, v. 4, p. 69–93.

—— , 1980, Cuenca Neuquina. In: Segundo Simposio de Geología Regional Argentina, p. 985–1032. Academia Nacional de Ciencias, Córdoba.

Digregorio, R. E., Gulisano, C. A., Gutiérrez Pleimling, A. R., and Minniti, S. A., 1984, Esquema de la Evolución Geodinámica de la cuenca Neuquina y sus implicancias paleogeográficas. Noveno Congreso Geológico Argentino, Actas 2, p. 147–162.

Dingman, R. J., 1963, Cuadrángulo Tulor, Provincia de Antofagasta. Instituto Investigaciones Geológicas Chile, Carta Geológica 11, p. 5–35.

Dingman, R. J., and Galli, C., 1965, Geology and ground water resources of the Pica area, Tarapaca Province, Chile. U.S. Geological Survey Bulletin 1189, p. 1–113.

Di Paola, E. C., and Marchese, H. G., 1970, Relaciones litoestratigráficas entre las formaciones Rayoso, Candeleros y Huincul, Provincia de Neuquén, República Argentina. Asociación Geológica Argentina, Revista 25(1), p. 111–120.

Doello Jurado, M., 1922, Una especie de "Viviparus" del Cretáceo superior de Río Negro. Physis, v. 5, p. 328–330.

—— , 1927, Noticia preliminar sobre los moluscos fósiles de agua dulce, mencionados en el precedente estudio de R. Wichmann "Sobre la Facies lacustre senoniana de los estratos con Dinosaurios y su fauna." Academia Nacional de Ciencias, Córdoba, Boletín 30, p. 407–416.

Dolgopol de Sáez, M., 1957, Crocodiloideos Fósiles Argentinos. Un nuevo crocodilo del Mesozoico argentino. Ameghiniana, 1(1–2), p. 48–50.

Dos Santos, E. L., Da Silva, L. C., Filho, V. O., Coutinho, M.G.N., Roisenberg, A., Ramalho, R., and Hartmann, L. A., 1984, Os Escudos Sul-Riograndense e Catarinense e a Bacia do Paraná. In: Schobbenhaus, C., Almeida, D. de, Derze, G. R., and Asmus, H. E., eds., Geologia do Brasil, p. 331–355. Ministério das Minas e Energía, Brasilia.

Dostal, J., Zentilli, M., Caelles, J. C., and Clark, A. H., 1977, Geochemistry and origin of volcanic rocks of the Andes (26°-28°S). Contributions to Mineralogy and Petrology, v. 63, p. 113–128.

Dott, R. H., Winn, R. D., De Witt, M. J., and Bruhn, R. L., 1977, Tectonic and sedimentary significance of Cretaceous Tekenika beds of Tierra del Fuego. Nature, v. 266(5603), p. 620–622.

Douvillé, R., 1910, Céphalopodes Argentins. Société Géologique de France, Mémoire 17(4), p. 5–24.

Drake, R., Curtiss, G., and Vergara, M., 1976, Potassium argon dating of igneous activity in the central chilean andes—latitude 33° S. Journal of Volcanological and Geothermal Research, v. 1, p. 265–284.

Drake, R. E., Charrier, R., Thiele, R., Munizaga, F., Padilla, H., and Vergara, M., 1982, Distribución y Edades K/Ar de Volcanitas post-Neocomianas en la Cordillera Principal entre 32° y 36° L.S: Implicaciones estratigráficas y tectónicas para el Meso-Cenozoico de Chile Central. III Congreso Geológico Chileno, Actas 2, p. D41–D78.

Eckel, E. B., 1959, Geology and mineral resources of Paraguay, A reconnaisance. U.S. Geological Survey Professional Paper 327, p. 1–110.

Eick, N. C., Gamermann, N., and Carraro, C. C., 1973, A discordância pré-Formação Serra Geral. Pesquisas, v. 2, p. 73–77.

Ellison, R. A., 1985, Nuevos aspectos de la Estratigrafía cretácica en la región del lago Titicaca del sur del Perú. Sociedad Geológica del Perú, Boletin 75:51-63.

Enay, R., 1972, Palaeobiogeographie des Ammonites du Jurassique terminal (Tithonique/Volgien/Portlandien s.l.) et Mobilité Continentale. Geobios, v. 5(4), p. 355–407.

d'Erasmo, G., 1935. Sopra alcuni avanzi di vertebrati fossili della Patagonia raccolti dal Dott. E. Feruglio. Accademia delle Science Fisiche e Matematiche Napoli, Atti (2) 20 (8), p. 1–26.

Ernesto, M., Hiodo, F. Y., and Pacca, I. G., 1979, Estudo Paleomagnético de Seqüência de Derrames Basálticos da Formação Serra Geral em Santa Catarina. Academia Brasileira de Ciências, Anais 51(2), p. 327–332.

Estes, R., and Price, L. I., 1973, Iguanid lizard from the Upper Cretaceous of Brazil. Science, v. 180(4087), p. 748–751.

Estes, R., Frazzetta, T. H., and Williams, E. E., 1970, Studies on the fossil snake *Dinilysia patagonica* Woodward, pt. 1. Cranial Morphology. Harvard University, Museum of Comparative Zoology Bulletin, v. 140(2), p. 25–74.

Evernden, J., Kríz, S. J., and Cherroni, C., 1966. Potassium-argon ages of some Bolivian rocks. Economic Geology, v. 72, p. 1042–1061.

Ewing, M., Ludwig, W. J., and Ewing, J., 1963, Geophysical investigations in the submerged Argentine coastal plain. 1, Buenos Aires to Peninsula Valdés. Geological Society of America Bulletin, v. 74(3), p. 275–292.

Falconer, J. D., 1931, Terrenos Gondwánicos del Departamento de Tacuarembó. Instituto Geología y Perforaciones Uruguay, Boletín 15, p. 3–17.

—— , 1937a, Formación de Gondwana en el Nordeste del Uruguay, con referencia especial a los Terrenos eo-gondwánicos. Instituto Geología y Perforaciones Uruguay, Boletín 23b, p. 1–122.

—— , 1937b, The Gondwana system (Permian and Triassic?) of northeastern Uruguay. Instituto Geología y Perforaciones Uruguay, Boletín 23a, p. 1–112.

Farquharson, G. W., 1982, Late Mesozoic sedimentation in the northern Antarctic Peninsula and its relationship to the southern Andes. Journal of the Geological Society of London, v. 139, p. 721–727.

Farrar, E., Clark, A. H., Haynes, S. J., Quirt, G. S., and Zentilli, M., 1970, K-Ar evidence for the post-Paleozoic migration of granitic intrusion foci in the Andes of northern Chile. Earth and Planetary Science Letters, v. 10(1), p. 60–66.

Favre, F., 1908, Die Ammoniten der unteren Kreide Patagoniens. Neues Jahrbuch für Mineralogie, Geologie und Paläontologie, v. 25, p. 601–647.

Ferello, R., and Lesta, P., 1973, Acerca de la existencia de una dorsal interior en el sector central de la Serranía de San Bernardo (Chubut). Quinto Congreso Geológico Argentino, Actas 4, p. 19–26.

Fernández, J., 1976, Hallazgo de peces pulmonados fósiles en la Puna Jujeña. Sociedad Científica Argentina, Anales 201(1-6), p. 13–18.

Fernández, J., Bondesio, P., and Pascual, R., 1973, Restos de *Lepidosiren paradoxa* (Osteichthyes, Dipnoi) de la Formación Lumbrera (Eogeno, Eoceno?) de Jujuy, Consideraciones estratigráficas, paleoecológicas y paleozoogeográficas. Ameghiniana, v. 10(2), p. 152–172.

Fernández Garrasino, C. A., 1977, Contribución a la Estratigrafía de la zona comprendida entre Estancia Ferrarotti, Cerro Colorado y Cerrito Negro–Departamento de Tehuelches–Provincia del Chubut–Argentina. Asociación Geológica Argentina, Revista 32(2), p. 130–144.

Ferrando, L., and Fernández, A., 1971, Esquema tectónico cronoestratigráfico del Pre-Devoniano en Uruguay. XXV Congresso Brasileiro Geologia, Anais, p. 199–210.

Ferraris, F., 1976, Esquema Estratigráfico de la Provincia de Antofagasta, Chile.

Segundo Congreso Latinoamericano de Geología, Memorias 2, p. 741–753.

——, 1978, Hoja Tocopilla, Región de Antofagasta. Instituto de Investigaciones Geológicas Chile. Mapas Geológicos Preliminares 3, p. 1–32.

——, 1981, Hoja Los Angeles-Angol (Escala 1:250,000), Región del Biobio. Mapas Geológicos Preliminares de Chile 5, p. 1–26.

Ferraris, F., and Di Biase, F., 1978, Hoja Antofagasta, Región de Antofagasta. Instituto de Investigaciones Geológicas Chile, Carta Geológica 30, p. 1–48.

Feruglio, E., 1935, Relaciones Estratigráficas y Faunísticas entre los estratos cretáceos y terciarios en la región del lago Argentino y en la del golfo San Jorge (Patagonia). Boletín Informaciones Petroleras 128, p. 69–93; 130, p. 65–100.

——, 1936a, Nota preliminar sobre algunas nuevas especies de moluscos del Supracretáceo y Terciario de la Patagonia. Museo La Plata, Paleontología, Notas 6, p. 277–300.

——, 1936b, Nuevas especies de moluscos supracretáceos y terciarios de la Patagonia. Boletín Informaciones Petroleras 139, p. 121–136.

——, 1936–1937, Palaeontographia Patagonica. Instituto Geologico della Universita di Padova, Memorie 11, p. 1–384.

——, 1938, El Cretáceo superior del lago San Martín y de las regiones adyacentes. Physis, v. 12, p. 293–342.

——, 1944, Estudios Geológicos y Glaciológicos en la región del lago Argentino (Patagonia). Academia Nacional de Ciencias, Córdoba, Boletín 37(1), p. 3–255.

——, 1949–1950, Descripción Geológica de la Patagonia, v. 1, 2, 3. Yacimientos Petrolíferos Fiscales, Buenos Aires.

Figueiredo, P. M., and Bortoluzzi, C. A., 1975, Lexico Estratigráfico de Río Grande do Sul. Pesquisas, v. 6, p. 9–74.

Fletcher, C.J.N., and Litherland, N., 1981, The Geology and tectonic setting of the Velasco Alkaline Province, Eastern Bolivia. Journal of the Geological Society, v. 138(5), p. 541–548.

Flint, S., Clemmey, H., and Turner, P., 1986a, The Lower Cretaceous Way Group of northern Chile: an alluvial fan-fan delta complex. Sedimentary Geology, v. 46, p. 1–22.

——, 1986b, Conglomerate-hosted copper mineralization in Cretaceous Andean molasse: The Coloso Formation of northern Chile. Geological Magazine, v. 123(5), p. 525–536.

Flores, M. A., 1969, El bolsón de Las Salinas en la Provincia de San Luis. Cuartas Jornadas Geológicas Argentinas, Actas 1, p. 311–327.

——, 1979, Cuenca de San Luis. Segundo Simposio de Geología Regional Argentina. Academia Nacional Ciencias, Córdoba, v. 1, p.745–769.

Flores, M. A., and Criado Roque, P., 1972, Cuenca de San Luis. In: Leanza, A. F., ed., Geología Regional Argentina, p. 567–579. Academia Nacional Ciencias, Córdoba.

Flores, M. A., Malumián, N., Masiuk, V., and Riggi, J. C., 1973, Estratigrafía Cretácica del Subsuelo de Tierra del Fuego. Asociación Geológica Argentina, Revista 28(4), p. 407–437.

Fodor, R. V., and Vetter, S. K., 1984, Rift-zone magmatism: Petrology of basaltic rocks transitional from CFB to MORB, southeastern Brazil margin. Contributions to Mineralogy and Petrology 88(4), p. 307–321.

Forbes, E., 1846, Descriptions of secondary fossil shells from South America. In: Darwin, C., Geological Observations on South America, p. 265–268. Smith, Elder and Co., London.

Fossa Mancini, E., 1937, Las investigaciones geológicas de YPF en la provincia de Mendoza y algunos problemas de estratigrafía regional. Boletín Informaciones Petroleras 14(154), p. 51–118.

Franchi, M. R., and Page, R.F.N., 1980, Los Basaltos Cretácicos y la evolución magmática del Chubut Occidental. Asociación Geológica Argentina, Revista 35(2), p. 208–229.

Francis, J. C., 1975, Esquema Bioestratigráfico Regional de la República Oriental del Uruguay. Primer Congreso Argentino de Paleontología y Bioestratigrafía, Actas 2, p. 539–568.

Francisconi, O., and Kowsmann, R. O., 1976, Preliminary structural study of the south Brazilian continental margin. In: International Symposium on Continental Margins of Atlantic Type. Academia Brasileira de Ciências, Anais 48

(suppl.), p. 25–26.

Freile, C., 1972, Estudio Palinológico de la Formación Cerro Dorotea (Maestrichtiano-Paleoceno) de la provincia de Santa Cruz, I. Museo La Plata, Paleontología, Revista 6(38), p. 39–63.

Freitas, R. O., 1955, Sedimentação, Estratigrafia e Tectónica da Serie Bauru. Universidade São Paulo, Faculdade de filosofia, ciências e letras, Geologia, Boletim 194(14), p. 1–185.

——, 1964a, Eruptivas Alcálicas. In: Geologia do Estado de São Paulo. Boletim Instituto Geografico e Geologico, v. 41, p. 101–120.

——, 1964b, Grupo Bauru. In: Geologia do Estado de São Paulo. Boletim Instituto Geografico e Geologico, v. 41, p. 126–147.

——, 1973, Geologia e Petrologia da Formação Caiuá no Estado de São Paulo. Boletim Instituto Geografico e Geologico, v. 50, p. 13–122.

——, 1976, Definição petrológica, estrutural e geotectónica das cintas orogénicas antigas do litoral norte do Estado de São Paulo. Instituto Geologico São Paulo, Boletim 1, p. 5–176.

Freneix, S., 1981, Faunes de Bivalves du Sénonien de Nouvelle-Calédonie. Analyses Paléobiogéographique, Biostratigraphique. Annales de Paléontologie (Invertébrés), v. 67(1), p. 13–32.

Frenguelli, J., 1927, Sobre la posición estratigráfica y edad de los basaltos del Río Uruguay. Sociedad Argentina Estudios Geográficos, Anales 2(3), p. 403–424.

——, 1930a, Apuntes de Geología Uruguaya. Instituto Geología y Perforaciones Uruguay, Boletín 11, p. 5–47.

——, 1930b, Apuntes de Geología Patagónica. Sobre restos de vegetales procedentes del Chubutiano de la Sierra de San Bernardo en el Chubut. Sociedad Científica de Santa Fe, Anales 2, p. 29–39.

——, 1935, "Ptilophyllum hislopi" (Oldham) en los "Mayer River Beds" del lago San Martín. Museo La Plata, Paleontología, Notas 1(3), p. 71–83.

——, 1937, Investigaciones geológicas en la zona salteña del Valle de Santa María. Museo La Plata, Obra Cincuentenario, v. 2, p. 215–572.

——, 1944, *Stomechinus pulchellus* n. sp., nuevo equinodermo del Titonense del Neuquén. Museo La Plata, Paleontología, Notas 9(61), p. 1–11.

——, 1945, Moluscos continentales en el Paleozoico superior y en el Triásico de la Argentina. Museo La Plata, Paleontología, Notas 10(83), p. 181–204.

——, 1946, Un nido de Esfégido del Cretácico Superior del Uruguay. Museo La Plata, Paleontología, Notas 11(90), p. 259–267.

——, 1948, Estratigrafía del llamado Rético en la Argentina. Sociedad Argentina Estudios Geográficos, Anales 8(2), p. 159–309.

——, 1953a, La flora fósil de la región del alto río Chalía en Santa Cruz (Patagonia). Museo La Plata, Paleontología, Notas 16(98), p. 239–257.

——, 1953b, Recientes progresos en el conocimiento de la Geología y Paleogeografía de Patagonia basados sobre el estudio de sus plantas fósiles. Museo La Plata, Geología, Revista 4(27), p. 321–341.

Fricke, W., and Voges, A., 1968, Beitrag zur Kenntnis der andinen Sedimentationsräume in Bolivien, Süd-Peru und NordChile. Geologisches Jahrbuch, v. 85, p. 941–972.

Fricke, W., Samtleben, C., Schmidt-Kaler, H., Uribe, H., and Voges, A., 1965, Geologische Untersuchungen im zentralen Teil des bolivianischen Hochlandes nord-westlich Oruro. Geologisches Jahrbuch, v. 83, p. 1–29.

Fritzsche, C. H., 1919, Eine Fauna aus schichten der Kreide-Tertiärgrenze in der argentinische Cordillere des südlichen Mendoza. Zentralblatt für Mineralogie, Geologie und Paläontologie, p. 359–369.

——, 1924, Neue Kreidefaunen aus Südamerika. Neues Jahrbuch für Mineralogie, Geologie und Paläontologie, B.Bd. 50, p. 1–56, 313–334.

Frutos, J., 1980, Andean tectonics as a consequence of sea-floor spreading. Tectonophysics, v. 72, p. T21–T32.

Frutos, J., Mencarini, P., Pincheira, M., Bourret, Y., and Alfaro, G., 1982, Geología de la Isla Quiriquina. III Congreso Geológico Chileno, Actas 3, p. F307–F338.

Fuenzalida, H., 1964, El Geosinclinal Andino y el Geosinclinal de Magallanes. Universidad de Chile. Escuela de Geología, Comunicaciones 5, p. 1–27.

Fuenzalida, R., 1968, Reconocimiento Geológico de Alto Palena (Chiloe Continental). Universidad de Chile, Facultad Ciencias Físicas y Naturales, Anales

22-23, p. 91–158.

Fúlfaro, V. J., Saad, A. R., Santos, M. V., and Vianna, R. B., 1982, Compartimentação e Evolução tectônica da Bacia do Paraná. Revista Brasileira de Geociências, v. 12(4), p. 590–610.

Fúlfaro, V. J., and Suguio, K., 1967, Campos de diques de Diabásio da Bacia do Paraná. Sociedad Brasileira Geologia, Boletim 16(2), p. 23–37.

Furque, G., 1966, Algunos aspectos de la Geología de Bahía Aguirre, Tierra del Fuego. Asociación Geológica Argentina, Revista 21(1), p. 61–66.

Furque, G., and Camacho, H. H., 1949, El Cretácico superior de la Costa Atlántica de Tierra del Fuego. Asociación Geológica Argentina, Revista 4(4), p. 263–297.

Gajardo, A., 1981, Hoja Concepción-Chillán (escala 1:250,000), Región del Biobio. Mapas Geológicos Preliminares de Chile 4, p. 1–32.

Galli, C., 1957, Las formaciones geológicas en el borde occidental de la Puna de Atacama, sector de Pica, Tarapacá. Minerales, v. 56, p. 3–15.

——, 1968, Cuadrángulo Juan de Morales, Provincia de Tarapacá. Instituto Investigaciones Geológicas Chile, Carta Geológica 18, p. 5–53.

Galli, C., and Dingman, R. J., 1962, Cuadrángulos Pica, Alca, Matilla y Chacarilla, Provincia de Tarapacá. Instituto Investigaciones Geológicas Chile, Carta Geológica 3(2-5), p. 7–125.

Galván, A. F., and Ruiz Huidobro, O. J., 1965, Geología del valle de Santa María. Estratigrafía de las formaciones mesozoico-Terciarias. Segundas Jornadas Geológicas Argentinas, Actas 3, p. 217–230.

Gamermann, N., Coulon, F. K., Carraro, C. C., and Eick, N. C., 1974, Conglomerado Bom Retiro do Sul. Pesquisas, v. 3(1), p. 7–16.

Gamerro, J. C., 1965a, Morfología del Polen de la Conífera *Trisacocladus tigrensis* Archang. de la Formación Baqueró, Provincia de Santa Cruz. Ameghiniana, v. 4(1), p. 31–38.

——, 1965b, Morfología del Polen de *Apterocladus lanceolatus* Archang. (Coniferae) de la Formación Baqueró, Provincias de Santa Cruz. Ameghiniana, v. 4(4), p. 133–138.

——, 1975a, Megasporas del Cretácico de Patagonia I, Ultraarquitectura de la pared megasporal en *Hughesisporites patagonicus* Archang. y *Horstisporites feruglioi* Archang. Ameghiniana, v. 12(1), p. 97–108.

——, 1975b, Megasporas del Cretácico de Patagonia II. Megasporas petrificadas de la Formación La Amarga, Cretácico inferior, Provincia Neuquén. Primer Congreso Argentino Paleontología y Bioestratigrafía, Actas 2, p. 11–28.

——, 1977, Megasporas del Cretácico de Patagonia III. Megasporas Petrificadas del "Chubutense," Prov. de Chubut, Argentina. Ameghiniana, v. 14(1-4), p. 100–116.

Gamerro, J. C., and Archangelsky, S., 1981, Palinozonas Neocretácicas y Terciarias de la Plataforma Continental Argentina en la Cuenca del Colorado. Revista Española de Micropaleontología, v. 13(1), p. 119–140.

Gansser, A., 1973, Facts and theories on the Andes. Geological Society of London Journal, v. 129, p. 93–131.

García, E. R. de, and Camacho, H. H., 1965, Descripción de fósiles procedentes de una perforación efectuada en la provincia de Santa Cruz (Argentina). Ameghiniana, v. 4(3), p. 71–73.

García, E. R. de, and Leanza, H. A., 1976, *Leanzacythere* gen. nov. (Ostracoda) del Berriasiano de Neuquén, República Argentina. Ameghiniana, v. 12(4), p. 315–321.

García, F., 1967, Geología del Norte Grande de Chile. Simposio Geosinclinal Andino (1962). ENAP, Santiago.

Gasparini, Z. B. de, 1973, Revisión de ?*Purranisaurus potens* Rusconi 1948 (Crocodilia, Thalattosuchia). Los Thalattosuchia como un nuevo Infraorden de los Crocodilia. Quinto Congreso Geológico Argentino, Actas 3, p. 423–431.

——, 1979, Comentarios críticos sobre los Vertebrados Mesozoicos de Chile. Segundo Congreso Geológico Chileno, Actas 3, p. H15–H32.

——, 1982, Una nueva familia de Cocodrilos Zifodontes Cretácicos de America del Sur. Quinto Congreso Latinoamericano de Geología (Argentina, 1982), Actas 4, p. 317–329.

Gasparini, Z. B. de, and Biró-Bagoczky, L., 1986, *Osteopygis* sp. (Reptilia, Testudines, Toxochelydae) tortuga fósil de la Formación Quiriquina, Cretácico

superior, Sur de Chile. Revista Geológica de Chile, v. 27, p. 27–32.

Gasparini, Z. B. de, and Buffetaut, E., 1980, *Dolichochampsa minima* n.g.n. sp., a representative of a new family of eusuchian crocodiles from the Late Cretaceous of northern Argentina. Neues Jahrbuch für Geologie und Paläontologie, Monatshefte 5, p. 257–271.

Gasparini, Z. B. de, and Dellapé, D., 1976, Un nuevo Cocodrilo marino (Thalattosuchia, Metriorhynchidae) de la Formación Vaca Muerta (Jurásico, Tithoniano) de la Provincia de Neuquén (República Argentina). Primer Congreso Geológico Chileno, Actas 1, p. C1–C19.

Gasparini, Z., Goñi, R., and Molina, O., 1982, Un plesiosaurio (Reptilia) Tithoniano en Cerro Lotena, Neuquén, Argentina. Quinto Congreso Latinoamericano de Geología (Argentina, 1982), Actas 5, p. 33–47.

Gayet, M., 1982a, Découverte dans le Crétacé supérieur de Bolivie des plus anciens Characiformes connus. Comptes Rendus des Séances de l'Académie des Sciences (II) 294 (16), p. 1037–1040.

——, 1982b, Nouvelle extension géographique et stratigraphique du genre *Lepidotes*. Comptes Rendus des Séances de l'Académie des Sciences (II) 294 (23), p. 1387–1390.

——, 1982c, Cypriniformes crétacés en Amérique du Sud. Comptes Rendus des Séances de l'Académie des Sciences (III) 12, p. 661–664.

Gentili, C. A., and Rimoldi, H. V., 1979, Mesopotamia. In: Segundo Simposio de Geología Regional Argentina, v. 1, p. 185–223, Academia Nacional Ciencias, Córdoba.

Gerth, H., 1914, Stratigraphie und Bau der argentinischen Kordillere zwischen dem Río Grande und Río Diamante. Deutsche Geologische Gesellschaft, Zeitschrift 65, p. 568–575.

——, 1925, La Fauna Neocomiana de la Cordillera Argentina en la parte meridional de la provincia de Mendoza. Academia Nacional Ciencias, Córdoba, Actas 9(2), p. 57–132.

——, 1932, Geologie Südamerikas. Gebrüder, Berlin.

——, 1955, Die Geologische Bau der Südamerikanischen Kordillere. Gebrüder Borntraeger, Berlin.

——, 1960, Die Entwicklung der Orogene der Südamerikanischen Kordillere während des Mesozoikums. Geologische Rundschau, v. 50, p. 619–630.

Giovine, A.T.Y., 1950, Algunos cefalópodos del Hauteriviense del Neuquén. Asociación Geológica Argentina, Revista 5(2), p. 35–76.

——, 1952, Sobre una nueva especie de Crioceras. Asociación Geológica Argentina, Revista 7(1), p. 71–75.

Giudiccini, G., and Campos, J. O., 1968, Notas sobre a morfogênese dos derrames basálticos. Boletim Sociedade Brasileira Geologia, v. 17(1), p. 15–28.

Godoy, E., 1981, Sobre la discordancia Intrasenoniana y el origen de los depósitos de caolin de Montenegro, región Metropolitana, Chile. Octavo Congreso Geológico Argentino, Actas 3, p. 733–741.

——, 1982, Geología del área Montenegro-Cuesta de Chacabuco, Region Metropolitana: el "problema" de la formación Lo Valle. III Congreso Geológico Chileno, v. 1, p. A124–A146.

Goin, F. J., Carlini, A. A., and Pascual, R., 1986, Un probable Marsupial del Cretácico tardío del Norte de Patagonia, Argentina. IV Congreso Argentino de Paleontología y Bioestratigrafía, Actas 2, p. 43–47.

Gómez Omil, R., De Muro, D., Hernaez, S., and Navarro, C., 1984, Reconocimiento de Islas Barrera en la Formación Yacoraite en el Flanco Suroriental de la Subcuenca de Lomas de Olmedo, Provincia de Salta. Noveno Congreso Geológico Argentino, Actas 5, p. 102–110.

Gonçalves, A., Oliveira, M.A.M. de, and Oliveira Mota, S., 1979, Geología da Bacia de Pelotas e da Plataforma de Florianópolis. Boletin Técnico Petrobras, v. 22(3):157–174.

Gonzaga de Campos, L. F., 1889, Relatorio da Commissão Geografica e Geológica de S. Paulo, Secção Geológica, Annexo, Sao Pãulo, p. 21–34.

——, 1905, Reconnhecimento da zona comprendida entre Bauru e Itapura. E. F. Noroeste do Brasil. Tipografía Ideal, São Paulo, p. 1–40.

González, O., and Vergara, M., 1962, Reconocimiento geológico de la Cordillera de los Andes entre los paralelos 35° y 38° Lat. S. Universidad de Chile, Instituto de Geología, Publicación 24, p. 1–121.

González, R. R., and Aceñolaza, F. G., 1972, La Cuenca de deposición Neopaleozoica-Mesozoica del Oeste de Argentina. Instituto Miguel Lillo, Miscelánea 40, p. 629–641.

González Díaz, E. F., 1982, Chronological zonation of granitic plutonism in the northern Patagonian Andes of Argentina: The migration of intrusive cycles. Earth Sciences Reviews, v. 18, p. 365–392.

González Díaz, E. F., and Nullo, F. E., 1980, Cordillera Neuquína. In: Segundo Simposio Geología Regional Argentina, v. 2, p. 1099–1147. Academia Nacional Ciencias, Córdoba.

González Díaz, E. F., and Valvano, J., 1979, Plutonitas Graniticas Cretácicas y Neoterciarias entre el Sector Norte del lago Nahuel Huapi y el lago Traful (Provincia del Neuquén). Séptimo Congreso Geológico Argentino, Actas 1, p. 227–242.

Goñi, J. C., and Delaney, P.J.V., 1961, Estudo estatistico dos minerais pesados da Formação Botucatu, Río Grande do Sul (Brasil) e Uruguay. Universidade Federal do Parana, Instituto Geologia, Boletim 6, p. 1–27.

Goñi, J. C., and Hoffstetter, R., 1964, Uruguay. In: Lexique Stratigraphique International V9a, p. 3–200. Commission International de Stratigraphie.

Gordillo, C. E., and Lencinas, A. N., 1967, Geología y Petrología del Extremo Norte de la Sierra de los Cóndores, Córdoba. Academia Nacional Ciencias, Córdoba, Boletín 46(1), p. 73–108.

——, 1970, Sierras Pampeanas de Córdoba y San Luis. In: Leanza, A. F., ed., Geología Regional Argentina, p. 1-39. Academia Nacional Ciencias, Córdoba.

——, 1972, Petrografía y composición química de los basaltos de la Sierra de Las Quijadas (San Luis) y sus relaciones con los basaltos Cretácicos de Córdoba. Asociación Geológica Córdoba Boletín 1(3-4), p. 1.

——, 1979, Sierras Pampeanas de Córdoba y San Luis. In: Segundo Simposio de Geología Regional Argentina, v. 1, p. 557–650. Academia Nacional Ciencias, Córdoba.

Gordon, M., Jr., 1947, Classification of the Gondwanic Rocks of Paraná, Santa Catarina and Río Grande do Sul. Ministerio de Agricultura, Departamento Nacional Produção Mineral, Divisão Geologia e Mineralogía, Notas Preliminares Estudos 38, p. 1–19.

Groeber, P., 1929, Líneas fundamentales de la geología del Neuquén, sur de Mendoza y regiones adyacentes. Dirección Nacional Geología y Minería, Publicación 58, p. 1–109.

——, 1933, Confluencia de los ríos Grande y Barrancas (Mendoza y Neuquén). Dirección Nacional Geología y Minería, Boletín 38, p. 1–72.

——, 1942, Rasgos geológicos generales de la región ubicada entre los paralelos 41° a 44° y entre los meridianos 69° y 71°. Primer Congreso Panamericano Ingeniería Minas y Geología, Anales 2, p. 368–379.

——, 1946, Observaciones Geológicas a lo largo del Meridiano 70. I, Hoja Chos Malal. Asociación Geológica Argentina, Revista 1, p. 177–208.

——, 1951, La Alta Cordillera entre las latitudes 34° y 29° 30′. Museo Argentino Ciencias Naturales, Geología, Revista 1(5), p. 235–352.

——, 1953, Andico. In: Geografía de la República Argentina, v. 2(1), Mesozoico, p. 349–536. Sociedad Argentina de Estudios Geográficos, Buenos Aires.

——, 1955, Anotaciones sobre Cretácico, Supracretácico, Paleoceno, Eoceno y Cuartario. Asociación Geológica Argentina, Revista 10(4), p. 234–261.

——, 1959, Supracretácico. In: Geografía de la República Argentina 2(2), p. 1–165. Sociedad Argentina de Estudios Geográficos, Buenos Aires.

——, 1963, La Cordillera entre las latitudes 22°20′ y 40° S. Academia Nacional Ciencias, Córdoba, Boletín 43(2-4), p. 111–175.

Grossi, J. H., Cardoso, R. N., and Da Costa, H. T., 1971, Formações cretacicas em Minas Gerais; uma revisão. Revista Brasileira Geociências, v. 1(1), p. 2–13.

Gulisano, C. A., Gutiérrez Pleimling, A. R., and Digregorio, R. E., 1984, Análisis Estratigráfico del intervalo Tithoniano-Valanginiano (Formaciones Vaca Muerta, Quintuco y Mulichinco) en el Suroeste de la Provincia del Neuquén. Noveno Congreso Geológico Argentino, Actas 1, p. 221–235.

Hallam, A., 1967, The bearing of certain paleogeographic data on continental drift. Palaeogeography, Palaeoclimatology, Palaeoecology, v. 3(2), p. 201–241.

Hallam, A., Biró-Bagóczky, L., and Pérez, E., 1986, Facies analysis of the Lo Valdés Formation (Tithonian-Hauterivian) of the High Cordillera of central Chile, and the palaeogeographic evolution of the Andean Basin. Geological Magazine, v. 123(4), p. 425–435.

Halle, T. G., 1913, Some mesozoic plant-bearing deposits in Patagonia and Tierra del Fuego and their floras. Svenska vetenskap-akademien, Stockholm, Handlingar 51(3), p. 3–58.

Haller, M. J., 1985, El Magmatismo Mesozoico en Trevelin, Cordillera Patagónica, Argentina. IV Congreso Geológico Chileno, Actas 3, p. 215–234.

Haller, M. J., and Lapido, O. R., 1980, El Mesozoico de la Cordillera Patagónica Central. Asociación Geológica Argentina, Revista 35(2), p. 230–247.

Haller, M. J., Lapido, O. R., Lizuain, A., and Page, R.F.N., 1981, El Mar Tithono-Neocomiano en la Evolución de la Cordillera Nordpatagónica. In: Cuencas Sedimentarias del Jurásico y Cretácico de América del Sur, v. 1, p. 221–237. Buenos Aires.

Halpern, M., 1962, Potassium-argon dating of plutonic bodies in Palmer Peninsula and southern Chile. Science, v. 138(3546), p. 1261–1262.

——, 1972, Rb-Sr and K-Ar dating of rocks from southern Chile and west Antarctica. Antarctic Journal of the United States, v. 7(5), p. 149–150.

——, 1973, Regional geochronology of Chile south of 50° latitude. Geological Society of America Bulletin, v. 84, p. 2407–2422.

Halpern, M., and Carlin, G. M., 1971, Radiometric chronology of crystalline rocks from southern Chile. Antarctic Journal of the United States, v. 6(5), p. 191–193.

Halpern, M., and Latorre, C. O., 1973, Estudio Geocronológico inicial de rocas del noroeste de la República Argentina. Asociación Geológica Argentina, Revista 28(2), p. 195–205.

Halpern, M., and Rex, D. C., 1972, Time of folding of the Yaghan Formation and age of the Tekenika beds, southern Chile, South America. Geological Society of America Bulletin, v. 83, p. 1881–1886.

Halpern, M., Cordani, U. G., and Berenholc, M., 1974, Variations in strontium isotopic composition of Paraná Basin volcanic rocks of Brazil. Revista Brasileira Geociências, v. 4(4), p. 223–7.

Halpern, M., Stipanicic, P. N., and Toubes, R. O., 1975, Geocronología (Rb/Sr) en los Andes Australes Argentinos. Asociación Geológica Argentina, Revista 30(2), p. 180–192.

Harrington, H. J., 1950, Geología del Paraguay Oriental. Universidad Buenos Aires, Facultad de Ciencias Exactas, Físicas y Naturales. Serie E. Geología 1, p. 9–82.

——, 1956a, Paraguay. In: Jenks, W. F., ed., Handbook of South American Geology. Geological Society of America Memoir 65, p. 103–114.

——, 1956b, Uruguay. In: Jenks, W. F., ed., Handbook of South American Geology. Geological Society of America Memoir 65, p. 119–128.

——, 1956c, Argentina. In: Jenks, W. F., ed., Handbook of South American Geology. Geological Society of America Memoir 65, p. 131–165.

——, 1961, Geology of parts of Antofagasta and Atacama Provinces, northern Chile. American Association of Petroleum Geologists Bulletin, v. 45(2), p. 169–197.

——, 1962, Paleogeographic development of South America. American Association of Petroleum Geologists Bulletin, v. 46(10), p. 1773–1814.

Hasui, Y., and Cordani, U. G., 1968, Idades potássio-argônio de rochas eruptivas mesozóicas do este mineiro e sul de Goiás. XXII Congresso Brasileiro Geologia, Anais, p. 139–143.

Hasui, Y., Sadowski, G. R., Suguio, K., and Fuck, G. F., 1975, The Phanerozoic tectonic evolution of the western Minas Gerais State. Academia Brasileira Ciências, Anais 47(3/4), p. 431–438.

Hatcher, J. B., 1897, On the geology of southern Patagonia. American Journal of Science, (4) 4 (23), p. 327–354.

——, 1900, Sedimentary rocks of southern Patagonia. American Journal of Science, (4) 9 (50), p. 85–108.

Haupt, O., 1907, Beiträge zur fauna des oberen Malm und der unteren Kreide in der Argentinischen Kordillere. Neues Jahrbuch für Geologie und Paläontologie, Abhandlungen 23, p. 187–236.

Heald, K. C., and Mather, K. F., 1922, Reconaissance of the eastern Andes between Cochabamba and Santa Cruz, Bolivia. Geological Society of America Bulletin 33(3), p. 553–570.

Heim, A., 1940, Geological observations in the Patagonian Cordillera. Eclogae Geologicae Helvetiae, v. 33(1), p. 25–51.

Heinz, R., 1928, Uber die Oberkreide-Inoceramen Süd Amerikas und ihre Beziehungen zu denen Europas und anderer Gebiete. Hamburg, Geologisches Staatsinstitut Mitteilungen, v. 10, p. 41–110.

Herbst, R., 1971a, Palaeophytologia Kurtziana III. 7, Revisión de las especies argentinas del género *Cladophlebis*. Ameghiniana, v. 8(3-4), p. 265–281.

——, 1971b, Esquema estratigráfico de la provincia de Corrientes, República Argentina. Asociación Geológica Argentina, Revista 26(2), p. 221–243.

——, 1980, Consideraciones estratigráficas y litológicas sobre la Formación Fray Bentos (Oligoceno inf.-medio) de Argentina y Uruguay. Asociación Geológica Argentina, Revista 35(3), p. 308–317.

Herm, D., 1967, Zur Mikrofazies kalkiger Sedimenteinschaltungen in Vulkaniten der andinen Geosynklinale Mittelchiles. Geologisches Rundschau, v. 56(2), p. 657–669.

Herrero Ducloux, A., 1946, Contribución al conocimiento geológico del Neuquén extraandino. Boletín Informaciones Petroleras 266, p. 245–281.

——, 1947, Los depósitos terrestres del Cretácico medio y superior del Neuquén y sur de Mendoza. Boletín Informaciones Petroleras 271, p. 171–178.

——, 1963, The Andes of western Argentina. In: The Backbone of the Americas—Tectonic History from Pole to Pole, A Symposium. American Association Petroleum Geologists Memoir 2, p. 16–28.

Hervé, F., Moreno, H., and Parada, M. A., 1974, Granitoids of the Andean Range of Valdivia Province, Chile. Pacific Geology, v. 8, p. 39–45.

Hervé, F., Nelson, E., Kawashita, K., and Suárez, M., 1981, New isotopic ages and the timing of orogenic events in the Cordillera Darwin, southernmost Chilean Andes. Earth and Planetary Science Letters, v. 55(2), p. 257–265.

Hervé, M., Marinovic, N., Mpodozis, C., and Pérez de Arce, C., 1985a, Geocronología K-Ar de la Cordillera de la Costa entre los 24° y 25° Latitud Sur. Antecedentes Preliminares. IV Congreso Geológico Chileno, Resúmenes, p. 158.

Hervé, M., Puig, A., Rivano, S., and Sepúlveda, P., 1985b, Geocronología K-Ar de las Rocas Intrusivas entre los 31° a 32° Latitud Sur, Chile. IV Congreso Geológico Chileno, Resúmenes, p. 159.

Herz, N., 1977, Timing of spreading in the South Atlantic: Information from Brazilian alkalic rocks. Geological Society of America Bulletin 88, p. 101–112.

Hinterwimmer, G. A., Meissinger, V. E., and Soave, L. A., 1984, Análisis de Facies, Porosidad y Diagénesis de una Secuencia de Playa–Formación Springhill-en el Sondeo Puesto Barros, Provincia de Santa Cruz. Noveno Congreso Geológico Argentino, Actas 5, p. 136–145.

Hoffstetter, R., and Ahlfeld, F., 1958, Paraguay. In: Lexique Stratigraphique International V9b, p. 5–30. Commission International de Stratigraphie.

Hoffstetter, R., Fuenzalida, H., and Cecioni, G., 1957, Chile-Chili. In: Lexique Stratigraphique International V7, p. 1–444. Commission International de Stratigraphie.

Huene, F. v., 1927, Beitrag zur Kenntnis mariner mesozoischer Wirbeltiere in Argentinien. Zentralblatt für Mineralogie, Geologie und Paläontologie B(1), p. 22–29.

——, 1929a, Terrestrische Oberkreide in Uruguay. Zentralblatt für Mineralogie, Geologie und Paläontologie B(4), p. 107–112.

——, 1929b, Los Saurisquios y Ornitisquios del Cretáceo Argentino. Museo La Plata, Paleontología, Anales (2)3, p. 1–196.

——, 1931, Verschiedene mesozoiche wirbeltierreste aus sudamerika. Neues Jahrbuch für Mineralogie, Geologie und Paläontologie B66, p. 181–196.

——, 1934a, Nuevos dientes de Saurios del cretáceo del Uruguay. Instituto Geología y Perforaciónes Uruguay, Boletín 21, p. 13–20.

——, 1934b, Neue Saurier-Zähne aus der Kreide von Uruguay. Zentralblatt für Mineralogie, Geologie und Paläontologie B(4), p. 182–189.

Huete, C., Maksaev, V., Moscoso, R., Ulriksen, C., and Vergara, H., 1977, Antecedentes Geocronológicos de rocas intrusivas y volcánicas en la Cordillera

de los Andes comprendida entre la Sierra Moreno y el río Loa, y los 21° y 22° Lat. Sur, II Región, Chile. Revista Geológica Chile, v. 4, p. 35–41.

Hünicken, M., 1965, Algunos Cefalópodos Supracretácicos de Río Turbio (Santa Cruz). Universidad Nacional Córdoba, Facultad Ciencias Exactas, Físicas y Naturales, Serie Ciencias Naturales, Revista 26(1-2), p. 49–99.

——, 1967, Flora Terciaria de los Estratos de Río Turbio, Santa Cruz (Niveles plantíferos del Arroyo Santa Flavia). Universidad Nacional Córdoba, Facultad Ciencias Exactas, Físicas y Naturales, Serie Ciencias Naturales, Revista, 27(3-4), p. 139–227.

——, 1971a, Palaeophytologia Kurtziana III. 4. Atlas de la Flora Fósil de Cerro Guido (Cretácico superior), Ultima Esperanza, Chile (Especímenes examinados por F. Kurtz). Ameghiniana, v. 8(3-4), p. 231–250.

——, 1971b, Palaeophytologia Kurtziana III, 5, Referencias estratigráficas sobre las Colecciones de Plantas Fósiles de R. Hauthal (Patagonia sudoccidental 1898–1899). Ameghiniana, v. 8(3-4), p. 251–258.

Hünicken, M., and Covacevich, V., 1975, Baculitidae en el Cretácico superior de la isla Quiriquina, Chile y consideraciones paleontológicas y estratigráficas. Primer Congreso Argentino de Paleontología y Bioestratigrafía, Actas 2, p. 141–172.

Hünicken, M., Charrier, R., and Lahsen, A., 1975, *Baculites* (Lytoceratina) de la Provincia de Magallanes, Chile. Primer Congreso Argentino de Paleontología y Bioestratigrafía, Actas 2, p. 115–140.

——, 1980, *Baculites* (Lytoceratina) de la base de la Formación Fuentes (Campaniano medio-superior) de la isla Riesco, Provincia de Magallanes, Chile. Academia Nacional de Ciencias, Córdoba, Boletín 53(3-4), p. 221–235.

Ibañez, M. A., 1960, Informe preliminar sobre el hallazgo de anuros en las "Areniscas Inferiores" de la quebrada del río Las Conchas, Provincia de Salta. Acta Geológica Lilloana, v. 3, p. 173–180.

Ihering, H. v., 1899, Descripción de la Ostra guaranitica. Sociedad Científica Argentina, Anales 47, p. 63–64.

——, 1907, Les Mollusques fossiles du Tertiaire et du Crétacé supérieur de l'Argentine. Museo Nacional Buenos Aires, Anales (3)7, p. 1–611.

——, 1913, Pleiodon priscus sp. n. In: Exploração do rio Grande e seus afluentes. Relatorio da Commissão Geografica e Geológica de S. Paulo, p. 39.

Ihering, R., 1909, Fosseis de S. José do Rio Preto. Revista Museu Paulista, v. 8, p. 141–146.

Indans, J., 1954, Eine Ammonitenfauna aus dem Untertithon der Argentinischen Kordillere in Süd-Mendoza. Palaeontographica, v. A105(3-6), p. 96–132.

Introcaso, A., and Ramos, V. A., 1984, La Cuenca del Salado: un modelo de Evolución Aulacogénica. Noveno Congreso Geológico Argentino, Actas 3, p. 27–46.

Issler, R. S., 1968, Aspectos de variacão dos parametros quimicos do vulcanismo da Serra Geral. XXII Congresso Sociedade Brasileira Geologia, Resumos Comunicações, p. 14.

——, 1969a, Caracteres magmaticos regionais do vulcanismo da Bacia do Parana. XXIII Congresso Sociedade Brasileira Geologia, Resumos Comunicações, Publicação Especial 1, p. 49.

——, 1969b, Problema dos Toleitos e a Série Toleitica da Bacia do Paraná: XXIII Congresso Sociedade Brasileira Geologia, Resumos Comunicações, Publicação Especial 1, p. 49–50.

——, 1970, O Problema dos Toleiitos, Series Toleiiticas e a Serie Toleiitica da Bacia do Parana. Escola Geologia, Universidade Rio Grande do Sul, Boletim 18, p. 1–26.

James, D. E., 1971, Plate tectonic model for the evolution of the central Andes. Geological Society of America Bulletin, v. 82(12), p. 3325–3346.

——, 1973, Plate tectonic model for the evolution of the central Andes: Reply. Geological Society of America Bulletin, v. 84, p. 1497–1500.

Jarŏs, J., and Zelman, J., 1969, La Relación estructural entre las formaciones Abanico y Farellones en la Cordillera del Meson, Provincia de Aconcagua, Chile. Universidad de Chile, Escuela de Geología, Publicación 34, p. 5–18.

Jeletzky, J. A., 1960, Youngest marine rocks in western interior of North America and the age of the Triceratops beds; With remarks on comparable dinosaur-bearing beds outside North America. XXI International Geological Congress, 5, p. 25–40.

—— 1977, Causes of Cretaceous oscillations of sea level in western and Arctic Canada and some general geotectonic implications. Palaeontological Society of Japan Special Papers, v. 21, p. 233–246.

—— , 1978, Causes of Cretaceous oscillations of sea level in Western and Arctic Canada and some general geotectonic implications. Geological Survey of Canada Paper 77-18, p. 1–44.

Jensen, O., 1979, Comparación entre el Tectonismo Larámico de los Rocky Mountains (EE.UU.) y el de los Andes Chileno-Argentinos. Segundo Congreso Geológico Chileno, Actas 1, p. B167–B186.

Jensen, O. L., and Vicente, J. C., 1977, Estudio Geológico del Area de "Las Juntas" del Río Copiapó. Asociación Geológica Argentina, Revista 31(3), p. 145–173.

Jones, G., 1956, Some deep Mesozoic basins recently discovered in southern Uruguay. XX International Geological Congress II, p. 53–72.

Jurgan, H., 1974, Die Marine Kalkfolge der Unterkreide in der Quebrada El Way, Antofagasta, Chile. Geologische Rundschau, v. 63, p. 490–516.

—— , 1977a, Zur Gliederung der Unterkreide-Serien in der Provinz Atacama, Chile. Geologische Rundschau, v. 66(2), p. 404–434.

—— , 1977b, Strukturelle und lithofazielle Entwicklung des andinen Unterkreide-Beckens im Norden Chiles (Provinz Atakama). Geotektonische Forschungen, v. 52(1-2), p. 1–138.

Kaaschieter, J.P.H., 1963, Geology of the Colorado Basin. Tulsa Geological Society Digest, v. 31, p. 177–187.

—— , 1965. Geología de la Cuenca del Colorado. Acta Geológica Lilloana, v. 7, p. 251–271.

Katz, H. R., 1963, Revision of Cretaceous stratigraphy in Patagonian Cordillera of Ultima Esperanza, Magallanes Province, Chile. American Association of Petroleum Geologists Bulletin, v. 47(3), p. 506–524.

—— , 1973, Contrasts in tectonic evolution of orogenic belts in the southeast Pacific. Royal Society of New Zealand Journal 3(3), p. 333–362.

Kielbowicz, A. A., Ronchi, D. I., and Stach, N. H., 1984, Foraminíferos y ostrácodos Valanginianos de la Formación Springhill. Asociación Geológica Argentina, Revista, v. 38(3-4), p. 313–339.

Klohn, C., 1960, Geología de la Cordillera de los Andes de Chile Central, Provincias de Santiago, O'Higgins, Colchagua y Curicó. Instituto Investigaciones Geológicas Chile, Boletín 8, p. 1–95.

Kraglievich, L., 1932, Nuevos apuntes para la Geología y Paleontología uruguayas. Museo Historia Natural Montevideo, Anales (2)3, p. 257–321.

Kranck, E. H., 1932, Geological Investigations in the Cordillera of Tierra del Fuego. Societas Geographica Fenniae, Acta Geographica, v. 4(2), p. 1–231.

Krantz, F., 1928, La Fauna del Titono superior y medio en la parte meridional de la provincia de Mendoza. Academia Nacional Ciencias, Córdoba, Actas 10, p. 9–57.

Krumbein, W. C., and Sloss, L. L., 1963, Stratigraphy and Sedimentation, 2nd ed. W. H. Freeman and Co., San Francisco.

Kumar, N., and Gambôa, L.A.P., 1979, Evolution of the São Paulo Plateau (southeastern Brazilian margin) and implications for the early history of the south Atlantic. Geological Society of America Bulletin, v. 90(1), p. 281–293.

Kurtz, F., 1902, Contribuciones a la Palaeophytologia Argentina. 3, Sobre la existencia de una Dakota flora en la Patagonia Austrooccidental (Cerro Guido, Gobernación de Santa Cruz). Museo La Plata, Revista 10, p. 43–60.

Lahee, F. H., 1927, The petroliferous belt of central-western Mendoza Province, Argentina. American Association of Petroleum Geologists Bulletin, v. 11(3), p. 261–278.

Lahsen, A., and Charrier, R., 1972, Late Cretaceous ammonites from Seno Skyring, Strait of Magallan Area, Magallanes Province, Chile. Journal of Paleontology, v. 46(4), p. 520–532.

Lamb, W. C., and Truitt, P., 1963, The Petroleum Geology of the Santa Cruz area, Bolivia, Sixth World Petroleum Congress, Paper I(40), p. 573–594.

Lambert, R., 1939, Note explicative d'une carte géologique des Terrains sédimentaires et des roches effusives du Département de Durazno. Instituto Geológico Uruguay, Boletín 25a, p. 1–37.

—— , 1940, Notice Explicative d'une Carte géologique de Reconnaissance du Département de Paysandú et des environs de Salto. Instituto Geológico Uruguay, Boletín 27a, p. 9–41.

—— , 1941, Estado Actual de Nuestros conocimientos sobre la geología de la República Oriental del Uruguay. Instituto Geológico Uruguay, Boletín 29, p. 3–89.

—— , 1944, Algunas trigonias del Neuquén. Museo La Plata, Paleontología, Revista 2(14), p. 357–397.

Lambrecht, K., 1929, *Neogaeornis wetzeli* n.g.n.sp., der reste Kreidevogel der südlichen Hemisphäre. Paläontologische Zeitschrift, v. 11, p. 121–129.

Landim, P.M.B., and Fúlfaro, V. J., 1972, Nota sobra a gênese da Formação Caiuá. XXV Congresso Brasileiro de Geología Anais 2, p. 277–280.

Langguth, V. B. de, 1978, Nidos de Insectos Fósiles del Cretácico superior del Uruguay. Museo Argentino de Ciencias Naturales "B. Rivadavia," Revista (Paleontología), v. 2 (4), p. 69–75.

Lapido, O. R., Lizuaín, A., and Núñez, E., 1984, La cobertura sedimentaria Mesozoica. Noveno Congreso Argentino, Relatorio 1(6), p. 139–162.

Larrain, A. P., 1975, Los equinoideos regulares fósiles y recientes de Chile. Gayana (Zoología), v. 35, p. 1–189.

—— , 1985, A new, early *Hemiaster* (Echinodermata: Echinoidea) from the Lower Cretaceous of Antofagasta, northern Chile. Journal of Paleontology, v. 59(6), p. 1401–1408.

Larson, R. L., and Ladd, J. W., 1973, Evidence for the opening of the South Atlantic in the Early Cretaceous. Nature, v. 246, p. 209–212.

Laubacher, G., 1978, Géologie des Andes Péruviennes. ORSTOM, Travaux et Documents, v. 95, p. 1–217.

Leanza, A. F., 1944, Las Apófisis yugales de *Holcostephanus*. Museo La Plata, Paleontología, Notas 9, p. 13–22.

—— , 1945, Ammonites del Jurásico superior y del Cretácico inferior de la Sierra Azul, en la parte meridional de la provincia de Mendoza. Museo La Plata, Anales, n.s. 1, p. 1–99.

—— , 1947, Upper limit of the Jurassic system. Geological Society of America Bulletin, v. 58, p. 833–842.

—— , 1949, Sobre *Windhauseniceras humphreyi* n. sp. del Titoniano de Neuquén. Asociación Geológica Argentina, Revista, v. 4(3), p. 239–242.

—— , 1957, Acerca de la existencia de "Simbirskites" en el Neocomiano argentino. Asociación Geológica Argentina, Revista, v. 12(1), p. 5–17.

—— , 1963, *Patagoniceras* gen. nov. (Binneyitidae) y otros ammonites del Cretácico superior de Chile Meridional con notas acerca de su posición estratigráfica. Academia Nacional Ciencias, Córdoba, Boletín 43, p. 203–225.

—— , 1964, Los Estratos con "Baculites" de Elcaín (Rio Negro, Argentina) y sus relaciones con otros terrenos supracretácicos argentinos. Universidad de Córdoba, Facultad Ciencias Exactas, Físicas y Naturales, Revista, v. 25(3-4), p. 93–107.

—— , 1965, *Parabinneyites,* nuevo nombre genérico para *Patagoniceras* Leanza, 1963, non Wetzel 1960. Asociación Geológica Argentina, Revista 19(2), p. 84.

—— , 1967a, Descripción de la Fauna de *Placenticeras* del Cretácico superior de Patagonia Austral, con consideraciones acerca de su posición estratigráfica. Academia Nacional Ciencias, Córdoba, Boletín 46(1), p. 5–28.

—— , 1967b, Los *Baculites* de la provincia de La Pampa con notas acerca de la edad del Piso Rocanense. Academia Nacional Ciencias, Córdoba, Boletín 44(1), p. 49–58.

—— , 1967c, Descripción de la especie tipo de *Mimetostreon* Bonarelli 1921 emend. Leanza 1963 (Moll. Pel.) del Cretácico de Santa Cruz (Patagonia Austral). Academia Nacional Ciencias, Córdoba, Boletín 46(1), p. 61–70.

—— , 1968, Anotaciones sobre los fósiles Jurásico-Cretácicos de Patagonia Austral (Colección Feruglio) conservados en la Universidad de Bologna. Acta Geológica Lilloana, v. 9, p. 121–186.

—— , 1969a, Sistema de Salta. Su edad, sus peces voladores, su asincronismo con el Horizonte Calcáreo Dolomítico y con las Calizas de Miraflores y la hibridez del Sistema Subandino. Asociación Geológica Argentina, Revista 24(4), p. 393–407.

—— , 1969b, Sobre el Descubrimiento del Piso Coniaciano en Patagonia Austral y descripción de una nueva especie de Ammonites (*Peroniceras santacrucense* n. sp.). Academia Nacional Ciencias, Córdoba, Boletín 47(1), p. 5–20.

——, 1970, Ammonites nuevos o poco conocidos del Aptiano, Albiano y Cenomaniano de los Andes Australes con notas acerca de su posición estratigráfica. Asociación Geológica Argentina, Revista, v. 25(2), p. 197–261.

——, 1972, Andes Patagónicos Australes. In: Leanza, A. F., ed., Geología Regional Argentina, p. 689–706. Academia Nacional Ciencias, Córdoba.

Leanza, A. F., and Castellaro, H. A., 1955, Algunos fósiles Cretácicos de Chile. Asociación Geológica Argentina, Revista, v. 10(3), p. 179–213.

Leanza, A. F., and Giovine, A., 1949, Leopoldias nuevas en el Supravalanginiano de Neuquén. Asociación Geológica Argentina, Revista, v. 4(4), p. 255–262.

Leanza, A. F., and Hünicken, M., 1970, Sobre la presencia del género *Roudaireia* en el Cretácico superior del Salitral de La Amarga (Dpto. Chicalco), Provincia de La Pampa, República Argentina. Asociación Geológica Argentina, Revista, v. 25(4), p. 489–494.

Leanza, A. F., and Leanza, H. A., 1973, *Pseudofavrella* gen. nov. (Ammonitina) del Hauteriviano de Neuquén, sus diferencias con *Favrella* R. Douvillé 1909, del Aptiano de Patagonia Austral y una comparación entre el Geosinclinal Andino y el Geosinclinal Magallánico. Academia Nacional Ciencias, Córdoba, Boletín 50(127-145).

Leanza, H. A., 1972, *Acantholissonia,* nuevo género de ammonitas del Valanginiano de Neuquén, República Argentina, y su posición estratigráfica. Asociación Geológica Argentina, Revista, v. 27(1), p. 63–70.

——, 1973, Estudio sobre los cambios faciales de los Estratos limítrofes Jurásico-Cretácicos entre Loncopué y Picún Leufú, Provincia de Neuquén, República Argentina. Asociación Geológica Argentina, Revista, v. 28(2), p. 97–132.

——, 1975, *Himalayites andinus* n. sp. (Ammonitina) del Tithoniano superior de Neuquén, Argentina. Primer Congreso Argentino de Paleontología y Bioestratigrafía, Actas 1, p. 581–588.

——, 1980, The Lower und Middle Tithonian ammonite fauna from Cerro Lotena, Province of Neuquén, Argentina. Zitteliana, v. 5, p. 3–49.

——, 1981a, Faunas de Ammonites del Jurásico superior y del Cretácico inferior de América del Sur, con especial consideración de la Argentina. In: Cuencas sedimentarias del Jurásico y Cretácico de América del Sur, v. 2, p. 559–597.

——, 1981b, Una nueva especie de *Myophorella* (Trigoniidae-Bivalvia) del Cretácico inferior de Neuquén, Argentina. Ameghiniana, v. 18(1-2), p. 1–9.

——, 1981c, The Jurassic-Cretaceous boundary beds in west-central Argentina and their ammonite zones. Neues Jahrbuch für Geologie und Paläontologie, Abhandlungen 161(1), p. 62–92.

——, 1985, *Maputrigonia,* un nuevo género de Trigoniidae (Bivalvia) del Berriasiano del Neuquén, Argentina. Academia Nacional de Ciencias, Córdoba, Boletín 56(3-4), p. 275–285.

——, 1986, Un nuevo Mortoniceratido (Cephalopoda Ammonoidea) del Albiano superior de Santa Cruz, Argentina, Ameghiniana, v. 22(3-4), p. 249–254.

Leanza, H. A., and Garate Zubillaga, J. I., 1983, Una nueva subespecie de *Trigonia carinata* Ag. (Bivalvia) del Cretácico inferior de Neuquén, Argentina. Ameghiniana, v. 20(1-2), p. 105–110.

——, 1986, Faunas de Trigonias (Bivalvia) del Jurásico y Cretácico inferior de la Provincia del Neuquén, Argentina, conservadas en el Museo Juan Olsacher de Zapala. In: Volkheimer, W., ed., Bioestratigrafía de los Sistemas Regionales del Jurásico y Cretácico de América del Sur, vol. 1, p. 201–229. Mendoza, Argentina.

Leanza, H. A., and Hugo, C. A., 1978, Sucesión de ammonites y edad de la Formación Vaca Muerta y sincrónicas entre los Paralelos 35° y 40° l.s., Cuenca Neuquina-Mendocina. Asociación Geológica Argentina, Revista, v. 32(4), p. 248–264.

Leanza, H. A., and Wiedmann, J., 1980, Ammoniten des Valangin und Hauterive (Unterkreide) von Neuquén und Mendoza, Argentinien. Eclogae Geologicae Helvetiae, v. 73(3), p. 941–981.

Leanza, H. A., Marchese, H. G., and Riggi, J. C., 1978, Estratigrafía del Grupo Mendoza con especial referencia a la Formación Vaca Muerta entre los paralelos 35 y 40° l.s., Cuenca neuquina-mendocina. Asociación Geológica Argentina, Revista, v. 32(3), p. 190–208.

Legarreta, L., 1986, Litogénesis de las secuencias Deposicionales carbonáticas de la F. Huitrin (Cretácico inferior), Provincia de Mendoza. Primera Reunión Argentina de Sedimentología, Resúmenes Expandidos, p. 173–176.

Legarreta, L., and Kozlowski, E., 1981, Estratigrafía y sedimentología de la Formación Chachao, Provincia de Mendoza. Octavo Congreso Geológico Argentino, Actas 2, p. 521–543.

——, 1984, Secciones condensadas del Jurásico-Cretácico de los Andes del Sur de Mendoza. Estratigrafía y significado Tectosedimentario. Noveno Congreso Geológico Argentino, Actas 1, p. 286–297.

Legarreta, L., Kozlowski, E., and Boll, A., 1981, Esquema estratigráfico y distribución de facies del Grupo Mendoza en el ámbito submendocino de la Cuenca Neuquina. Octavo Congreso Geológico Argentino, Actas 3, p. 389–409.

Legarreta, L., Kokogian, D., and Boggetti, D., 1986, Secuencias Deposicionales del Grupo Malargüe (Cretácico superior-Terciario inferior), Cuenca Neuquina, Argentina. IV Congreso Argentino de Paleontología y Bioestratigrafía, Actas (in press).

Leinz, V., 1949, Contribucão a Geologia dos Derrames Basalticos do sul do Brasil. Universidade São Paulo, Facultade Filosofia, Ciências e Letras, Geologia, Boletim 5, p. 1–61.

Leinz, V., Bartorelli, A., Sadowski, G. R., and Isotta, C.A.L., 1966, Sôbre o comportamento espacial do Trapp Basáltico da Bacia do Paraná. Sociedade Brasileira de Geologia, Boletim 15(4), p. 79–91.

Leinz, V., Bartorelli, A., and Isotta, C.A.L., 1968, Contribução do estudo do magmatismo basáltico mesozóico da Bacia do Paraná. Academia Brasileira Ciências, Anais 40 (Suppl.), p. 167–181.

Lencinas, A. N., and Salfity, J. A., 1973, Algunas Características de la Formación Yacoraite en el Oeste de la Cuenca Andina, Provincia de Salta y Jujuy, República Argentina. Quinto Congreso Geológico Argentino, Actas 3, p. 253–267.

Leonardi, G., 1980, On the discovery of an abundant ichno-fauna (vertebrates and invertebrates) in the Botucatu Formation s.s. in Araraquara, São Paulo, Brazil. Academia Brasileira Ciências, Anais 52(3), p. 559–567.

——, 1981, Novo Icnogênero de Tetrápode Mesozóico da Formação Botucatu, Araraquara, SP. Academia Brasileira Ciências, Anais, v. 53(4), p. 793–805.

Lesta, P. J., 1968, Estratigrafía de la Cuenca del Golfo San Jorge. Terceras Jornadas Geológicas Argentinas, Actas 1, p. 251–289.

——, 1969, Algunas nuevas comprobaciones en la Geología de la Patagonia. Cuartas Jornadas Geológicas Argentinas, Actas 2, p. 187–194.

Lesta, P. J., and Ferello, R., 1972, Región Extraandina de Chubut y Norte de Santa Cruz. In: Leanza, A. F., ed., Geología Regional Argentina, p. 601–653. Academia Nacional Ciencias, Córdoba.

Lesta, P. J., Ferello, R., and Bianchi, J. L., 1973, Constitución Geológica de la Porción actualmente sumergida de la Cuenca del Golfo San Jorge. Quinto Congreso Geológico Argentino, Actas 4, p. 69–74.

Lesta, P. J., Turic, M. A., and Mainardi, E., 1979, Actualización de la Información Estratigráfica en la Cuenca del Colorado. Séptimo Congreso Geológico Argentino, Actas 1, p. 701–713.

Lesta, P. J., Ferello, R., and Chebli, G. A., 1980, Geología del Chubut Extraandino. In: Segundo Simposio Geología Regional Argentina, v. 2, p. 1307–1387. Academia Nacional Ciencias. Córdoba.

Leterrier, J., Roche, H.d.l., and Rüegg, N. R., 1972, Composition chimique et parenté tholéiitique des roches basaltiques du Basin du Parana. Académie des Sciences, Comptes Rendus, Série D, 274(12), p. 1772–1775.

Levi, B., 1960, Estratigrafía del Jurásico y Cretácico inferior de la Cordillera de la Costa entre las latitudes 32°40′ y 33°40′. Universidad de Chile, Departamento de Geología, Publicación 16, p. 223–269.

——, 1969, Burial metamorphism of a Cretaceous volcanic sequence west from Santiago, Chile. Contributions to Mineralogy and Petrology, v. 24, p. 30–49.

——, 1970, Burial metamorphic episodes in the Andean geosyncline, central Chile. Geologische Rundschau, v. 59(3), p. 994–1012.

——, 1973, Eastward shift of Mesozoic and Early Tertiary volcanic centers in the Coast Range of central Chile. Geological Society of America Bulletin, 84(12), p. 3901–3910.

Levi, B., and Aguirre, L., 1960, El Conglomerado de Algarrobo y su relación con las formaciones del Cretácico superior de Chile Central. Instituto Investigaciones Geológicas Chile, Apartado 17, p. 417–431.

——, 1962, El Conglomerado de Algarrobo y su relación con las formaciones del Cretácico superior de Chile Central. Primeras Jornadas Geológicas Argentinas, Actas 2, p. 417–431.

Levi, B., Mehech, S., and Munizaga, F., 1963, Edades radiométricas y petrografía de Granitos chilenos. Instituto Investigaciones Geológicas Chile, Boletín 12, p. 1–42.

Levy, R., 1967a, Revisión de las Trigonias de Argentina. Parte 3: Los Pterotrigoniinae de Argentina. Ameghiniana, v. 5(3), p. 101–107.

——, 1967b, Revisión de las Trigonias de Argentina. Parte 4. Los Megatrigoniinae de Argentina y su relación con *Anditrigonia* gen.nov. Ameghiniana, v. 5(4), p. 135–144.

——, 1985, Dos Nuevas Especies de Trigoniidae (Mollusca, Bivalvia) en la Formación Puesto El Alamo, Cretácico superior (Santa Cruz, Argentina). Ameghiniana, v. 22(1-2), p. 57–61.

Lima, M. R. de, 1983, Paleoclimatic reconstruction of the Brazilian Cretaceous based on Palynological data. Revista Brasileira de Geociências, v. 13(4), p. 223–228.

Linares, E., and Latorre, C. O., 1975, La Edad del Granito de Aguilar, Provincia de Jujuy, Argentina. Segundo Congreso Iberoamericano de Geologia Económica, v. 1, p. 91–98.

Linares, E., and Valencio, D., 1974, Edades Potasio-Argón y Paleomagnetismo de los Diques ultrabásicos del Río de los Molinos, Córdoba, República Argentina. Asociación Geológica Argentina, Revista, v. 29(3), p. 341–8.

——, 1975, Paleomagnetism and K-Ar ages of some Trachybasaltic dykes from Río de los Molinos, Province of Córdoba, República Argentina. Journal of Geophysical Research, v. 80, p. 3315–3321.

Ljungner, E., 1930, Geologische Aufnahmen in der Patagonischen Kordillera. University of Uppsala, Geological Institution, Bulletin 23, p. 203–242.

Llambías, E. J., and Brogioni, N., 1981, Magmatismo Mesozoico y Cenozoico. In: Geología y Recursos Minerales de la Provincia de San Luis, Relatorio del Octavo Congreso Geológico Argentino, p. 101–115.

Lohmann, H. H., 1970, Outline of tectonic history of Bolivian Andes. American Association of Petroleum Geologists Bulletin, v. 54(5), p. 737–757.

Lohmann, H. H., and Braniša, L., 1962, Estratigrafía y Paleontología del Grupo Puca en el Sinclinal de Miraflores, Potosí. Petroleo Boliviano, Boletín 4(2), p. 9–16.

Longobucco, M. I., Azcuy, C. L., and Aguirre Urreta, B., 1985, Plantas de la Formación Kachaike, Cretácico de la Provincia de Santa Cruz. Ameghiniana, v. 21(2-4), p. 305–315.

Lucero, H. N., 1979, Sierras Pampeanas del norte de Córdoba, sur de Santiago del Estero, borde oriental de Catamarca y ángulo sudeste de Tucumán. In: Segundo Simposio Geología Regional Argentina, v. 1, p. 293–347. Academia Nacional Ciencias, Córdoba.

Ludwig, W. J., Ewing, J. I., and Ewing, M., 1965, Structure of the Argentine continental margin. American Association of Petroleum Geologists Bulletin, v. 52(12), p. 2337–2368.

Lydekker, R., 1893, Contribuciones al conocimiento de los vertebrados fósiles de la Argentina. I, Los Dinosaurios de Patagonia. Museo La Plata, Paleontología Argentina, Anales 2, p. 1–14.

Maack, R., 1947, Breves notícias sôbre a geologia dos Estados do Parana e Santa Catarina. Instituto de Biologia e Pesquisas Tecnologicas, Arquivos de Biologia e Tecnologia 2, p. 63–154. Curitiba.

Macellari, C. E., 1979, La presencia del género *Aucellina* (Bivalvia, Cretácico) en la Formación Hito XIX (Tierra del Fuego, Argentina). Ameghiniana, v. 16(1-2), p. 143–172.

——, 1985a, Paleobiogeografía y Edad de la Fauna de *Maorites-Gunnarites* (Ammonoidea) del Cretácico superior de la Antártida y Patagonia. Ameghiniana, v. 21(2-4), p. 223–242.

——, 1985b, El límite Cretácico-Terciario en la Península Antártica y en el sur de Sudamérica: Evidencias Macropaleontológicas. Sexto Congreso Latinoamericano de Geología (Colombia, Oct. 1985), p. 1–12.

——, 1987, Progressive Endemism in the Late Cretaceous Ammonite Family Kossmaticeratidae and the Breakup of Gondwanaland. American Geophysical Union, Geophysical Monograph 41, p. 85–92.

Maeda, S., and Urdininea, M. H., 1976, Note on the Cretaceous system in Bolivia. Journal of Geography, v. 85(5), p. 19–33.

Maeda, S., Chisaka, T., Hamada, T., Kimura, T., and Tazuke, H., 1972, Geological and palaeontological researches to the Andes. Journal of Geography, v. 81(1), p. 1–14.

Maksaev, V., 1978, Cuadrángulo Chitigua y sector occidental del Cuadrángulo Cerro Palpana, Región de Antofagasta. Instituto Investigaciones Geológicas Chile, Carta Geológica 31, p. 1–55.

——, 1979, Las fases Incaica y Quechua en la Cordillera de los Andes del Norte Grande de Chile. Segundo Congreso Geológico Chileno, Actas 1, p. B63–B77.

——, 1984, Mesozoico a Paleogeno de la Región de Antofagasta. In: Seminario Actualización de la Geología de Chile. Servicio Nacional Geología y Minería, Chile, Miscelánea 4, p. C1–C20.

Malizzia, D. C., and Limeres, M. H., 1984, Estudio de Facies de la Formación Quebrada del Medano en el Perfil de la Quebrada epónima, Provincia de la Rioja. Noveno Congreso Geológico Argentino, Actas 1, p. 310–321.

Malumián, N., 1968, Foraminíferos del Cretácico superior y Terciario del Subsuelo de la Provincia de Santa Cruz, Argentina. Ameghiniana, v. 5(6), p. 191–227.

——, 1979, Aspectos paleoecológicos de los Foraminíferos del Cretácico de la Cuenca Austral. Ameghiniana, v. 15(1-2), p. 149–160.

Malumián, N., and Báez, A. M., 1978, Outline of Cretaceous stratigraphy of Argentina. Muséum d'Histoire Naturelle de Nice, Annales 4, p. 1–10.

Malumián, N., and Masiuk, V., 1973, Asociaciones foraminiferológicas fósiles de la República Argentina. Quinto Congreso Geológico Argentino, Actas 3, p. 433–453.

——, 1975, Foraminíferos de la Formación Pampa Rincón (Cretácico inferior), Tierra del Fuego. Revista Española de Micropaleontología, v. 7(3), p. 579–600.

——, 1976a, Foraminíferos Característicos de las Formaciones Nueva Argentina, Arroyo Alfa, Cretácico inferior, Tierra del Fuego, Argentina. Sexto Congreso Geológico Argentino, Actas 1, p. 393–411.

——, 1976b, Foraminíferos de la Formación Cabeza de León. Asociación Geológica Argentina, Revista, v. 31(3), p. 180–202.

——, 1978, Foraminíferos planctónicos del Cretácico de Tierra del Fuego. Asociación Geológica Argentina, Revista, v. 33(1), p. 36–51.

Malumián, N., and Nañez, C., 1983, Foraminíferos de Ambiente Anóxico de la Formación Rio Mayer (Cretácico inferior), Provincia de Santa Cruz. Ameghiniana, v. 20(3-4), p. 367–393.

Malumián, N., and Proserpio, C., 1979, Foraminíferos Aglutinados del Cretácico de Cuenca Austral. Su significado Geológico-Ambiental. Séptimo Congreso Geológico Argentino, Actas 2, p. 431–447.

Malumián, N., and Ramos, V. A., 1984, Magmatic intervals, transgression-regression cycles and oceanic events in the Cretaceous and Tertiary of southern South America. Earth and Planetary Science Letters, v. 67, p. 228–237.

Malumián, N., Masiuk, V., and Riggi, J. C., 1971, Micropaleontología y sedimentología de la Perforación SC-1, Provincia de Santa Cruz, República Argentina. Asociación Geológica Argentina, Revista, v. 26(2), p. 175–208.

Malumián, N., Masiuk, V., and García, E. R. de, 1972, Microfósiles del Cretácico superior de la Perforación SC-1, Provincia de Santa Cruz, Argentina. Asociación Geológica Argentina, Revista, v. 27(3), p. 265–272.

Malumián, N., Nullo, F. E., and Ramos, V. A., 1983, The Cretaceous of Argentina, Chile, Paraguay and Uruguay. In: Moullade, M., and Nairn, A.E.M., eds., The Phanerozoic Geology of the World II, Mesozoic B, p. 265–304. Elsevier Scientific Publishers, Amsterdam.

Malumián, N., Echeverría, A., Martínez Machiavello, J. C., and Nañez, C., 1984, Los Microfósiles. Noveno Congreso Geológico Argentino, Relatorio II(7),

p. 485–526.

Manceñido, M. O., and Damborenea, S. E., 1984, Megafauna de Invertebrados Paleozoicos y Mesozoicos. Noveno Congreso Geológico Argentino. Relatorio II(5), p. 413–465.

Marchese, H. G., 1971, Litoestratigrafía y variaciones faciales de las sedimentitas Mesozoicas de la Cuenca Neuquina, Provincia de Neuquén, República Argentina, Revista, v. 26(3), p. 343–410.

Marini, O. J., Fuck, R. A., and Trein, E., 1967, Intrusives básicas cretáceas do Paraná. Boletim Paranaense Geociências, Resumos Comunicações, 26, p. 73.

Marinovic, N., and Lahsen, A., 1984, Hoja Calama, Región de Antofagasta. Instituto Investigaciones Geológicas Chile, Carta Geológica 58:1–140.

Marshall, P. A., Olivero, E. B., and Scasso, R. A., 1984, El vulcanismo del Cerro Grande y su significado en la estratigrafía del Cretácico inferior de la comarca del Lago Fontana, Chubut. Noveno Congreso Geológico Argentino, Actas 1, p. 322–336.

Martínez, A., 1968, Microthyriales (Fungi, Ascomycetes) fósiles del Cretácico inferior de la Provincia de Santa Cruz, Argentina. Ameghiniana, v. 5(7), p. 257–263.

Martínez, C., Suáres, R., and Subieta, T., 1971, La Cadena hercínica en la parte septentrional de la Cordillera Oriental de los Andes bolivianos (Perfil La Paz–Alto Beni). Servicio Geológico de Bolivia, Boletín 15, p. 26–35.

Martínez, R., and Osorio, R., 1963, Consideraciones preliminares sobre la presencia de carofitas fósiles en la Formación Colimapu. Revista Minerales, v. 82, p. 27–43.

Martínez, R., Giménez, O., Rodríguez, J., and Bochatey, G., 1986, *Xenotarsosaurus bonapartei* nov. gen. et sp. (Carnosauria, Abelisauridae), un nuevo Theropoda de la Formación Bajo Barreal, Chubut, Argentina. IV Congreso Argentino de Paleontología y Bioestratigrafía, Actas 2, p. 23–31.

Martínez Pardo, R., 1965, *Bolivinoides dracodorreeni* Finlay from the Magellan Basin, Chile. Micropaleontology, v. 11, p. 360–364.

Masiuk, V., and Viña, F. J., 1986, Estratigrafía de la Formación Agrio de la Cuenca Neuquina. Boletín Informaciones Petroleras (3a época), 3(6), p. 2–38.

Mather, K. F., 1922, Front Range of the Andes between Santa Cruz, Bolivia, and Embarcación, Argentina. Geological Society of America Bulletin, v. 33(4), p. 703–764.

Matsumoto, T., 1977, On the so-called Cretaceous transgressions. Palaeontological Society of Japan Special Papers, v. 21, p. 75–84.

——, 1980, Inter-regional correlation of transgressions and regressions in the Cretaceous period. Cretaceous Research, v. 1, p. 359–373.

McDougall, I., and Rüegg, N. R., 1966, Potassium-argon dates on the Serra Geral Formation of South America. Geochimica et Cosmochimica Acta, v. 30, p. 191–195.

McNutt, R. H., Crocket, J. H., Clark, A. H., Caelles, J. C., Farrar, E., Haynes, S. J., and Zentilli, M., 1975, Initial $^{87}Sr/^{86}Sr$ ratios of plutonic and volcanic rocks of the central Andes between latitudes 26° and 29° south. Earth and Planetary Science Letters, v. 27(2), p. 305–313.

Melfi, A. J., 1967, Potassium-argon ages from core samples of basaltic rocks from southern Brasil. Geochimica et Cosmochimica Acta, v. 31, p. 1079–1089.

Melinossi, R., 1935, Su di un echinide della Patagonia. Societa Toscana Scienze Naturali Atti, v. 44, p. 32–39.

Mendes, J. C., 1961, Algumas considerações sôbre a estratigrafia da Bacía do Paraná. Boletim Paranaense Geografia, v. 4/5, p. 3–33.

——, 1971, As camadas Gondwanicas do Brasil e seus Problemas. Academia Brasileira Ciências, Anais 43 (Suppl.), p. 187–196.

Mendes, J. C., and Frakes, L., 1964, Tálus Fóssil na Formação Botucatu (Neomesozoico). Sociedad Brasileira Geologia, Boletim 13(1-2), p. 67–71.

Mendes, J. C., and Petri, S., 1971, Geologia do Brasil. Ministerio da Educação e Cultura, Brasil.

Méndez, I., and Viviers, M. C., 1973, Estudio micropaleontológico de sedimentitas de la Formación Yacoraite (Provincias de Salta y Jujuy). Quinto Congreso Geológico Argentino, Actas 3, p. 467–470.

Méndez, V., 1975, Estructuras de las Provincias de Salta y Jujuy a partir del meridiano 65°30′ oeste, hasta el límite con las Repúblicas de Bolivia y Chile. Asociación Geológica Argentina, Revista, v. 29(4), p. 391–424.

Méndez, V., Turner, J.C.M., Navarini, A., Amengual, R., and Viera, V., 1979, Geología de la Región Noroeste, Provincias de Salta y Jujuy. Dirección General de Fabricaciones Militares, Buenos Aires.

Mendía, J. E., 1978, Paleomagnetism of alkaline lava flows from El Salto–Almafuerte, Córdoba Province, Argentina. Royal Astronomical Society Geophysical Journal, v. 54(3), p. 539–546.

Mendiberri, H., 1985, Estratigrafía de la sección inferior de la Fm. Agrio. Boletín Informaciones Petroleras (3a época), 2(3), p. 35–51.

Menegotto, E., Sartori, P. L., and Maciel, F.C.L., 1968, Nova sequencia sedimentar sobre a Serra Geral no Rio Grande do Sul. Instituto de Solos e Culturas da Universidade Federal do Santa Maria, Publicação Especial 1, p. 1–19.

Menéndez, C. A., 1959, Flora cretácica de la Serie del Castillo al sur del cerro Cachetaman. Asociación Geológica Argentina, Revista, v. 14(3-4), p. 219–238.

——, 1965a, Microplancton fósil de sedimentos Terciarios y Cretácicos del Norte de Tierra del Fuego (Argentina). Ameghiniana, v. 4(1), p. 7–15.

——, 1965b, *Sueria rectinervis* n. gen. et sp. de la flora fósil de Ticó, Provincia de Santa Cruz. Ameghiniana, v. 4(3), p. 75–83.

——, 1966a, Fossil Bennettitales from the Ticó flora, Santa Cruz province, Argentina. British Museum (Natural History) Bulletin 12(1), p. 3–42.

——, 1966b, La presencia de *Thyrsopteris* en el Cretácico superior de Cerro Guido, Chile. Ameghiniana, v. 4(8), p. 299–302.

——, 1969, Datos Palinológicos de las Floras Preterciarias de la Argentina. Gondwana Stratigraphy, IUGS Symposium, Proceedings, p. 55–69. Buenos Aires.

——, 1972a, Palaeophytologia Kurtziana III. 8, La Flora del Cretácico superior de Cerro Guido, Chile (1-2). Ameghiniana, v. 9(3), p. 209–212.

——, 1972b, Palaeophytologia Kurtziana III. 9, La Flora del Cretácico superior de Cerro Guido, Chile (3-7). Ameghiniana, v. 9(4), p. 289–297.

——, 1973, Acerca de la validez de *Ptilophyllum longipinnatum* Menendez. Ameghiniana, v. 10(2), p. 196–7.

Menéndez, C. A., and Caccavari de Filice, M. A., 1975, Las especies de *Nothofagidites* (Polen fósil de *Nothofagus*) de sedimentos Terciarios y Cretácicos de Estancia La Sara, Norte de Tierra del Fuego, Argentina. Ameghiniana, v. 12(2), p. 165–183.

Mercado, M., 1978, Hojas Chañaral y Potrerillos. Instituto de Investigaciones Geológicas Chile, Mapas Geológicos Preliminares 2, p. 1–24.

——, 1981, Jurassic-Neocomian paleogeography of the Atacama Region, Chile. Zentralblatt Geologie Paläontologie, v. 1, (3/4), p. 356–358.

——, 1982, Hoja Laguna del Negro Francisco, Región de Atacama. Carta Geológica de Chile (Escala 1:100,000) 56, p. 1–73.

Mezzalira, S., 1964, Formação Caiuá. Instituto Geografico e Geologico São Paulo, Boletim 41, p. 120–125.

——, 1966, Os fósseis do Estado de São Paulo. Instituto Geografico e Geologico São Paulo, Boletim 45, p. 5–128.

——, 1974, Contribução ao conhecimento da Estratigrafia e Paleontologia do Arenito Baurú. Instituto Geografico e Geologico São Paulo, Boletim 51, p. 3–163.

——, 1981, Léxico Estratigráfico do Estado de São Paulo. Governo do Estado de São Paulo, Secretaria da Agricultura e Abastecimento, Coordenadoria da Pesquisa de Recursos Naturais, Instituto geológico, Boletim 5, p. 5–161.

Mezzalira, S., and Arruda, M. R., 1965, Observaçoès geológicas na Região do Pontal do Paranapanema, Estado de São Paulo. Academia Brasileira Ciências, Anais 37(1), p. 69–77.

Millan, J. H., 1976, O Gondwana Brasileiro. Sem. Estudios Geologicos, Universidade Federal Rural Rio Janeiro, Itaguaí (1976) I, II, III, p. 173–188.

Mingramm, A., and Russo, A., 1970, Sierras Subandinas y Chaco Salteño. In: Leanza, A. F., ed., Geología Regional Argentina, p. 184–211. Academia Nacional Ciencias, Córdoba.

Mingramm, A., Russo, A., Pozzo, A., and Cazau, L., 1979, Sierras Subandinas. In: Segundo Simposio Geología Regional Argentina, v. 1, p. 95–137.

Academia Nacional Ciencias, Córdoba.

Minioli, B., 1971, Determinações potássio-argônio em rochas localizadas no litoral norte do Estado de São Paulo. Academia Brasileira Ciências, Anais 43(2), p. 443–448.

Minioli, B., Ponçano, W. L. de, and Oliveira, S.M.B., 1971, Extensão Geografica do Volcanismo Basáltico do Brasil Meridional. Academia Brasileira Ciências, Anais 43(2), p. 433–437.

Mitchum, R. M., and Uliana, M. A., 1985, Seismic stratigraphy of carbonate depositional sequences, Upper Jurassic–Lower Cretaceous, Neuquén Basin, Argentina. American Association Petroleum Geologists Memoir 39, p. 255–274.

Mitchum, R. M., Vail, P. R., and Thompson, S., 1977, Seismic stratigraphy and global changes of sea level, Pt. 2: The depositional sequence as a basic unit for stratigraphic analysis. American Association Petroleum Geologists Memoir 26, p. 53–62.

Mombru, C. A., Uliana, M. A., and Bercowski, F., 1979, Estratigrafía y Sedimentología de las Acumulaciones Biocarbonáticas del Cretácico inferior surmendocino. Séptimo Congreso Geológico Argentino, Actas 1, p. 685–700.

Mon, R., 1971, Estructura Geológica del extremo austral de las Sierras Subandinas, Prov. de Salta y Tucumán, República Argentina. Asociación Geológica Argentina, Revista, v. 26(2), p. 209–220.

Mon, R., and Urdaneta, A., 1972, Introducción a la Geología de Tucumán, República Argentina. Asociación Geológica Argentina, Revista, v. 27(3), p. 309–329.

——, 1976, Geología del borde Oriental de los Andes en las Provincias de Tucumán y Santiago del Estero, República Argentina. Segundo Congreso Latinoamericano de Geología Memorias 2 (Caracas), p. 609–631.

Mones, A., 1980, Nuevos elementos de la Paleoherpetofauna del Uruguay (Crocodilia y Dinosauria). Segundo Congreso Argentino de Paleontología y Bioestratigrafía, Actas 1, p. 265–274.

Montanelli, S. B., 1987. Presencia de Pterosauria (Reptilia) en la Formación La Amarga (Hauteriviano-Barremiano) Neuquén, Argentina. Ameghiniana, v. 24(1-2), p. 109–113.

Montecinos, P., 1979, Plutonismo durante el Ciclo Tectónico Andino en el Norte de Chile, entre los 18°-29° Lat. Sur. Segundo Congreso Geológico Chileno, Actas 3, p. E89–E108.

——, 1985, Petrología de los cuerpos intrusivos que rodean al Yacimiento de Fierro "El Algarrobo" Ascensión Sincrónica de 2 magmas. III Región, Chile. IV Congreso Geológico Chileno, Actas 3, p. 335–356.

Montecinos, P., and Helle, S., 1985, Comportamiento del Rb y del Sr en las rocas intrusivas que rodean el yacimiento de Fierro "El Algarrobo": Polaridad del Sr en la Costa Pacífica, III Región, Chile. IV Congreso Geológico Chileno, Actas 3, p. 314–334.

Montecinos, P., and Oyarzún, R., 1976, Distribución de Cr, Ni, Co, Cu, Pb, Zn, Rb, Sr, en las rocas intrusivas Cretácicas y Terciarias de la Cordillera Domeyko entre los 23°00-24°00 Lat. S y 68°45′ - 69°00 Long. W. Primer Congreso Geológico Chileno, Actas 2, p. F72–F91.

Moraes Rego, L. F., 1935, Camadas Cretáceas do Sul do Brasil. Escola Politécnica São Paulo, Anuario (2) 4, p. 231–234.

Moraga, A., 1977, Cuadrángulo Quebrada Desierto (Quebrada Salitrosa), III Región. Instituto Investigaciones Geológicas Chile, Carta Geológica 25, p. 5–12.

Moraga, A., Chong, G., Fortt, M. A., and Henriquez, H., 1974, Estudio Geológico del Salar de Atacama, Provincia de Antofagasta. Instituto Investigaciones Geológicas Chile, Boletin 29, p. 1–56.

Moreno, H., and Parada, M. A., 1976, Esquema Geológico de la Cordillera de los Andes entre los Paralelos 39°00′ y 41°30′ S. Primer Congreso Geológico Chileno, Actas 1, p. A213–A226.

Moreno, J. A., 1970, Estratigrafía y Paleogeografía del Cretácico superior en la Cuenca del Noroeste Argentino, con especial mención de los Subgrupos Balbuena y Santa Barbara. Asociación Geológica Argentina, Revista, v. 25(1), p. 9–44.

Möricke, W., 1895, Die Gastropoden und Bivalven der Quiriquina-Schichten. Neues Jahrbuch für Mineralogie, Geologie und Paläontologie, v. 10, p. 95–114.,

Moroni, A. M., 1982, Correlación palinológica en las Formaciones Olmedo y Yacoraite. Cuenca del Noroeste Chileno. Tercer Congreso Geológico Chileno, Actas 3, p. F339–F349.

Moscoso, R., 1976, Antecedentes sobre un engrane volcánico-sedimentario marino del Neocomiano en el área de tres Cruces, IV Región, Chile. Primer Congreso Geológico Chileno, Actas 1, p. A155–A167.

——, 1984, El Mesozoico superior–Cenozoico inferior del sector occidental de las Regiones de Atacama y Coquimbo (27°-30°S). In: Seminario Actualización de la geología de Chile. Servicio Nacional Geología y Minería, Chile, Miscelánea 4, p. G1–G21.

Moscoso, R., Padilla, H., and Rivano, S., 1982, Hoja Los Andes, Región de Valparaiso. Carta Geológica de Chile (Escala 1:250,000) 52: p. 1–67.

Moya, M. C., and Salfity, J. A., 1982, Los ciclos magmáticos en el Noroeste Argentino. Quinto Congreso Latinoamericano de Geología (Argentina, 1982), Actas 3, p. 523–536.

Mpodozis, A. C., and Rivano, S., 1976, Evidencias de Tectogénesis en el límite Jurásico-Cretácico en la Alta Cordillera de Ovalle (Provincia de Coquimbo). Primer Congreso Geológico Chileno, Actas 1, p. B57–B68.

Mpodozis, S., Rivano, S., and Vicente, J. C., 1973, Resultados preliminares del Estudio Geológico de la Alta Cordillera de Ovalle entre los ríos Grande y Los Molles (Prov. de Coquimbo, Chile). Quinto Congreso Geológico Argentino, Actas 4, p. 117–132.

Muhlmann, P., 1937, Algunas observaciones preliminares sobre los "Estratos de Malargüe." Boletín Informaciones Petroleras 153, p. 43–56.

Munizaga, F., and Vicente, J. C., 1982, Acerca de la zonación plutónica y del Volcanismo Miocénico en los Andes de Aconcagua (Lat. 32-33° S): datos Radimétricos K-Ar. Revista Geológica de Chile, v. 16, p. 3–21.

Munizaga, F., Hervé, F., and Drake, R., 1984, Geocronología K-Ar del extremo septentrional del Batolito Patagónico en la Región de los Lagos, Chile. Noveno Congreso Geológico Argentino, Actas 3, p. 133–145.

Muñoz Cristi, J., 1956, Chile. In: Jenks, W. F., ed., Handbook of South American Geology. Geological Society of America Memoir 65, p. 131–165.

Murature de Sureda, F., and Alonso, R. N., 1980, Nuevos hallazgos de insectos fósiles en la Formación Lumbrera (Grupo Salta, Cretácico-Terciario) en la provincia de Salta, República Argentina. Segundo Congreso Argentino de Paleontología y Bioestratigrafía, Actas 2, p. 127–129.

Musacchio, E., 1970, Ostrácodos de la Superfamilia Cytheracea y Darwinulacea de la Formación La Amarga (Cretácico inferior) en la Provincia de Neuquén, República Argentina. Ameghiniana, v. 7(4), p. 301–316.

——, 1971, Hallazgo del género *Cypridea* (Ostracoda) en Argentina y consideraciones estratigráficas sobre la Formación La Amarga (Cretácico inferior) en la Provincia de Neuquén. Ameghiniana, v. 8(2), p. 105–125.

——, 1972a, Charophytas de la Formación Yacoraite en Tres Cruces y Yavi Chico, Jujuy, Argentina. Ameghiniana, v. 9(3), p. 223–237.

——, 1972b, Charophytas del Cretácico inferior en sedimentitas "Chubutenses" al Este de La Herrería, Chubut. Ameghiniana, v. 9(4), p. 354–356.

——, 1973, Charophytas y Ostrácodos no marinos del Grupo Neuquén (Cretácico superior) en algunos afloramientos de las provincias de Río Negro y Neuquén, República Argentina. Museo La Plata, Paleontología, Revista 8(48), p. 1–32.

——, 1978, Microfauna del Jurásico y del Cretácico inferior. Relatorio Geología y Recursos Naturales del Neuquén, Séptimo Congreso Geológico Argentino, p. 147–161.

——, 1979a, Ostrácodos del Cretácico inferior en el Grupo Mendoza, Cuenca del Neuquén, Argentina. Séptimo Congreso Geológico Argentino, Actas 2, p. 459–473.

——, 1979b, Datos Paleobiogeográficos de algunas asociaciones de foraminíferos, ostrácodos y carofitas del Jurásico medio y el Cretácico inferior de Argentina. Ameghiniana, v. 16(3-4), p. 247–271.

——, 1980, Algunos microfósiles calcáreos, marinos y continentales, del Jurásico y Cretácico inferior de la República Argentina. Segundo Congreso Argentino de Paleontología y Bioestratigrafía, Actas 5, p. 67–76.

——, 1981, South American Jurassic and Cretaceous foraminifera, ostracoda and charophyta of Andean and sub-Andean regions. In: Cuencas Sedimentarias Jurásico y Cretácico de América del Sur, v. 2, p. 461–498.

Musacchio, E. A., and Abrahamovich, A. H., 1984, Early Cretaceous Platycopina (marine ostracods) from the Andean Neuquén Basin, Argentina. Neues Jahrbuch für Geologie und Paläontologie, Abhandlungen 167(2), p. 251–274.

Musacchio, E. A., and Chebli, G. A., 1975, Ostrácodos no marinos y carofitas del Cretácico inferior en las Provincias de Chubut y Neuquén, Argentina. Ameghiniana, v. 12(1), p. 70–96.

Musacchio, E. A., and Palamarczuk, S. C., 1976, Microfósiles calcáreos de la Formación Ranquiles (Cretácico inferior) en la Provincia de Neuquén, Argentina. Ameghiniana, v. 12(4), p. 306–321.

Naranjo, J. A., 1978, Zona Interior de la Cordillera de la Costa entre los 26°00′ y 26°20′, Región de Atacama. Instituto Investigaciones Geológicas Chile, Carta Geológica 34, p. 1–48.

Naranjo, J. A., 1981, Evolución Geológica de los Andes de Antofagasta Meridional. Octavo Congreso Geológico Argentino, Actas 3, p. 457–470.

Naranjo, J. A., and Covacevich, V., 1979, Nuevos antecedentes sobre la Geología de la Cordillera de Domeyko en el área de Sierras Vaquillas Altas, Región de Antofagasta. Segundo Congreso Geológico Chileno, Actas 1, p. A45–A64.

Naranjo, J. A., and Puig, A., 1984, Hojas Taltal y Chañaral, Región de Antofagasta y Atacama. Instituto Investigaciones Geológicas Chile, Carta Geológica 62-63, p. 1-140.

Nasi, C., and Thiele, R., 1982, Estratigrafía del Jurásico y Cretácico de la Cordillera de la Costa, al Sur del río Maipo, entre Melipilla y Laguna de Aculeo (Chile Central). Revista Geológica de Chile, v. 16, p. 81–99.

Natland, M. L., González, E., Cañón, A., and Ernst, M., 1974, A system of stages for correlation of Magallanes Basin sediments. Geological Society of America Memoir 139, p. 1–126.

Navarro García, L. F., 1984, Estratigrafía de la región comprendida entre los paralelos de 26°00′ a 27°15′ de latitud sur y los meridianos de 66° a 67°00′ de longitud oeste, Provincia de Catamarca. Noveno Congreso Geológico Argentino, Actas I, p. 353–383.

Nelson, E. P., Dalziel, I.W.D., and Milnes, A. G., 1980, Structural geology of Cordillera Darwin—Collisional-style orogenesis in the southernmost Chilean Andes. Eclogae Geologicae Helvetiae, v. 73(3), p. 727–751.

Newell, N. D., 1949, Geology of the Lake Titicaca region, Peru and Bolivia. Geological Society of America Memoir, v. 36, p. 1–111.

Niemeyer, H., Skarmeta, J., Fuenzalida, R., and Espinosa, W., 1984, Hojas Península de Taitao y Puerto Aisén. Instituto Investigaciones Geológicas Chile, Carta Geológica (Esc. 1:500,000) N° 60-61:1-80.

Nordenskjold, E., 1905, Die Krystallinen Gesteine der Magallansländer. Wissenschaftliche Ergebnisse der Schwedischen Expedition nach den Magallansländer, v. 1(6), p. 175–240.

Nullo, G. B. de, Nullo, F., and Proserpio, C., 1981, Santoniano-Campaniano: Estratigrafía y contenido ammonitífero. Cuenca Austral. Asociación Geológica Argentina, Revista, v. 35(4), p. 467–493.

Nullo, F., Proserpio, C. A., and Nullo, G. B. de, 1981a, El Cretácico de la Cuenca Austral entre el Lago San Martín y Río Turbio. In: Cuencas Sedimentarias del Jurásico y Cretácico de América del Sur, v. 1, p. 181–220.

——, 1981b, Estratigrafía del Cretácico superior en el C° Indice y alrededores, Provincia de Santa Cruz. Octavo Congreso Geológico Argentino, Actas 3, p. 373–387.

Oblitas, D., Salinas, C., and Davila, J., 1978, Reseña de la Geología Petrolera de Bolivia. Segundo Congreso Latinoamericano de Geología, Memoria 5, p. 4161–4203.

Odin, G. S., and Kennedy, W. J., 1982, Mise a jour de l'échelle des temps mésozoiques. Comptes Rendus Académie des Sciences (II) 94, p. 383–386.

Ojeda, H.A.O., 1982, Structural framework, stratigraphy and evolution of Brazilian marginal basins. American Association of Petroleum Geologists Bulletin, v. 66(6), p. 732–749.

Olivares, A., Rivano, S., Sepúlveda, P., and Vicente, J. C., 1985, Sobre la presencia del Neocomiano volcánico en la Cordillera de Illapel–Los Andes entre los 31° a 33° Latitud Sur, Chile. Cuarto Congreso Geológico Chileno,

Resúmenes, p. 40.

Oliveira, A. I., 1956, Brazil. In: Jenks, W. F., ed., Handbook of South American Geology. Geological Society of America Memoir 65, p. 5–62.

Oliveira, A. I., and Leonardos, O. H., 1943, Geologia do Brasil. Serviço de Informação Agricola. Río de Janeiro.

Oliver Schneider, C., 1923, Contribución a la paleontología chilena. Apuntes sobre *Cimoliasaurus andium* Deecke. Revista Chilena de Historia Natural, v. 25, p. 89–95.

Olivero, E. B., 1983, Ammonoideos y Bivalvos Berriasianos de la Cantera Tres Lagunas, Chubut. Ameghiniana, v. 20(1-2), p. 11–20.

Orbigny, A. d', 1842, Voyage dans l'Amerique méridionale, 1826-1833. 3(3). Géologie, p. 1–289; 4, Paléontologie, p. 1–188.

——, 1848, Atlas Paléontologique. In: Grange, M. J., ed., Géologie, Minéralogie et Géographie Physique du Voyage. Voyage au Pole Sud et dans l'Océanie sur les corvettes l'Astrolabe et la Zélée pendant les annees 1837-38-39-40 sous le commandement de M. J. Dumont d'Urville.

Ott, E., and Volkheimer, W., 1972, *Palaeospongilla chubutensis* n.g. et n.sp. ein Süsswasser-schwamm aus der Kreide Patagoniens. Neues Jahrbuch für Geologie und Paläontologie, Abhandlungen 140(1), p. 49–63.

Oyarzún, J., and Villalobos, J., 1969, Recopilación de análisis químicos de rocas chilenas. Universidad de Chile, Departamento de Geología, Publicación 33, p. 1–47.

Pacca, I. G., and Hiodo, F. Y., 1976, Paleomagnetic analysis of Mesozoic Serra Geral lava flows in southern Brazil. In: Symposium on Continental Margins of Atlantic Type. Academia Brasileira Ciências, Anais 48 (Suppl.), p. 207–214.

Pacheco, J. A., 1931, Notas sobre a geologia do vale do Rio Grande a partir da fóz do Rio Pardo até a sua confluencia com o Rio Parnaiba. In: Exploração do Rio Grande seus afluentes. Comissão Geográfica e Geológica, São Paulo.

Padula, E. L., 1972, Subsuelo de la Mesopotamia y regiones adyacentes. In: Leanza, A. F., ed., Geología Regional Argentina, p. 213–235. Academia Nacional Ciencias, Córdoba.

Padula, E. L., and Mingramm, A., 1968, Estratigrafía, distribución y cuadro geotectónico-sedimentario del "Triásico" en el subsuelo de la llanura Chaco-Paranense. Terceras Jornadas Geológicas Argentinas, Actas 1, p. 291–331.

——, 1969, Sub-surface Mesozoic redbeds of the Chaco-Mesopotamian region, Argentina, and their relatives in Uruguay and Brazil. Gondwana Stratigraphy, IUGS Symposium Proceedings, p. 1053–1071. Buenos Aires.

Palacios C., and Espinoza, S., 1982, Geología y Petrología del complejo plutónico de la Cordillera de la Costa entre Tocopilla y río Loa, norte de Chile. Tercer Congreso Geológico Chileno, Actas 2, p. D154–D171.

Palacios, O., and Ellison, R., 1986, El Sistema Cretácico en la Región del Lago Titicaca (Perú). In: Cretácico de América Latina, IGCP 242, Primer Simposio (La Paz, 1986), p. 32–48.

Palmer, H. C., Hayatsu, A., and MacDonald, W. D., 1980, Palaeomagnetic and K-Ar age studies of a 6 km-thick Cretaceous section from the Chilean Andes. Royal Astronomical Society, Geophysical Journal, v. 62, p. 133–153.

Palmer, H. S., 1914, Geological notes on the Andes of northwestern Argentina. American Journal of Science, v. 38(4), p. 309–330.

Pareja, J., Vargas, C., Suárez, R., Ballon, R., Carrasco, R., and Villarroel, C., 1978, Mapa Geológico de Bolivia, Memoria Explicativa. Yacimientos Petrolíferos Fiscales Bolivianos-Servicio Geológico de Bolivia, La Paz.

Parker, G., 1974, Posición estratigráfica del "Famatinense" y sus correlaciones. Asociación Geológica Argentina, Revista, v. 29(2), p. 231–247.

Parodi Bustos, R., 1962, Los anuros cretácicos de Puente Morales (Salta) y sus vinculaciones con *Shelania pasquali* Casamiquela (Chubut) y *E. reuningi* Haughton de Africa del Sur. Facultad de Ciencias Naturales de Salta, Revista 1(3), p. 81–85.

Parodi Bustos, R., and Kraglievich, J. L., 1960, A propósito de los anuros cretácicos descubiertos en la Provincia de Salta. Facultad de Ciencias Naturales de Salta, Revista 1(1), p. 37–40.

Parodi Bustos, R., Figueroa Caprini, M., Kraglievich, J. L., and Del Corro, G., 1960, Noticia preliminar acerca del yacimiento de anuros extinguidos de

Puente Morales (Dep. Guachipas, Salta). Facultad de Ciencias Naturales de Salta, Revista 1(1), p. 1–27.

Parodiz, J. J., 1969, The Tertiary non-marine Mollusca of South America. Carnegie Museum Annals, v. 40, p. 1–242.

Pascual, R., 1980, Nuevos y singulares tipos ecológicos de Marsupiales extinguidos de América del Sur (Paleoceno tardío o Eoceno temprano) del Noroeste Argentino. Segundo Congreso Argentino Paleontología y Bioestratigrafía, Actas 2, p. 151–171.

Pascual, R., and Bondesio, P., 1976, Notas sobre vertebrados de la frontera Cretácica-Terciaria III: Ceratodontidae (peces Osteichthyes, Dipnoi) de la Formación Coli-Toro y de otras unidades del Cretácico tardío de Patagonia y sur de Mendoza. Sus implicancias paleobiogeográficas. Sexto Congreso Geológico Argentino, Actas 1, p. 565–576.

Pascual, R., and Odreman, O. E., 1973, Las unidades estratigráficas del Terciario portadoras de mamíferos. Su distribución y sus relaciones con los acontecimientos diastróficos. Quinto Congreso Geológico Argentino, Actas 3, p. 293–338.

Pascual, R., Bondesio, P., Scillato Yane, G. J., Vucetich, M. G., and Gasparini, Z. B. de, 1978, Vertebrados. In: Relatorio Geología y Recursos Naturales del Neuquén, Séptimo Congreso Geológico Argentino, p. 177–185.

Pascual, R., Vucetich, M. G., and Fernández, J., 1979, Los Primeros mamíferos (Notoungulata, Henricosborniidae) de la Formación Mealla (Grupo Salta, Subgrupo Santa Bárbara). Sus implicancias filogenéticas, taxonómicas y cronológicas. Ameghiniana, v. 15(3-4), p. 366–390.

Paulcke, W., 1903, Ueber die Kreideformation in Südamerika und ihre Beziehungen zu anderen Gebieten. Neues Jahrbuch für Mineralogie, Geologie und Paläontologie, v. 17, p. 252–312.

——, 1907, Die Cephalopoden der oberen Kreide Südpatagoniens. Naturforschungen Gesellschaft Freiburg, Berichte 15, p. 167–248.

Pérez, E., and Reyes, R., 1978, Las Trigonias del Cretácico superior de Chile y su valor cronoestratigráfico. Instituto Investigaciones Geológicas Chile, Boletín 34, p. 1–67.

——, 1980, *Buchotrigonia (Buchotrigonia) topocalmensis* sp. nov. (Trigoniidae; Bivalvia) del Cretácico superior de Chile. Revista Geológica de Chile, v. 9, p. 37–48.

——, 1983a, Las especies del género *Anditrigonia* Levy, 1967, en la Colección Philippi. Revista Geológica de Chile, v. 18, p. 15–41.

——, 1983b, *Paranditrigonia* subgénero nuevo de *Anditrigonia* Levy (Mollusca; Bivalvia). Revista Geológica de Chile, v. 19-20, p. 57–79.

Pérez, E., Reyes, R., and Pérez, V., 1981, Clave de la especies y subespecies sudamericanas del género *Steinmanella* Crickmay, 1930. (Trigoniidae, Bivalvia). Revista Geológica de Chile, v. 13-14, p. 103–106.

Pérez Leytón, M., 1987, Datos palinológicos del Cretácico superior de la sección estratigráfica de Carata, Potosí, Bolivia. IV Congreso Latinoamericano de Paleontología (Bolivia, 1987), Memorias v. 2, p. 739–756.

Perry, L. D., 1963, Flora Formation (Upper Cretaceous) of Northern Bolivia. American Association Petroleum Geologists, Bulletin, v. 47(10), p. 1855–1860.

Pesce, A. H., 1979a, Estratigrafía de la Cordillera Patagónica entre los 43°30′ y 44° de Latitud Sur y sus áreas mineralizadas. Séptimo Congreso Geológico Argentino, Actas 1, p. 257–270.

——, 1979b, Estratigrafía del Arroyo Perdido en su tramo medio e inferior, Provincia de Chubut. Séptimo Congreso Geológico Argentino, Actas 1, p. 315–333.

Petersen, C. S., 1946, Estudios Geológicos en la región del Río Chubut Medio. Dirección General de Minas y Geología, Boletín 59, p. 5–137.

Petri, S., 1955, Charophyta cretácica de S. Paulo (Formação Bauru). Sociedad Brasileira de Geologia, Revista 4(1), p. 67–72.

——, 1983, Brazilian Cretaceous Paleoclimates: Evidence from clay-minerals, sedimentary structures and palynomorphs. Revista Brasileira de Geociências, v. 13(4), p. 215–222.

——, 1987, Cretaceous Paleogeographic Maps of Brazil. Palaeogeography, Palaeoclimatology, Palaeoecology, v. 59, p. 117–168.

Petri, S., and Fúlfaro, V. J., 1983, Geologia do Brasil. Editora da Universidade de São Paulo, São Paulo.

Philippi, R. A., 1899, Los Fósiles Secundarios de Chile. F. A. Brockhaus, Leipzig, Santiago.

Piatnitzky, A., 1936, Estudio Geológico de la región del río Chubut y del río Genua. Boletín Informaciones Petroleras 137, p. 83–118.

——, 1938, Observaciones geológicas en el oeste de Santa Cruz (Patagonia). Boletín Informaciones Petroleras 165, p. 45–85.

Pichowiak, S., Damm, K. W., and Zeil, W., 1982, Origin and structurally controlled development of coastal batholiths in northern Chile. Quinto Congreso Latinoamericano de Geología (Argentina 1982), Actas 1, p. 547–554.

Pilsbry, H. A., 1939, Freshwater mollusca and crustacea from near El Molino, Bolivia. Johns Hopkins University Studies in Geology, v. 13, p. 69–72.

Pincheira, M., and Thiele, R., 1982, El Neocomiano de la Cordillera de la Costa al NW de Vallenar (28°15′ a 28°30′ L.S.): Situación tectónica del Borde Occidental de la Cuenca Marina Neocomiana Tras-Arco. Tercer Congreso Geológico Chileno, Actas 1, p. A236–A261.

Piraces, R., 1976, Estratigrafía de la Cordillera de la Costa entre la Cuesta El Melón y Limache, Provincia de Valparaíso, Chile. Primer Congreso Geológico Chileno, Actas 1, p. A65–A82.

——, 1979, Velocidades de depositación de las secuencias Mesozoicas en los Andes a la Latitud de Chile Central: Su relación con los eventos magmáticos y Tectónicos. Segundo Congreso Geológico Chileno. Actas 1, p. F83–F106.

Pires, F.R.M., 1982, Formação Bauru: Controvérsias. Academia Brasileira de Ciências, Anais 54(2), p. 369–393.

Ponce de León, V., 1966, Estudio Geológico del Sinclinal de Camargo (Parte Norte). Instituto Boliviano del Petróleo 6(1), p. 33–53.

Ponte, F. C., and Asmus, H. E., 1976, The Brazilian marginal basins: current stage of knowledge. In: Continental Margins of Atlantic Type. Academia Brasileira de Ciências, Anais 48 (Suppl.), p. 215–239.

——, 1978, Geological framework of the Brazilian Continental margin. Geologische rundschau, v. 67(1), p. 201–235.

Porto, J. C., and Danieli, C. A., 1979, La extensión Noroeste de la Cuenca del Grupo Salta en la Provincia de Tucumán. Séptimo Congreso Geológico Argentino, Actas 1, p. 563–576.

Porto, J. C., Danieli, C. A., and Ruiz Huidobro, O. J., 1982, El Grupo Salta en la Provincia de Tucumán, Argentina. Quinto Congreso Latinoamericano de Geología (Argentina 1982), Actas 4, p. 253–264.

Pothe de Baldis, E. D., 1987, Dinoflagelados de la Facies de Mar Abierto del Santoniano-Campaniano del Sur de Lago Viedma, Provincia de Santa Cruz, Argentina. Ameghiniana, v. 23 (3-4), p. 167–183.

Powell, J. E., 1979, Sobre una asociación de dinosaurios y otras evidencias de vertebrados del Cretácico superior de la región de La Candelaria. Prov. de Salta, Argentina. Ameghiniana, v. 16(1-2), p. 191–204.

Powell, J. E., and Palma, R. M., 1981, Primer hallazgo de Mamíferos en la Formación Río Loro, Provincia de Tucumán y su significado cronológico. Asociación Geológica Argentina, Revista, v. 36(2), p. 208–212.

Price, L. I., 1945, A new reptile from the Cretaceous of Brazil. Departamento Nacional Produção Mineral, Divisão Geologia e Mineralogia. Notas Preliminares e Estudos 25, p. 1–8.

——, 1950a, On a new crocodilian *Sphagesaurus* from the Cretaceous of the state of São Paulo, Brazil. Academia Brasileira de Ciências, Anais 22(1), p. 77–83.

——, 1950b, Os crocodilídeos da fauna da formação Bauru do cretáceo terrestre do Brasil Meridional. Academia Brasileira de Ciências, Anais 22(4), p. 473–490.

——, 1951, Um ovo de dinossaurio na formação Bauru do cretáceo do Estado de Minas Gerais. Departamento Nacional Produção Mineral, Divisão Geologia e Mineralogia, Notas Preliminares e Estudos 53, p. 1–5.

——, 1953, Os quelônios da Formação Bauru, Cretáceo terrestre do Brasil Meridional. Departamento Nacional Produção Mineral, Divisão Geologia e Mineralogia. Boletim 147, p. 1–34.

——, 1954, Os Crocodilios da Formação Bauru. Academia Brasileira de Ciências, Anais 22, p. 473–490.

——, 1955, Novos crocodilídeos dos arenitos de série Bauru cretáceo do Estado de Minas Gerais. Academia Brasileira de Ciências, Anais 27(4), p. 487–498.

—— , 1959, Sôbre un crocodilídeo notusúquio do Cretáceo brasileiro. Departamento Nacional Produção Mineral, Divisão Geologia e Mineralogia, Boletim 188, p. 1–55.

Putzer, H., 1962, Geologie von Paraguay. Gebrüder Borntraeger, Berlin.

Quattrocchio, M., 1978, Contribución al conocimiento de la palinología estratigráfica de la Formación Lumbrera (Terciario inferior, Grupo Salta). Ameghiniana, v. 15(3-4), p. 285–300.

—— , 1980, Estudio palinológico preliminar de la Formación Lumbrera (Grupo Salta), Localidad Pampa Grande, Provincia de Salta, República Argentina. Segundo Congreso Argentino de Paleontología y Bioestratigrafía, Actas 2, p. 131–144.

—— , 1984, Sobre el posible significado paleoclimático de los quistes de Dinoflagelados en el Jurásico y Cretácico inferior de la Cuenca Neuquina. Tercer Congreso Argentino de Paleontología y Bioestratigrafía (Corrientes 1982), Actas, p. 107–113.

Quattrocchio, M. E., and Volkheimer, W., 1985, Estudio Palinológico del Berriasiano en la Localidad Mallin Quemado, Provincia de Neuquén, Argentina. Ameghiniana, v. 21(2-4), p. 187–204.

Quensel, P.D., 1911, Geologisch-petrographische Studien in der patagonischen Cordillera. University of Uppsala, Geological Institutionen, Bulletin 9, p. 1–113.

—— , 1913, Die Quarzporphyr und Porphyroidformation in Südpatagonien und Feuerland. University of Uppsala, Geological Institutionen, Bulletin 12, p. 9–40.

Radelli, L., 1964, Ensayo de reconstrucción de la cronoestratigrafía y de la paleogeografía del Altiplano y de las cordilleras orientales de Bolivia despues del Paleozoico superior. Rivista Italiana di Paleontologia e Stratigrafía, v. 70(4), p. 833–868.

Ramírez, C. F., and Gardeweg, M., 1982, Hoja Toconao, Región Antofagasta. Carta Geológica de Chile (Escala 1:250,000) 54, p. 1–121.

Ramos, E. D., and Ramos, V. A., 1979, Los ciclos magmáticos de la República Argentina. Séptimo Congreso Geológico Argentino, Actas 1, p. 771–786.

Ramos, V. A., 1976, Estratigrafía de los lagos La Plata y Fontana, Provincia del Chubut, República Argentina. Primer Congreso Geológico Chileno, Actas 1, p. A43–A64.

—— , 1979, El vulcanismo del Cretácico inferior de la Cordillera Patagónica. Séptimo Congreso Geológico Argentino. Actas 1, p. 423–435.

—— , 1982, Geología de la Región del Lago Cardiel, Provincia de Santa Cruz. Asociación Geológica Argentina, Revista, v. 37(1), p. 23–49.

—— , 1985a, El Mesozoico de la Alta Cordillera de Mendoza: Facies y Desarrollo estratigráfico-Argentina. Cuarto Congreso Geológico Chileno, Actas 1(1), p. 492–513.

—— , 1985b, El Mesozoico de la Alta Cordillera de Mendoza: Reconstrucción Tectónica de sus facies, Argentina. Cuarto Congreso Geológico Chileno. Actas 1(2), p. 104–118.

—— , 1986, The Tectonics of the Central Andes: 30-33° S Latitude. Geological Society of America Special Paper 218 (in press).

Ramos, V. A., and Palma, M. A., 1981, El Batolito Granítico del Monte San Lorenzo. Cordillera Patagónica (Provincia de Santa Cruz). Octavo Congreso Geológico Argentino, Actas 3, p. 257–280.

—— , 1983, Las Lutitas Pizarreñas fosilíferas del Cerro Dedo y su evolución Tectónica; Lago La Plata, Provincia de Chubut. Asociación Geológica Argentina. Revista, v. 38(2), p. 148–160.

Ramos, V. A., Niemeyer, H., Skarmeta, J., and Muñoz, J., 1982. Magmatic evolution of the Austral Patagonian Andes. Earth Science Reviews, v. 18, p. 411–443.

Rassmuss, J. E., 1957, Zur Erdölgeologie Südamerikas, insbesondere des pazifischen Raumes, Geologische Rundschau, v. 45(3), p. 686–707.

Rawson, P. F., 1971, Lower Cretaceous ammonites from north-east England: The Hauterivian genus *Simbirskites*. British Museum (Natural History) Bulletin, v. 20(2), p. 25–86.

Regairaz, A. C., 1970, Contribución al conocimiento de las discordancias en el área de las Huayquerias (Mendoza, Argentina). Cuartas Jornadas Geológicas Argentinas, Actas 2, p. 243–254.

Reig, O. A., 1959, Primeros datos descriptivos sobre los anuros del eocretáceo de la Provincia de Salta (República Argentina). Ameghiniana, v. 1(4), p. 3–7.

Reutter, K. J., 1974, Entwicklung und Bauplan der Chilenischen Hochkordillere im Bereich 29° Südlicher Breite. Neues Jahrbuch für Geologie und Paläontologie, Abhandlungen 146 (2), p. 153–178.

Reyes, F. C., 1972, Correlaciones en el Cretácico de la cuenca Andina de Bolivia, Perú, Chile. Yacimientos Petrolíferos Fiscales Bolivianos, Revista Técnica 1(2-3), p. 101–144.

—— , 1974, Consideraciones sobre el Cretácico de la Cuenca Subandina de Bolivia. Revista Brasileira Geociências, v. 4(2), p. 95–113.

Reyes, F. C., and Salfity, J. A., 1973, Consideraciones sobre la estratigrafía del Cretácico (Subgrupo Pirgua) del Noroeste argentino. Quinto Congreso Geológico Argentino, Actas 3, p. 355–385.

Reyes, F. C., Salfity, J. A., Viramonte, J. G., and Gutiérrez, W., 1976, Consideraciones sobre el Vulcanismo del Subgrupo Pirgua (Cretácico) en el Norte Argentino. Sexto Congreso Geológico Argentino, Actas 1, p. 205–223.

Reyes, R., 1970, La Fauna de Trigonias de Aisén. Instituto de Investigaciones Geológicas, Chile, Boletín 26, p. 5–31.

Reyes, R., and Pérez, E., 1978, Las Trigonias del Titoniano y Cretácico inferior de la Cuenca Andina de Chile y su valor cronoestratigráfico. Instituto Investigaciones Geológicas Chile, Boletín 32, p. 1–105.

—— , 1979, Estado actual del conocimiento de la Familia Trigoniidae (Mollusca, Bivalvia) en Chile. Revista Geológica de Chile, v. 8, p. 13–64.

—— , 1985, *Myophorella (Myophorella) hillebrandti* sp. nov. (Trigoniidae; Bivalvia) del Neocomiano, Norte de Chile. Revista Geológica de Chile, v. 24, p. 93–100.

Reyes, R., Pérez, E., and Serey, I., 1981, Estudio sistemático y filogenético de las especies sudamericanas del género *Steinmanella* (Trigoniidae, Bivalvia). Revista Geológica de Chile, v. 12, p. 25–47.

Reyes C., R., 1960, Informe sobre la geología de la zona petrolífera del Noroeste. Instituto Boliviano del Petróleo, Boletín 1(2), p. 9–31.

Reyment, R. A., 1972, Cretaceous history of the South Atlantic. In: Tarling, D. H., and Runcorn, S. K., eds., Implications of Continental Drift to the Earth Sciences, v. 2, p. 805–814.

—— , 1977, Las transgresiones del Cretácico Medio en el Atlántico Sur. Asociación Geológica Argentina, Revista 32 (4), p. 291–299.

—— , 1979, Events of the Mid-Cretaceous in South America. Academia Brasileira de Ciências, Anais 51(3), p. 489–500.

—— , 1980, Paleo-oceanology and paleobiogeography of the Cretaceous South Atlantic Ocean. Oceanologica Acta, v. 3 (1), p. 127–133.

Reyment, R., and Mörner, N. A., 1977, Cretaceous transgressions and regressions exemplified by the South Atlantic. Palaeontological Society of Japan Special Papers, v. 21, p. 247–261.

Reyment, R. A., and Tait, E. A., 1972, Biostratigraphical dating of the early history of the South Atlantic. Royal Society of London Philosophical Transactions, v. B264, p. 55–95.

Ribeiro, M., 1971, Uma província alcalina no Rio Grande do Sul. Iheringia, v. 4, p. 59–71.

Riccardi, A. C., 1968, Estratigrafía de la región oriental de la Bahía de la Lancha, Lago San Martín, Santa Cruz. Universidad Nacional La Plata, Facultad Ciencias Naturales, Ph.D. Thesis (unpublished).

—— , 1970, *Favrella* R. Douvillé, 1909 (Ammonitina, Cretácico inferior): Edad y Distribución. Ameghiniana, v. 7(2), p. 119–138.

—— , 1971, Estratigrafía en el Oriente de la Bahía de la Lancha, lago San Martín, Santa Cruz, Argentina. Museo La Plata, Geología, Revista, v. 7(61), p. 245–318.

—— , 1975, *Eubaculites* Spath (Ammonoidea) del Cretácico superior de Argentina. Ameghiniana, v. 11(4), p. 379–399.

—— , 1976, Paleontología y Edad de la Formación Springhill. Primer Congreso Geológico Chileno, Actas 1, p. C41–C56.

—— , 1977, Berriasian invertebrate fauna from the Springhill Formation of southern Patagonia. Neues Jahrbuch für Geologie und Paläontologie, Abhandlungen 155(2), p. 216–252.

——, 1979, El género *Calycoceras* Hyatt (Ammonitina, Cretácico superior) en Patagonia Austral. Museo La Plata, Obra Centenario, v. 5, p. 63–72.

——, 1983a, The Jurassic of Argentina and Chile. In: Moullade, M., and Nairn, A.E.M., eds., The Phanerozoic Geology of the World II, Mesozoic B8, p. 201–263. Elsevier Scientific Publishers, Amsterdam.

——, 1983b, Kossmaticeratidae (Ammonitina) y Nomenclatura Estratigráfica del Cretácico Tardío en Lago Argentino, Santa Cruz, Argentina. Ameghiniana, v. 20(3-4), p. 317–345.

——, 1984a, Las Asociaciones de Amonitas del Jurásico y Cretácico de la Argentina. Noveno Congreso Geológico Argentino Actas 4, p. 559–595.

——, 1984b, Las Zonas de Amonitas del Cretácico de la Patagonia (Argentina y Chile). III Congreso Latinoamericano de Paleontología (México 1984), Memoria, p. 396–405.

——, 1987, Cretaceous paleogeography of southern South America. Palaeogeography, Palaeoclimatology, Palaeoecology v. 59(1-3), p. 169–195.

Riccardi, A. C., and Rolleri, E. O., 1980, Cordillera Patagónica Austral. In: Segundo Simposio de Geología Regional Argentina, v. 2, p. 1173–1306. Academia Nacional de Ciencias, Córdoba.

——, 1984, Kossmaticeratidae (Ammonitina) Campaniano de la márgen sur del Lago Argentino, Santa Cruz, Argentina. Academia Nacional de Ciencias, Córdoba, Miscelánea 70, p. 1–18.

Riccardi, A. C., Westermann, G.E.G., and Levy, R., 1971, The Lower Cretaceous Ammonitina *Olcostephanus, Leopoldia* and *Favrella* from west-central Argentina. Palaeontographica, v. A136, p. 83–121.

Riccardi, A. C., Aguirre Urreta, M. B., and Medina, F., 1987, Aconeceratidae (Ammonitina) from the Hauterivian-Albian of Southern Patagonia. Palaeontographica, v. A 196, p. 105–185.

Richter, M., 1925, Beiträge zur Kenntnis der Kreide in Feuerland. Neues Jahrbuch für Mineralogie, Geologie und Paläontologie, v. 52, p. 524–568.

Rivano, S., 1980, Cuadrángulos D86, Las Ramadas, Carrizal y Paso Río Negro, Región de Coquimbo. Instituto Investigaciones Geológicas Chile, Carta Geológica 41-44, p. 1–68.

——, 1984, Geología del Meso-Cenozoico entre los 31° y 33° Lat. Sur. In: Seminario Actualización de la Geología de Chile. Servicio Nacional Geología y Mineria, Chile, Miscelánea 4, p. G1–G21.

Rivano, S., Sepúlveda, P., Hervé, M., and Puig, A., 1985, Geocronología K-Ar de las rocas intrusivas entre los 31-32° Latitud Sur, Chile. Revista Geológica de Chile v. 24, p. 63–74.

Rivas, S., and Carrasco, R., 1968, Geología y Yacimientos Minerales de la región de Potosí, 1, Parte Geológica. Servicio Geológico de Bolivia, Boletín 11, p. 1–95.

Rivera, S., 1985, Una discusión sobre la edad de las Formaciones Cerrillos y Hornitos, Región de Atacama. Cuarto Congreso Geológico Chileno, Actas III, p. 489–501.

Robles, D., 1982, El desarrollo de la Formación Springhill en la Cuenca de Magallanes. Primer Congreso Nacional Hidrocarburos, Petróleo y gas, Trabajos Técnicos, p. 293–312. Instituto Argentino del Petróleo, Buenos Aires.

Robles, D. E., 1984, Los Depocentros de la Formación Springhill en el Norte de Tierra del Fuego. Noveno Congreso Geológico Argentino. Actas 1, p. 449–457.

Rodríguez Schelotto, M. L., Orchuela, I., Baliña, M., Blanco Ibañez, S., Ferraresi, P., Ambrosis, E., and Carozzi, A. V., 1981, Medios deposicionales y microfacies de la Formación Quintuco (Berriasiano-Valanginiano) en el Yacimiento Loma La Lata. Provincia de Neuquén. Octavo Congreso Geológico Argentino, Actas 2, p. 503–520.

Rolleri, E. O., and Criado Roque, P., 1968, La Cuenca Triásica del Norte de Mendoza. Terceras Jornadas Geológicas Argentinas, Actas 1, p. 1–76.

——, 1970, Geología de la Provincia de Mendoza. Cuartas Jornadas Geológicas Argentinas, Actas 2, p. 1–46.

Rolleri, E. O., and Fernández Garrasino, C. A., 1979, Comarca Septentrional de Mendoza. In: Segundo Simposio de Geología Regional Argentina, v. 1, p. 773–809. Academia Nacional de Ciencias, Córdoba.

Rolleri, E. O., Dellapé, D. A., and Manceñido, M. O., 1984a, Estudio geológico del valle del Rio Limay entre Piedra del Aguila y el Chocón (Provincias del Neuquén y Río Negro). Noveno Congreso Geológico Argentino, Actas 1, p. 478–497.

Rolleri, E. O., Manceñido, M. O., and Dellapé, D. A., 1984b, Relaciones Estratigráficas y correlación de la Formación Ortiz en el Sur de la Cuenca Neuquina. Noveno Congreso Geológico Argentino, Actas 1, p. 498–523.

Romero, E. J., 1973, Polen fósil de *"Nothofagus" ("Nothofagidites")* del Cretácico y Paleoceno de Patagonia. Museo de la Plata, Paleontología, Revista, n.s., 7(47), p. 291–303.

Romero, E. J., and Arguijo, M. H., 1981, Análisis Bioestratigráfico de las Tafofloras del Cretácico superior de Austrosudamerica. In: Cuencas Sedimentarias del Jurásico y Cretácico de América del Sur, v. 2, p. 393–406.

Roselli, F. L., 1939, Apuntes de paleontología y geología uruguayas y sobre insectos del Cretácico del Uruguay o descubrimientos de admirables instintos constructivos de esa época. Sociedad Amigos Ciencias Naturales "Kraglievich Font.," Boletín 1(2), p. 29–102.

Rossi de García, E., 1979, El género *Novocythere* (Ostrácoda) (Perforación SC-1), Santa Cruz, República Argentina. Ameghiniana, v. 14(1-4), p. 117–128.

Rossi de García, E., and Proserpio, C., 1980, Ostrácodos del Cretácico superior de Patagonia, República Argentina (Hoja 44e, Valle General Racedo), Chubut. Segundo Congreso Argentino de Paleontología y Bioestratigrafía, Actas 2, p. 15–30.

Roxo, M. G. de O., 1936, On a new species of fossil crocodilia from Brazil: *Goniopholis paulistanus* sp. n. Academia Brasileira Ciências, Anais 8(1), p. 33–34.

Rüegg, N. R., 1970, A composição quimica das rochas basálticas da Bacia do Paraná (America do Sul) e de outras provincias basalticas gondwânicas equivalentes. Museum e Laboratorio da Universidade de Coimbra e do Centro de Estudos Geológicos, Memorias e Noticias 70, p. 26–85.

——, 1976, Caracteristicas de Distribuição e Teor de Elementos Principais em Rochas Basálticas de Bacia do Paraná. Universidade São Paulo, Instituto de Geociências, Boletim 7, p. 81–106.

Rüegg, N. R., and Amaral, G., 1976, Variação da Composição quimica das Rochas Basálticas da Bacia do Paraná. Universidade São Paulo, Instituto de Geociências Boletim 7, p. 131–147.

Rüegg, N. R., and Dutra, C. V., 1965, Short note on the trace element content of undifferentiated basaltic rocks of the state of São Paulo, Brazil. Academia Brasileira Ciências, Anais 37(3/4), p. 491–496.

——, 1970, Variation in the content of some trace elements in basaltic rocks from the Paraná Basin. XXIV Congresso Brasileiro de Geologia, Anais, p. 219–226.

Rüegg, N. R., and Vandoros, P., 1965, O diabásio de Laranjal Paulista. Sociedad Brasileira Progresso Ciencia Reuniâo 17(2), p. 128–129. São Paulo.

Ruiz, C., Aguirre, L., Corvalán, J., Rose, H. J., Segerstrom. K., and Stern, T. W., 1961, Ages of batholithic intrusions of northern and central Chile. Geological Society of America Bulletin, v. 72(10), p. 1551–1560.

Ruiz, C., Aguirre, L., Corvalán, J., Klohn, C., Klohn, E., and Levi, B., 1965, Geología y Yacimientos Metalíferos de Chile. Instituto Investigaciones Geológicas Chile, Santiago.

Ruiz, L. C., 1984, Plantas Fósiles Cretácicas procedentes de la Zona del Lago Cardiel, Provincia de Santa Cruz. Noveno Congreso Geológico Argentino, Actas 4, p. 444–454.

Rusconi, C., 1933, Sobre reptiles Cretáceos del Uruguay (Uruguaysuchus Aznarezi, n.g.n.sp.) y sus relaciones con los notosúchidos de Patagonia. Instituto Geología y Perforaciones Uruguay, Boletín 19, p. 3–64.

Russo, A., and Rodrigo, L., 1965, Estratigrafía y Paleogeografía del Grupo Puca en Bolivia. Instituto Boliviano del Petróleo, Boletín 5(3-4), p. 5–51.

Russo, A., Ferello, R., and Chebli, G., 1979, Llanura Chaco Pampeana. In: Segundo Simposio de Geología Regional Argentina, v. 1, p. 139–183. Academia Nacional Ciencias, Córdoba.

Rutland, R.W.R., 1974, Andes: Antofagasta segment (20°-25°S). In: Mesozoic-Cenozoic Orogenic Belts. Geological Society of London Special Publication 4, p. 733–743.

Saavedra, A., 1970, Resumen de la Actividad Ignea en Bolivia. In: Conference on Solid Earth Problems, II: Symposium on the Results of Upper Mantle Investigations with Emphasis on Latin America. International Upper Mantle Project, Scientific Report 37-II, p. 375–380.

Salamuni, R., and Bigarella, J. J., 1967, The Botucatu Formation. In: Bigarella, J. J., Becker, R. D., and Pinto, I. D., eds., Problems in Brazilian Gondwana Geology, p. 197–206. Curitiba, Paraná.

Salas, R., Kast, R. F., Montecinos, F., and Salas, I., 1966, Geología y recursos minerales del Departamento de Arica, Provincia de Tarapacá. Instituto Investigaciones Geológicas Chile, Boletín 21, p. 7–114.

Salfity, J. A., 1979, Paleogeología de la Cuenca del Grupo Salta. Séptimo Congreso Geológico Argentino, Actas 1, p. 505–515.

——, 1980, Estratigrafía de la Formación Lecho (Cretácico) en la Cuenca Andina del Norte Argentino. Universidad Nacional Salta, Publicaciones Especiales, Tésis 1, p. 1–91.

——, 1982, Evolución Paleogeográfica del grupo Salta (Cretácico-Eogénico), Argentina. Quinto Congreso Latinoamericano de Geología (Argentina 1981), Actas 1, p. 11–26.

Salfity, J. A., and Marquillas, R. A., 1981, Las unidades estratigráficas cretácicas del Norte de la Argentina. In: Cuencas Sedimentarias Jurásicas y Cretácicas de América del Sur, v. 1, p. 303–317.

Salfity, J. A., Gorustovich, S. A., and Moya, M. C., 1984, Las Fases Diastróficas en los Andes del Norte Argentino. International Symposium Central-Andean Tectonics and Relations with Natural Resources (La Paz, Bolivia, 1984), p. 1–23.

Salfity, J., Marquillas, R., Gardeweg, M., Ramírez, C., and Davidson, J., 1985, Correlaciones en el Cretácico superior del Norte de Argentina y Chile. Cuarto Congreso Geológico Chileno, Actas 4, p. 654–667.

Salso, J. H., 1966, La Cuenca de Macachín, Provincia de La Pampa, Nota Preliminar. Asociación Geológica Argentina, Revista, v. 21(2), p. 107–117.

Sánchez, T. M., 1973, Redescripción del cráneo y mandíbulas de *Pterodaustro guinazui* Bonaparte (Pterodactyloidea, Pterodaustriidae). Ameghiniana, v. 10(4), p. 313–325.

Sanford, R. M., and Lange, F. W., 1960, Basin study approach to oil evaluation of Paraná miogeosyncline of south Brazil. American Association of Petroleum Geologists Bulletin, v. 44(8), p. 1316–1370.

Sanjines-Saucedo, G., 1982, Estratigrafía del Carbónico, Triásico y Cretácico Boliviano en el borde oriental de las Sierras Subandinas Centrales. Quinto Congreso Latinoamericano de Geología (Argentina 1982), Actas 1, p. 301–318.

Santa Cruz, N., 1980, Edades K-Ar de rocas del área de las cuencas de los ríos Quinto y Conlara, provincia de San Luis. Asociación Geológica Argentina, Revista, v. 35(3), p. 434–435.

——, 1981, Bases Hidrogeológicas de la Provincia de Corrientes. Octavo Congreso Geológico Argentino, Actas 4, p. 231–242.

Sartori, P. L., Filho, C. M., and Menegotto, E., 1975, Contribuição ao Estudo das Rochas Vulcânicas da Bacia do Paraná na Região de Santa María, RS. Revista Brasileira de Geociências, v. 5(3), p. 141–159.

Saunders, A. D., and Tarney, J., 1982, Igneous activity in the southern Andes and northern Antarctic Peninsula: A review. Journal of the Geological Society of London, v. 139, p. 691–700.

Schaeffer, B., 1963, Cretaceous fishes from Bolivia with comments on Pristid evolution. American Museum Novitates 2159, p. 1–20.

Schiller, W., 1922, Los sedimentos marinos del límite entre el Cretáceo y Terciario de Roca en la Patagonia septentrional. Museo La Plata, Revista 26, p. 256–280.

Schlagintweit, O., 1941, Correlación de las calizas de Miraflores en Bolivia con el Horizonte calcáreo-dolomítico del Norte Argentino. Museo La Plata, Geología, Notas 6(14), p. 337–354.

Schlatter, L. E., and Nederlof, M. H., 1966, Bosquejo de la Geología y Paleogeografía de Bolivia. Servicio Geológico Bolivia, Boletín 8, p. 1–49.

Schneider, A. W., 1964, Contribuição a petrologia dos derrames basálticos da bacia do Paraná. Escola Engenharia, Universidade Rio Grande do Sul, Publicações Avulsas 1, p. 1–76.

——, 1970, Vulcanismo Basaltico da Bacia do Paraná: Perfil Foz do Iguaçu-Serra da Esperança. XXIV Congresso Brasileiro Geologia, Anais, p. 211–217.

Schneider, R. L., Mühlmann, H., Tommasi, E., Madeiros, R. A., Daemon, R. F., and Nogueira, A. A., 1974, Revisão Estratigráfica da Bacia do Paraná. XXVIII Congresso Brasileiro Geologia, Anais, p. 41–65.

Schobbenhaus, C., Almeida, D. de, Derze, G. R., and Asmus, H. E., 1984, Geologia do Brasil. Ministerio das Minas e Energia, Brasilia.

Schultze, H. P., 1981, A Pycnodont dentition (*Paramicrodon volcanensis* n. sp.; Pisces, Actinopterygii) from the Lower Cretaceous of El Volcan region, southeast of Santiago, Chile, Revista Geológica Chile, v. 12, p. 87–93.

Schwab, K., 1973, Die Stratigraphie in der Umgebung des Salar de Cauchari (NW-Argentinien). Geotektonische Forschungen, v. 43(1-2), p. 1–168.

——, 1984, Contribución al conocimiento del sector occidental de la Cuenca sedimentaria del grupo Salta (Cretácico-Eogénico), en el Noroeste Argentino. Noveno Congreso Geológico Argentino, Actas 1, p. 586–604.

Sciutto, J. C., 1981, Geología del Codo del Río Senguerr, Chubut, Argentina. Octavo Congreso Geológico Argentino, Actas 3, p. 203–219.

Sclater, J. G., Hellinger, S., and Tapscott, C., 1977, The Paleobathymetry of the Atlantic Ocean from the Jurassic to the Present. Journal of Geology, v. 85, p. 509–552.

Scorza, E. P., 1952, Considerações sobre o Arenito Caiuá. Departamento Nacional Produção Mineral, Divisão Geologia e Mineralogia, Boletim 139, p. 1–62.

Scott, K. M., 1966, Sedimentology and dispersal pattern of a Cretaceous flysch sequence, Patagonian Andes, southern Chile. American Association of Petroleum Geologists Bulletin, v. 50(1), p. 72–107.

Secretan, S., 1972, Crustacés Décapodes nouveaux du Crétacé supérieur de Bolivie. Museum National d'Histoire Naturelle, Paris, Bulletin (3) 49 (7), p. 1–15.

Segerstrom, K., 1959, Cuadrángulo Los Loros, Provincia de Atacama. Instituto Investigaciones Geológicas Chile, Carta Geológica 1(1), p. 5–33.

——, 1960, Structural Geology of an Area East of Copiapó, Atacama Province, Chile. XXI International Geological Congress 18, p. 14–20.

——, 1961, Facies change in Neocomian rocks of the Teresita Chula area, Atacama Province, Chile. U.S. Geological Survey Research 1961, p. 219–223.

——, 1962, Regional geology of the Chañarcillo Silver Mining District and adjacent areas, Chile. Economic Geology, v. 57(8), p. 1247–1261.

——, 1967, Geology and ore deposits of central Atacama Province, Chile. Geological Society of America Bulletin, v. 78(3), p. 305–318.

——, 1968, Geología de las Hojas Copiapó y Ojos del Salado, Provincia de Atacama. Instituto Investigaciones Geológicas Chile, Boletín 24, p. 1–58.

Segerstrom, K., and Moraga, A. B., 1964, Cuadrángulo Chañarcillo, Provincia de Atacama. Instituto Investigaciones Geológicas Chile, Carta Geológica 13, 49 p.

Segerstrom, K., and Parker, R. L., 1959, Cuadrángulo Cerrilos. Instituto Investigaciones Geológicas Chile, Carta Geológica 1(2), p. 1–32.

Segerstrom, K., and Ruiz, C., 1962, Cuadrángulo Copiapó, Provincia de Atacama. Instituto Investigaciones Geológicas Chile, Carta Geológica 3(1), p. 1–112.

Segerstrom, K., Thomas, H., and Tilling, R. I., 1963, Cuadrángulo Pintadas, Provincia de Atacama. Instituto Investigaciones Geológicas Chile, Carta Geológica 12, p. 1–51.

Seiler, J., 1979, Paleomicroplancton del Cretácico inferior en el Subsuelo del Sudoeste de la Provincia del Chubut, Ameghiniana, v. 16(1-2), p. 183–190.

Seiler, J. and Moroni, A. M., 1984, Zonación Palinológica del Subsuelo en el oeste del golfo San Jorge. Correlación con Pozos de la misma zona. Tercer Congreso Argentino de Paleontología y Bioestratigrafía (Corrientes 1982), Actas, p. 115–121.

Sempere, Th., 1986, Estratigrafía secuencial del Mesozoico andino boliviano. Cretácico de América Latina, IGCP 242, Primer Simposio (La Paz 1986), p. 72–73.

Sepúlveda, E. G., and Viera, R. M., 1980, Geología y área de alteración en el Cerro Colorado y alrededores. Chubut Noroccidental. Asociación Geológica Argentina, Revista, v. 35(2), p. 195–202.

Sepúlveda, P., and Naranjo, J. A., 1982, Hoja Carrera Pinto, Región de Atacama. Carta Geológica de Chile (Escala 1:100,000) 53, p. 1–60.

Serra, N., 1945, Memoria Explicativa del mapa geológico del Departamento de Soriano. Instituto Geológico Uruguay, Boletín 32, p. 1–42.

Sigal, J., Grekoff, N., Singh, N. P., Cañón, A., and Ernst, M., 1970, Sur l'âge et les affinités "gondwaniennes" de microfaunes (Foraminiféres et Ostracodes) malgaches, indiennes, et chiliennes au sommet du Jurassique et à la base du Crétacé. Académie des Sciences Paris, Comptes Rendus (D) 271, p. 24–27.

Sillitoe, R., 1977, Permo-Carboniferous, Upper Cretaceous and Miocene porphyry copper-type mineralization in the Argentinian Andes. Economic Geology, v. 72(1), p. 99–103.

Simeoni, M., 1985, Foraminíferos del Cretácico inferior en los niveles basales de la Formación Agrio, Perfil el Marucho, Neuquén, Argentina. Ameghiniana, v. 21(2-4), p. 285–293.

Simeoni, M., and Musacchio, E. A., 1986, Ostrácodos no marinos y Carofitos hauterivianos de la Formación Agrio en la localidad El Marucho, Cuenca de Neuquén, Argentina. Ameghiniana, v. 23(1-2), p. 89–96.

Skarmeta, J., 1978, Región Continental de Aysén entre el lago General Carrera y la Cordillera Castillo. Instituto Investigaciones Geológicas Chile, Carta Geológica 29, p. 1–53.

Skarmeta, J., and Charrier, R., 1976, Geología del Sector fronterizo de Aysén entre los 45° y 46° de Latitud Sur. Sexto Congreso Geológico Argentino, Actas 1, p. 267–286.

Skarmeta, J., and Marinovic, N., 1981, Hoja Quillagua, Región de Antofagasta. Carta Geológica de Chile (Escala 1:250,000) 51, p. 1–63.

Sloss, L. L., Krumbein, W. C., and Dapples, E. C., 1949, Integrated facies analysis. Geological Society of America Memoir 39, p. 91–124.

Soares, P. C., 1975, Divisão Estratigráfica do Mesozóico no Estado de São Paulo. Revista Brasileira Geociências, v. 5(4), p. 229–251.

——, 1981, Estratigrafia das Formações Jurássico-Cretáceas na Bacia do Paraná-Brasil. In: Cuencas Sedimentarias del Jurásico y Cretácico de América del Sur, v. 1, p. 271–304.

Soares, P. C., and Landim, P.M.B., 1976, Comparison between the tectonic evolution of the intracratonic and marginal basins in south Brazil. In: International Symposium on Continental Margins of Atlantic Type. Academia Brasileira Ciências, Anais 48 (Suppl.), p. 313–323.

Soares, P. C., Landim, P.M.B., and Fúlfaro, V. J., 1978, Tectonic cycles and sedimentary sequences in the Brazilian intracratonic basins. Geological Society of America Bulletin, v. 89(2), p. 181–191.

Soares, P. C., Landim, P.M.B., Fúlfaro, V. J., and Neto, A.F.S., 1980, Ensaio de caracterização estratigráfica do Cretáceo no Estado de São Paulo: Grupo Bauru. Revista Brasileira Geociências, v. 10(3), p. 177–185.

Sokolov, D. N., 1946, Algunos fósiles Suprajurásicos de la República Argentina. Asociación Geológica Argentina, Revista, v. 1(1), p. 7–16.

Sonnenberg, F. P., 1963, Bolivia and the Andes. In: Backbone of the Americas, A Symposium. American Association of Petroleum Geologists Memoir 2, p. 36–46.

Souza, A., Sinelli, O., and Gonçalves, N.M.M., 1971, Nova ocorrência fossilifera na Formação Botucatu. XXV Congresso Brasileiro Geologia, Anais 2, p. 281–295.

Spencer, F. N., 1950, The geology of the Aguilar lead-zinc mine, Argentina. Economic Geology, v. 45(5), p. 405–433.

Sprechmann, P., Bossi, J., and Da Silva, J., 1981, Cuencas del Jurásico y Cretacico del Uruguay. In: Cuencas Sedimentarias del Jurasico y Cretácico de América del Sur, v. 1, p. 239–270.

Stach, A. K. de, and Angelozzi, G., 1984, Microfósiles calcáreos de la Formación Yacoraite en la Subcuenca Lomas de Olmedo, Provincia de Salta. Noveno Congreso Geológico Argentino. Actas 4, p. 508–522.

Staesche, K., 1937, *Podocnemis brasiliensis* n. sp. aus der oberen Kreide Brasiliens. Neues Jahrbuch für Mineralogie, Geologie und Paläontologie, v. 77, p. 291.

——, 1944, Una Tortuga do Cretáceo superior do Brasil. Departamento Nacional Produção Mineral, Divisão Geologia e Mineralogia, Boletim 114, p. 1–24.

Stanton, T. W., 1901, The Marine Cretaceous Invertebrates. Princeton University Expedition to Patagonia 1896–1899, Reports, v. 4(1), p. 1–43.

Steinmann, G., 1883, Reisenotizen aus Patagonien. Neues Jahrbuch für Mineralogie, Geologie und Paläontologie, v. 2, p. 255–288.

——, 1895, Die Cephalopoden der Quiriquina-Schichten. Neues Jahrbuch für Mineralogie, Geologie und Paläontologie, v. 10, p. 64–94.

——, 1906, Die Entstehung der Kupfererzlagerstätte von Corocoro und verwandter Vorkommnisse in Bolivia. In: Festschrift 70. Geb. H. Rosenbusch, p. 335–368. E. Schweizerbartsche Verb., Stuttgart.

——, 1929, Geologie von Peru. C. Winters, Heidelberg.

Steinmann, G., and Wilckens, O., 1908, Kreide und Tertiärfossilien aus den Magellansländern, gesammelt von der schwedischen Expedition 1895–1897. Svenska vetenskapakademien, Stockholm, Arkiv for Zoologi 4, 6, p. 1–103.

Steinmann, G., Deecke, W., and Möricke, W., 1895, Das Alter und die Fauna der Quiriquina-Schichten in Chile. Neues Jahrbuch für Mineralogie, Geologie und Paläontologie, v. 10, p. 1–118.

Stern, C. R., and Stroup, J. B., 1982, The petrochemistry of the Patagonian Batholith, Ultima Esperanza, Chile. In: Craddock, C., ed., Antarctic Geoscience, p. 135–142. University of Wisconsin Press, Madison.

Steuer, A., 1897, Beiträge zur Kenntniss der Geologie und Paläontologie der Argentinischen Anden. Palaeontologische Abhandlungen, N.F., v. 3, p. 127–222.

Stinnesbeck, W., 1986, Zur den Faunistischen und Palökologischen verhältnissen in der Quiriquina Formation (Maastrichtian), Zentral-Chiles. Palaeontographica, v. A194(4-6), p. 99–237.

Stipanicic, P. N., 1957, El Sistema Triásico en la Argentina. XX International Geological Congress 2, p. 73–111.

——, 1967, Consideraciones sobre las edades de algunas fases magmáticas del Neopaleozoico y Mesozoico. Asociación Geológica Argentina, Revista, v. 22(2), p. 101–133.

——, 1983, The Triassic of Argentina and Chile. In: Moullade, M., and Nairn, A.E.M., eds., The Phanerozoic Geology of the World II, The Mesozoic B, p. 181–199. Elsevier Scientific Publishers, Amsterdam.

Stipanicic, P. N., and Bonaparte, J. F., 1972, Cuenca triásica de Ischigualasto-Villa Unión (Provincias de San Juan y La Rioja). In: Leanza, A. F., ed., Geología Regional Argentina, p. 507–536. Academia Nacional Ciencias, Córdoba.

——, 1979, Cuenca triásica de Ischigualasto-Villa Unión (Provincias de La Rioja y San Juan). In: Segundo Simposio de Geología Regional Argentina, v. 1, p. 523–575. Academia Nacional Ciencias, Córdoba.

Stipanicic, P. N., and Linares, E., 1969, Edades Radimétricas determinadas para la República Argentina y su significado geológico. Academia Nacional Ciencias, Córdoba, Boletín 47(1), p. 51–96.

——, 1975, Catálogo de Edades Radimétricas determinadas para la República Argentina I, Años 1960–1974. Asociación Geológica Argentina, Publicación Especial B3, p. 1–42.

Stipanicic, P. N., and Methol, E. J., 1972, Macizo de Somuncurá. In: Leanza, A. F., ed., Geología Regional Argentina, p. 581–599. Academia Nacional Ciencias, Córdoba.

Stipanicic, P. N., and Rodrigo, F., 1970, Diastrofismo Eo- y Mesocretácico en Argentina y Chile, con referencia a los movimientos Jurásicos de la Patagonia. Cuartas Jornadas Geológicas Argentinas, Actas 2, p. 337–352.

Stipanicic, P. N., Rodrigo, F., Baulies, O. L., and Martínez, C. G., 1968, Las Formaciones Presenonianas en el denominado Macizo Nordpatagónico y regiones adyacentes. Asociación Geológica Argentina, Revista, v. 23(2), p. 67–98.

Stolley, E., 1912, Uber einige Cephalopoden aus der Unteren Kreide Patagoniens. Svenska vetenskaps-akademien, Stockholm, Arkiv for Zoologi 7(23), p. 13–18.

——, 1928, "Helicerus" Dana als Erhaltungszustand von "Belemnopsis." Neues Jahrbuch für Mineralogie, Geologie, und Paläontologie, v. 60, p. 315–323.

Suárez, J. M., 1969, Um quelônio da formação Bauru. Departamento Geografia, Faculdade Filosofia, Ciências e Letras, Presidente Prudente, Boletim 2, p. 35–54.

Suárez, M., 1978, Región al sur del Canal Beagle. Instituto Investigaciones Geológicas Chile, Carta Geológica 36, p. 1–48.

——— . 1979a, Evolución Geológica de los Andes del sur, Arco de Scotia y Península Antártica: una síntesis. Estudios Geológicos 35, p. 473–485. Madrid.

——— , 1979b, A Late Mesozoic island arc in the southern Andes, Chile. Geological Magazine, v. 116(3), p. 167–179.

Suárez, M., and Pettigrew, T. H., 1976, An Upper Mesozoic island-arc–back-arc system in the southern Andes and south Georgia. Geological Magazine 113(4), p. 305–400.

Suárez, M., Puig, A., and Hervé, M., 1985a, Depósitos de un abanico submarino de Trans-arco del mesozoico superior: Formación Yahgan, Islas Hoste y Navarino. Cuarto Congreso Geológico Chileno, Actas 1(1), p. 534–545.

Suárez, M., Hervé, M., and Puig, A., 1985b, Plutonismo diapírico del Cretácico en Isla Navarino. Cuarto Congreso Geológico Chileno, Actas 3, p. 549–563.

Suárez, M., Puig, A., and Hervé, M., 1986. K-Ar dates on granitoids from Archipiélago Cabo de Hornos, southernmost Chile. Geological Magazine, v. 123(5), p. 581–584.

Suguio, K., Coimbra, A. M., and Guardado, L. R., 1974, Correlação sedimentológica de arenitos da Bacia do Paraná. Universidade São Paulo, Instituto de Geociências, Boletim 5, p. 85–116.

Suguio, K., Berenholc, M., and Salati, E., 1975, Composição Quimica e isotópica dos Calcarios e Ambiente de sedimentação da Formação Bauru. Universidade Sao Paulo, Instituto de Geociências, Boletim 6, p. 55–75.

Szubert, E. C., 1979, Esquema Interpretativo da evolução Geológica das rochas vulcânicas mesozoicas da Bacía do Paraná. Acta Geologica Leopoldensia, v. 6, p. 113–124.

Takahashi, K., 1978, Upper Cretaceous Palynofossils from Quiriquina Island, Chile. Journal of Palynology, v. 14(1), p. 30–49.

——— , 1979, Phytoplankton from the Upper Cretaceous Quiriquina Formation, central Chile. Nagasaki Daigaku Faculty of Liberal Arts and Education Science Bulletin 19, p. 31–37.

Tasch, P., and Volkheimer, W., 1970, Jurassic Conchostracans from Patagonia. University of Kansas Paleontological Contributions, Paper 50, p. 1–23.

Tavera, J., 1956, Fauna del Cretáceo inferior de Copiapó. Universidad de Chile, Publicación 9, p. 205–216.

Teruggi, M. E., 1955, Los basaltos tholeíticos de Misiones. Museo La Plata, Geología, Notas 18(70), p. 257–278.

Thiede, J., and van Andel, T., 1977, The paleoenvironment of anaerobic sediments in the Late Mesozoic south Atlantic Ocean. Earth and Planetary Science Letters, v. 33, p. 301–309.

Thiele, R., 1964, Reconocimiento Geológico de la Alta Cordillera de Elqui. Universidad de Chile, Departamento de Geología, Publicación 27, p. 135–197.

——— , 1980, Hoja Santiago, Región Metropolitana. Instituto Investigaciones Geológicas Chile, Carta Geológica 39, p. 1–51.

Thiele, R., and Hein, R., 1979, Posición y Evolución tectónica de los Andes Nord-Patagónicos. Segundo Congreso Geológico Chileno, Actas 1, p. B33–B46.

Thiele, R., and Nasi, C., 1982, Evolución Tectónica de los Andes a la latitud 33° a 34° Sur (Chile Central) durante el Mesozoico-Cenozoico. Quinto Congreso Latinoamericano de Geología (Argentina, 1982), Actas 3, p. 403–426.

Thiele, R., Castillo, J. C., Hein, R., Romero, G., and Ulloa, M., 1979, Geología del sector fronterizo de Chile Continental entre los 43°00'–43°45' latitud sur (Comunas de Futaleufú y de Palena). Séptimo Congreso Geológico Argentino, Actas 1, p. 577–591.

Thiele, R., Bobenrieth, L., and Borič, R., 1980, Geología de los Cerros Renca, Ruiz y Colorado (Santiago): Contribución a la Estratigrafía de Chile Central. Universidad de Chile, Departamento de Geología, Comunicaciones 30, p. 1–14.

Thieuloy, J. P., 1977, Les ammonites boréales des formations néocomiennes du Sud-Est français, Province subméditerranée). Geobios, v. 10, p. 395–442.

Thomas, A., 1967, Cuadrángulo Mamiña, Provincia de Tarapacá. Instituto Investigaciones Geológicas Chile, Carta Geológica 17, p. 5–49.

——— , 1970, Cuadrángulos Iquique y Caleta Molle, Provincia de Tarapacá. Insti-

tuto Investigaciones Geológicas Chile, Carta Geológica 21-22, p. 5–52.

Thomas, C. R., 1949, Manantiales field, Magallanes Province, Chile. American Association of Petroleum Geologists Bulletin, v. 33(9), p. 1579–1589.

Thomas, H., 1958, Geología de la Cordillera de la Costa entre el Valle de La Ligua y la Cuesta de Barriga. Instituto Investigaciones Geológicas Chile Boletín 2, p. 5–86.

——— , 1967, Geología de la Hoja Ovalle, Provincia de Coquimbo. Instituto Investigaciones Geológicas Chile, Boletín 23, p. 5–58.

Thomson, M.R.A., 1981, Mesozoic ammonite faunas of Antarctica and the break-up of Gondwana. In: Cresswell, M. M., and Vela, P., eds., Gondwana Five, p. 269–275. A. A. Balkema, Rotterdam.

Thorpe, R. S., Francis, P. W., and Harmon, R. S., 1980, Andean Andesites and Crustal Growth. Revista Geológica de Chile, v. 10, p. 55–73.

Tilling, R. I., 1976, El Batolito Andino cerca de Copiapó, Provincia de Atacama, Geología y Petrología. Revista Geológica de Chile, v. 3, p. 1–24.

Tobar, A., Salas, I., and Kast, R. F., 1968, Cuadrángulos Camaraca y Azapa. Instituto Investigaciones Geológicas Chile, Carta Geológica 19-20, p. 5–20.

Torres, M. A., 1985, Estratigrafía de la ladera occidental del Cerro Amarillo y Quebrada de la Yesera, Departamento de Cafayate, Salta. Asociación Geológica Argentina, Revista, v. 40(3–4), p. 141–157.

Torres, T., 1982, Hallazgo de *Palmoxylon chilensis* n. sp., del Cretácico superior en Huechun, Región Metropolitana. Tercer Congreso Geológico Chileno, Actas 1, p. A302–A320.

Torres, T., and Rallo, M., 1981, Anatomía de troncos fósiles del Cretácico superior de Pichasca, en el norte de Chile. Segundo Congresso Latino-Americano de Paleontología, Anais, 1, p. 385–398.

Toubes, R. O., and Spikermann, P., 1974, Algunas edades K/Ar y Rb/Sr de Plutonitas de la Cordillera Patagónica entre los paralelos 40° y 44° de Latitud Sur. Asociación Geológica Argentina, Revista, v. 28(4), p. 382–396.

Traverso, N. E., 1966, *Brachyphyllum tigrense,* nueva conífera de la Formación Baqueró, Cretácico de Santa Cruz. Ameghiniana, v. 4(6), p. 189–194.

——— , 1968, *Brachyphyllum baqueroense,* otra nueva conífera de la Formación Baqueró, Cretácico de Santa Cruz. Ameghiniana, v. 5(10), p. 374–8.

Troncoso, A., and Doubinger, J., 1980, Dinoquistes (Dynophyceae) del límite Cretácico-Terciario del Pozo El Ganso No. 1 (Magallanes, Chile). Segundo Congreso Argentino de Paleontología y Bioestratigrafía, v. 2, p. 93–120.

Turner, J.C.M., 1959, Estratigrafía del cordón de Escaya y de la Sierra de Rinconada (Jujuy). Asociación Geológica Argentina, Revista, v. 13(1), p. 15–39.

——— , 1964, Descripción Geológica de la Hoja 15c, Vinchina (Provincia de La Rioja). Dirección Nacional de Geología y Minería, Boletín 100, p. 7–81. Buenos Aires.

——— , 1965, Estratigrafía de Aluminé y adyacencias (Provincia del Neuquén). Asociación Geológica Argentina, Revista, v. 20(2), p. 153–184.

——— , 1970, The Andes of northwestern Argentina. Geologische Rundschau, v. 59(3), p. 1028–1063.

——— , 1979, Perfil geológico entre los ríos Chubut y Tecka (Provincia del Chubut). Museo Argentino de Ciencias Naturales, Geología, Revista 8(3), p. 71–93.

Turner, J.C.M., and Méndez, V., 1979, Puna. In: Segundo Simposio Geología Regional Argentina, v. 1, p. 13–56. Academia Nacional Ciencias, Córdoba.

Turner, J.C.M., and Mon, R., 1979, Cordillera Oriental. In: Segundo Simposio Geología Regional Argentina, v. 1, p. 57–94. Academia Nacional Ciencias, Córdoba.

Turner, J.C.M., and Salfity, J. A., 1977, Perfil Geológico Humahuaca-Pueblo Abra Laite (Jujuy, República Argentina). Asociación Geológica Argentina, Revista, v. 32, p. 111–121.

Turner, J.C.M., Méndez, V., and Lurgo, C. S., 1979, Geología de la región noroeste, Provincias de Salta y Jujuy, República Argentina. Séptimo Congreso Geológico Argentino, Actas 1, p. 367–387.

Uhlig, V., 1911, Die marinen Reiche des Jura und der Unterkreide. Geologische Gesellschaft Wien, Mitteilangungen 4, p. 329–448.

Uliana, M. A., and Biddle, K. T., 1987, Permian to Late Cenozoic Evolution of Northern Patagonia: Main Tectonic Events, Magmatic Activity, and Depositional Trends. American Geophysical Union, Geophysical Monograph 40,

p. 271–286.

Uliana, M. A., and Dellapé, D. A., 1981, Estratigrafía y Evolución Paleoambiental de la Sucesión Maestrichtiano-Eoterciaria del Engolfamiento Neuquino (Patagonia Septentrional). Octavo Congreso Geológico Argentino, Actas 3, p. 673–711.

Uliana, M. A., and Musacchio, E. A., 1979, Microfósiles calcáreos no-marinos del Cretácico superior en Zampal, Provincia de Mendoza, Argentina. Ameghiniana, v. 15(1-2), p. 111–135.

Uliana, M. A., Dellapé, D. A., and Pando, G. A., 1975a, Distribución y génesis de las sedimentitas Rayosianas (Cretácico inferior de las Provincias de Neuquén y Mendoza, República Argentina). Segundo Congreso Ibero-Americano de Geología Económica, Actas 1, p. 151–176.

——, 1975b, Estratigrafía de las sedimentitas Rayosianas (Cretácico inferior de las Provincias de Neuquén y Mendoza). Segundo Congreso Ibero-Americano de Geología Económica, Actas 1, p. 177–196.

——, 1977, Análisis estratigráfico y Evaluación del Potencial Petrolífero de las Formaciones Mulichinco, Chachao y Agrio, Cretácico inferior de las Provincias de Neuquén y Mendoza. Petrotecnia, v. 16(2), p. 31–46; (3), p. 25–33.

Umpierre, M., and Halpern, M., 1971, Edades Estroncio-Rubidio en rocas cristalinas del sur de la República Oriental de Uruguay. Asociación Geológica Argentina, Revista, v. 26(2), p. 133–151.

Urien, C. M., Zambrano, J. J., 1973, The geology of the basins of the Argentine continental margin and Malvinas Plateau. In: Nairn, A.E.M., and Stehli, F. G., eds., The Ocean Basins and Margins, v. 1: The South Atlantic, p. 135–169. Plenum Publishing Corporation, New York.

Urien, C. M., Martins, L. R., and Zambrano, J. J., 1975, The geology and tectonic framework of southern Brazil, Uruguay and northern Argentina continental margin: Their behavior during the southern Atlantic opening. In: International Symposium on Continental Margins of Atlantic Type. Academia Brasileira Ciências, Anais 48 (Suppl.), p. 365–376.

Urien, C. M., Zambrano, J. J., and Martins, L. R., 1981, The basins of southeastern South America (southern Brazil, Uruguay and eastern Argentina) including the Malvinas Plateau and southern South Atlantic paleogeographic evolution. In: Cuencas Sedimentarias del Jurásico y Cretácico de América del Sur, v. 1, p. 45–125.

Vail, P. R., Mitchum, R. M., and Thompson, S., 1977, Seismic stratigraphy and global changes of sea level, Pt. 4: Global Cycles of Relative Changes of Sea Level. American Association of Petroleum Geologists Memoir 26, p. 83–97.

Valencio, D. A., 1972, Paleomagnetism of the Lower Cretaceous volcanites Cerro Colorado Formation of the Sierra de los Cóndores Group, Province of Córdoba, Argentina. Earth and Planetary Science Letters, v. 16(3), p. 370–378.

Valencio, D. A., Giudici, A., Mendía, J. E., and Gascón, J. O., 1976, Paleomagnetismo y edades K-Ar del Subgrupo Pirgua, Provincia de Salta, Argentina. Sexto Congreso Geológico Argentino, Actas 1, p. 527–537.

Valencio, D. A., Mendía, J. E., Giudici, A., and Gascón, J. O., 1977, Palaeomagnetism of the Cretaceous Pirgua Subgroup (Argentina) and the age of the opening of the South Atlantic. Royal Astronomical Society Geophysical Journal, v. 51(1), p. 47–58.

Vandoros, P., Rüegg, N. R., and Cordani, U. G., 1966, On potassium-argon age measurements of basaltic rocks from southern Brazil. Earth and Planetary Science Letters, v. 1(6), p. 449–452.

Vargas, E., 1970, Estudio geológico del área Llallagua. Servicio Geológico Bolivia, Boletín 12, p. 11–52.

Vergara, H., 1978a, Cuadrángulo Quehuita y Sector Occidental del Cuadrángulo Volcan Miño, Región de Tarapacá. Instituto Investigaciones Geológicas Chile, Carta Geológica 32, p. 1–44.

——, 1978b, Cuadrángulo Ujina, Región de Tarapacá. Instituto Investigaciones Geológicas Chile, Carta Geológica 33, p. 1–63.

Vergara, M., 1969, Rocas volcánicas y sedimentario-volcánicas mesozoicas y cenozoicas en la latitud 34°30′S. Universidad de Chile, Departamento de Geología, Publicación 32, p. 5–36.

Vergara, M., and Drake, R., 1978, Edades potasio-argón y su implicancia en la geología regional de Chile. Universidad de Chile, Departamento de Geología, Comunicaciones 23, p. 1–11.

——, 1979a, Edades K/Ar en secuencias volcánicas continentales Postneocomianas de Chile Central; su depositación en cuencas intermontanas restringidas. Asociación Geológica Argentina, Revista, v. 34(1), p. 42–52.

——, 1979b, Eventos Magmáticos—Plutónicos en los Andes de Chile Central. Segundo Congreso Geológico Chileno, Actas 1, p. F19–F30.

Vicente, J. C., Charrier, R., Davidson, J., Mpodozis, A., and Rivano, S., 1973, La Orogénesis Subhercínica: Fase Mayor de la Evolución Paleogeográfica y Estructural de los Andes Argentino-Chilenos Centrales. Quinto Congreso Geológico Argentino, Actas 5, p. 81–98.

Vilas, J. E., 1976, Palaeomagnetism of the Lower Cretaceous Sierra de los Cóndores Group, Province of Córdoba, Argentina. Royal Astronomical Society Geophysical Journal, v. 46, p. 295–305.

Vilela, C. R., 1951, Acerca del hallazgo del Horizonte calcáreo dolomítico en la Puna salto-jujeña y su significado geológico. Asociación Geológica Argentina, Revista, v. 6(2), p. 101–107.

Vilela, C. R., and Csaky, A., 1968, Las turbiditas en los sedimentos cretácicos de la región de lago Argentino (Provincia de Santa Cruz). Terceras Jornadas Geológicas Argentinas, Actas 1, p. 209–225.

Vistelius, A. B., Ivanov, D. N., Kuroda, Y., and Ruiz, C., 1970, Variations of modal composition of granitic rocks in some regions around the Pacific. Mathematic Geology, v. 2(1), p. 63–80.

Volkheimer, W., 1969, Palaeoclimatic evolution in Argentina and relations with other regions of Gondwana. Gondwana Stratigraphy, IUGS Symposium, Proceedings, p. 551–587. Buenos Aires.

——, 1970, Neuere Ergebnisse der Anden Stratigraphie von Süd-Mendoza (Argentinien) und benachbarter Gebiete und Bemerkungen zur Klimageschichte des Südlichen Andenraums. Geologischen Rundschau, v. 59(3), p. 1088–1124.

——, 1973, Observaciones geológicas en el área de Ingeniero Jacobacci y adyacencias (Provincia de Río Negro). Asociación Geológica Argentina, Revista, v. 28(1), p. 13–36.

——, 1978, Microfloras Fósiles. Relatorio Geología y Recursos Naturales del Neuquén. Séptimo Congreso Geológico Argentino, p. 193–207.

Volkheimer, W., and Baldis, D. P., de, 1975, Significado Estratigráfico de Microfloras Paleozoicas y Mesozoicas de la Argentina y países vecinos. Segundo Congreso Ibero-Americano de Geología Económica, v. 4, p. 403–424.

Volkheimer, W., and Ott, E., 1973, Esponjas de Agua Dulce del Cretácico de la Patagonia. Nuevos Datos acerca de su posición sistemática y su importancia paleobiogeográfica y paleoclimatológica. Quinto Congreso Geológico Argentino, Actas 3, p. 455–461.

Volkheimer, W., and Prámparo, M. B., 1984, Datos palinológicos del Cretácico inferior en el borde austral de la Cuenca Neuquina, localidad Estancia Santa Elena, Argentina. Parte I: Especies Terrestres. Tercer Congreso Latinoamericano de Paleontología, Memoria, p. 269–279.

Volkheimer, W., and Quattrocchio, M. E., 1981, Distribución Estratigráfica de los Palinomorfos Jurásicos y Cretácicos en la Faja Andina y Areas adyacentes de América del Sur, con especial consideración de la Cuenca Neuquina. In: Cuencas Sedimentarias del Jurásico y Cretácico de América del Sur, v. 2, p. 407–444.

Volkheimer, W., and Salas, A., 1975, Die Alteste Angiosperm-Palynoflora Argentiniens von der Typus-Lokalität der unterkretazieschen Huitrin-Folge des Neuquén-Beckens. Ihre mikrofloristische Association und biostratigraphische Bedeutung. Neues Jahrbuch für Mineralogie, Geologie und Paläontologie, Monatshefte 7, p. 424–436.

——, 1976. Estudio palinológico de la Formación Huitrín, Cretácico de la Cuence Neuquima, en su localidad tipo. Sexto Congreso Geológico Argentino, Actas 1, p. 433–456.

Volkheimer, W., and Sepúlveda, E., 1976, Biostratigraphische Bedeutung und microfloristische Assoziation von *Cyclusphaera psilata* n. sp., einer leitform aus der Unterkreide des Neuquén Beckens (Argentinien). Neues Jahrbuch für Geologie und Palaeontologie, Monatshefte 1976 (2), p. 97–108.

Volkheimer, W., Caccavari de Filice, M. A., and Sepúlveda, E., 1977, Datos palinológicos de la Formación Ortiz (Grupo La Amarga), Cretácico inferior de la Cuenca Neuquina (República Argentina). Ameghiniana, v. 14(1-4), p. 59–74.

Volkheimer, W., Quattrocchio, M., and Salfity, J., 1984, Datos Palinológicos de la Formación Maíz Gordo, Terciario inferior de la Cuenca de Salta. Noveno Congreso Geológico Argentino, Actas 4, p. 523–538.

Walker, C. A., 1981, New subclass of birds from the Cretaceous of South America. Nature, v. 292, p. 51–53.

Walther, K., 1919, Líneas Fundamentales de la estructura geológica de la República Oriental del Uruguay. Instituto Nacional Agronomía, Revista (2) 3, p. 1–186.

—— , 1927, Contribución al conocimiento de las rocas "basálticas" de la formación de Gondwana en la América del Sur. Instituto Geología y Perforaciones Uruguay, Boletín 9, p. 1–43.

—— , 1930, Sedimentos gelíticos y clastogelíticos del cretáceo superior y Terciario uruguayos. Instituto Geología y Perforaciones Uruguay, Boletín 13, p. 3–94.

—— , 1932, Restos de un Pez Ganoide de gran tamaño, proveniente del Gondwana Uruguayo. Instituto Geología y Perforaciones Uruguay, Boletín 19, p. 65–72.

Wanless, H. R., and Weller, J. M., 1932, Correlation and extent of Pennsylvanian cyclothems. Geological Society of America Bulletin, v. 43, p. 1003–1016.

Washburne, C. W., 1930a, Petroleum geology of the state of São Paulo, Brazil. Comissão Geográfica e Geologica do Estado de São Paulo, Boletim 22, p. 1–282.

—— , 1930b, Geologia do Estado de São Paulo. Comissão Geográfica e Geologica do Estado de São Paulo, Boletim 22, p. 1–282.

Waterhouse, J. B., and Riccardi, A. C., 1970, The Lower Cretaceous bivalve *Maccoyella* in Patagonia and its paleogeographic significance for continental drift. Ameghiniana, v. 7(3), p. 281–296.

Weaver, C., 1927, The Roca Formation in Argentina. American Journal of Science, v. 15(5), p. 417–434.

—— , 1931, Paleontology of the Jurassic and Cretaceous of West Central Argentina. University of Washington Memoir 1, p. 1–469. Seattle.

—— , 1942, A General Summary of the Mesozoic of South America and Central America. Eighth American Science Congress, v. 4, p. 149–193. Washington.

Weeks, L. G., 1947, Paleogeography of South America. American Association of Petroleum Geologists Bulletin, v. 31(7), p. 1194–1241.

Wenz, S., 1969, Note sur quelques Poissons Actinoptérygiens du Crétacé supérieur de Bolivie. Société Géologique de France, Bulletin (7) 11, p. 434–438.

Wenzel, O., Wathelet, J. C., Chávez, L., and Bonilla, R., 1975, La sedimentación cíclica Meso-Cenozoica en la región carbonífera de Arauco-Concepción, Chile. Segundo Congreso Ibero-Americano de Geología Económica, Actas 1, p. 215–237.

Wetzel, W., 1930, Die Quiriquina-Schichten als Sediment und paläontologisches Archiv. Paläontographica, v. 73, p. 49–106.

—— , 1960, Die Coyhaique-Schichten des patagonischen Neocoms und ihre Ammoniten. Neues Jahrbuch für Geologie und Paläontologie, Monatshefte 6, p. 246–254.

White, C., 1890, On certain Mesozoic fossils from the islands of St. Pauls and St. Peters in the Strait of Magellan. United States National Museum Proceedings, v. 13, p. 13–14.

White, I. C., 1908, Relatorio sobre os Coal Measures e rochas asociadas no sul do Brasil. Relatorio Final da Comissão de Estudos das Minas de Carvao de Pedra do Brasil. Pt. 1, p. 1–300. Rio de Janeiro.

Wichmann, R., 1924, Nuevas Observaciones Geológicas en la parte oriental de Neuquén y en el Territorio de Río Negro. Dirección General de Minas, Publicación 2, p. 3–22.

—— , 1927a, Los Estratos con Dinosaurios y su Techo en el Este del Territorio de Neuquén. Dirección Geología y Minas, Publicacion 32, p. 3–25.

—— , 1927b, Sobre la facies lacustre senoniana de los estratos con dinosaurios y su fauna (en los territorios del Río Negro y del Chubut). Academia Nacional Ciencias, Córdoba, Boletín 30, p. 383–405.

Wiedmann, J., 1980, Paläogeographie und Stratigraphie im Grenzbereich Jura/Kreide Südamerikas. Münstersche Forschungen zur Geologie und Paläontologie, v. 51, p. 27–61.

Wilckens, O., 1904, Uber Fossilien der oberen Kreide Süd-Patagoniens. Zentralblatt für Mineralogie, Geologie und Paläntologie, 1904. p. 597–599.

—— , 1905, Die Meeresablagerungen der Kreide- und Tertiär Formation in Patagonien. Neues Jahrbuch für Mineralogie, Geologie und Paläontologie, v. 21, p. 98–195.

—— , 1907a, Die Lamellibranchiaten, Gastropoden u.s.w. der oberen Kreide Südpatagoniens. Naturforschenden Gesellschaft Freiburg, v. 15, p. 97–166.

—— , 1907b, Erläuterungen zu R. Hauthals Geologischer Skizze des Gebietes zwischen dem Lago Argentino und dem Seno de la Ultima Esperanza (Südpatagonien). Naturforschenden Gesellschaft Freiburg, v. 15, p. 75–96.

—— , 1921, Beiträge zur Paläontologie von Patagonien. Neues Jahrbuch für Mineralogie, Geologie und Paläontologie, p. 1–14.

Williams, B. G., and Hubbard, R. J., 1984, Seismic stratigraphic framework and depositional sequences in the Santos Basin, Brazil. Marine and Petroleum Geology, v. 1, p. 90–104.

Wilson, T. J., and Dalziel, I.W.D., 1983, Geology of the Ultima Esperanza fold-thrust, southernmost Andes. U.S. Antarctic Journal, Annual Review, v. 18(5), p. 75–76.

Windhausen, A., 1914a, Einige Ergebnisse zweir Reisen in den Territorien Rio Negro und Neuquén. Neues Jahrbuch für Mineralogie, Geologie und Paläontologie, v. 38, p. 325–362.

—— , 1914b, Contribución al conocimiento geológico de los territorios del Río Negro y Neuquén. Ministerio de Agricultura, Sección Geología, Anales 10(1), p. 1–60.

—— , 1918a, The problem of the Cretaceous-Tertiary boundary in South America and the stratigraphic position of the San Jorge Formation in Patagonia. American Journal of Science 4, 45, p. 1–53.

—— , 1918b, Líneas Generales de la Estratigrafía del Neocomiano en la Cordillera Argentina. Academia Nacional Ciencias, Córdoba, Boletín 23, p. 97–127.

—— , 1931, Geología Argentina, v. 2, 645 p. Buenos Aires.

Winn, R. D., and Dott, R. H., 1977, Large-scale traction-produced structures in deep-water fan-channel conglomerates in southern Chile. Geology, v. 5, p. 41–44.

—— , 1978, Submarine-fan turbidites and resedimented conglomerates in a Mesozoic arc–rear marginal basin in southern South America. In: Stanley, D. J., and Kelling, G., eds., Sedimentation in Submarine Canyons, Fans & Trenches, p. 362–373. Dowden, Hutchinson & Ross, Stroudsburg, PA.

YPFB (Yacimientos Petroliferos Fiscales Bolivianos), 1972, Resumen de la Geología Petrolera de Bolivia. La Paz.

Yrigoyen, M. R., 1972, Cordillera Principal. In: Leanza, A. F., ed., Geología Regional Argentina, p. 345–364. Academia Nacional Ciencias, Córdoba.

—— , 1975a, La Edad Cretácica del Grupo Gigante (San Luis) y su relación con cuencas circunvecinas. Primer Congreso Argentino de Paleontología y Bioestratigrafía, Actas 2, p. 29–56.

—— , 1975b, Geología del Subsuelo y Plataforma Continental. In: Geología de la Provincia de Buenos Aires, Relatorio Sexto Congreso Geológico Argentino, p. 139–168.

—— , 1979, Cordillera Principal. In: Segundo Simposio Geología Regional Argentina, v. 1, p. 651–694. Academia Nacional Ciencias, Córdoba.

Zambrano, J. J., 1971, Las Cuencas Sedimentarias de la plataforma continental Argentina. Petrotecnia, v. 21(4), p. 29–37.

—— , 1972, Cuenca del Colorado. In: Leanza, A. F., ed., Geología Regional Argentina, p. 419–437. Academia Nacional Ciencias, Córdoba.

—— , 1974, Cuencas sedimentarias en el subsuelo de la Prov. de Buenos Aires y zonas adyacentes. Asociación Geológica Argentina, Revista, v. 29(4), p. 443–469.

—— , 1975, Perspectivas petrolíferas de la plataforma continental Argentina. Petrotecnia, v. 15 (7-8), p. 20–28; (9-10), p. 15–34.

—— , 1980a, Comarca de la Cuenca Cretácica de Colorado. In: Segundo Simposio Geología Regional Argentina, v. 2, p. 1033–1070. Academia Nacional

Ciencias, Córdoba.

—— , 1980b, Cuencas Sedimentarias de la parte Austral del Continente Sudamericano: Esquema preliminar de su evolución a fines del Jurásico y comienzos del Cretácico. Segundo Congreso Argentino de Paleontología y Bioestratigrafía, Actas 5, p. 15–39.

—— , 1981, Distribución y Evolución de las Cuencas Sedimentarias en el Continente Sudamericano durante el Jurásico y el Cretácico. In: Cuencas Sedimentarias del Jurásico y Cretácico de América del Sur, v. 1, p. 9–44.

—— , 1986, Las Cuencas Sedimentarias de América del Sur durante el Jurásico y Cretácico: su relación con la actividad tectónica y magmática. In: Volkheimer, W., ed., Bioestratigrafía de los Sistemas Regionales del Jurásico y Cretácico de América del Sur, v. 1, p. 1–48. Mendoza, Argentina.

Zambrano, J. J., and Urien, C. M., 1970, Geological outline of the basins in southern Argentina and their continuation off the Atlantic shore. Journal of Geophysical Research, v. 75(8), p. 1303–1396.

—— , 1975, Pre-Cretaceous basins in the Argentina continental shelf. In: Burk, C. A., and Drake, C. L., ed., The Geology of Continental Margins, p. 463–470, Springer Verlag, Berlin.

Zanettini, J.C.M., 1979, Geología de la Comarca de Campana Mahuida (Provincia del Neuquén). Asociación Geológica Argentina, Revista, v. 34(1),

p. 61–68.

Zeil, W., 1960, Zur Geologie der nordchilenischen Kordilleren. Geologische Rundschau, v. 50, p. 639–673.

—— , 1964, Geologie von Chile. Gebrüder Borntraeger, Berlin.

—— , 1979, The Andes. A Geological Review. Gebrüder Borntraeger, Berlin.

—— , 1980, Los plutones de los Andes. Academia Nacional Ciencias, Córdoba, Boletín, 53(1-2), p. 45–58.

Zeil, W., Damm, K. W., and Pichowiak, S., 1980, Los plutones de la Cordillera de la Costa al norte de Chile. In: Nuevos Resultados de la Investigación Geo-Científica Alemana en Latinoamérica, p. 112–122. Deutsche Forschungsgemeinschaft, Bonn.

Zilli, N., Orchuela, I., Dellapé, D., and Otaño, R., 1979, Análisis de las Formaciones Quintuco y Loma Montosa en el sector centro oriental de la Cuenca Neuquina. Séptimo Congreso Geológico Argentino, Actas 1, p. 609–615.

Manuscript Accepted by the Society February 28, 1987

This Paper is a Contribution to the International Correlation Program (IGCP) Project 191, Cretaceous Paleoclimatic Atlas Project.

Index

[Italicized page numbers indicate major references.]

Typeset by WESType Publishing Services, Inc., Boulder, Colorado
Printed in U.S.A. by Malloy Lithographing, Inc., Ann Arbor, Michigan